DEVELOPMENTAL AND CELL BIOLOGY SERIES

EDITORS

P.W. BARLOW   P.B. GREEN   C.C.WYLIE

The epigenetic nature of early
chordate development

THE EPIGENETIC NATURE OF EARLY CHORDATE DEVELOPMENT

Inductive interaction and competence

PIETER D. NIEUWKOOP

Hubrecht Laboratory, Utrecht, The Netherlands

A. G. JOHNEN & B. ALBERS

Zoological Institute, Cologne, West Germany

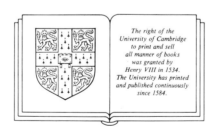

The right of the
University of Cambridge
to print and sell
all manner of books
was granted by
Henry VIII in 1534.
The University has printed
and published continuously
since 1584.

CAMBRIDGE UNIVERSITY PRESS

Cambridge

London New York New Rochelle

Melbourne Sydney

Published by the Press Syndicate of the University of Cambridge
The Pitt Building, Trumpington Street, Cambridge CB2 1RP
32 East 57th Street, New York, NY 10022, USA
10 Stamford Road, Oakleigh, Melbourne 3166, Australia

First published 1985

Printed in Great Britain by the University Press, Cambridge

Library of Congress catalogue card number: 84-23113

British Library cataloguing in publication data
Nieuwkoop, Pieter D.
The epigenetic nature of early chordate
development: inductive interaction and
competence

1. Embryology – Vertebrates
I. Title   II. Johnen, A. G.   III. Albers, B.
596′.033 QL959

ISBN 0 521 25107 9

Contents

Preface

The discovery of the so-called 'organisation centre' of the amphibian embryo by Spemann & H. Mangold in 1924 influenced embryological research to such an extent that we can actually speak of a milestone in the history of embryology. The work resulting in the concept of the organisation centre was summarised by Spemann (1936*) in the monograph *Experimentelle Beiträge zu einer Theorie der Entwicklung* (reprinted in 1968*). Spemann presented its main content in a series of lectures in New Haven, Connecticut, USA and published them (1938*) under the title *Embryonic Development and Induction* (reprinted in 1962*). Nearly 25 years later Saxén & Toivonen (1962*) devoted a monograph to the same subject, entitled *Primary Embryonic Induction*, in which the research results of the last decades were summarised. That monograph constitutes the starting point for the present book.

Spemann's epoch-making work induced his former pupil V. Hamburger, on the occasion of the 100th anniversary of Spemann's birthday in 1969, to dedicate a special article to Spemann as a teacher and a research worker under the title of 'Hans Spemann and the organiser concept'. Fifty years after the discovery of the organisation centre O. Nakamura and S. Toivonen took the initiative to edit a special monograph written by a number of Japanese and European embryologists, called *Organiser – a Milestone of a Half-century from Spemann* with, among others, an historical review of the organiser concept over the past 50 years by Nakamura, Hayashi & Asashima (1978*). It appeared in Japanese in 1976* and was translated into English in 1978*. These works mainly cover research on the amphibian embryo and deal primarily with the epigenetic interaction between the invaginating archenteron roof and the overlying ectoderm, called 'primary induction', an interaction which leads to the formation of the neural plate, the anlage of the central nervous system.

It is remarkable that just recently an attempt has been made by M. Jacobson (1982*) to propagate anew the old preformistic concept of development (worded by His, 1901), as applied to the development of the central nervous system, in modern terms of compartmentalisation, prepattern and cytoplasmic determinants. The author criticises the classical experiments of Spemann & H. Mangold (1924) with invalid arguments (see Chapter 12, p. 151) and simply ignores the extensive literature on the epigenetic nature of vertebrate development. The readers of this volume may judge for themselves whether they agree with Jacobson that another review on embryonic induction would be superfluous.

The enormous literature on inductive interactions between the various parts of the developing egg and embryo from the time of fertilisation up to the establishment of the three-dimensional pattern of embryonic organ anlagen has never been comprehensively summarised. A partial treatment of the subject was given by Saxén & Kohonen in 1968*, and more recently by Saxén *et al.* in 1976*. We feel that the entire sequence of inductive interactions during early embryogenesis ought to be reviewed for two main reasons: (1) in order to bring out the fundamental role of inductive interactions in the establishment of the three-dimensional pattern of organ anlagen, demonstrating the essentially epigenetic nature of vertebrate development, and (2) in order to summarise our present knowledge of the nature of inductive interactions during embryogenesis and to determine what common features may be discerned which can help us in planning further analysis of this fundamental aspect of animal development.

The development of the cephalochordates (and, to some extent, also that of the urochordates) shows pronounced similarities with that of vertebrate development. We therefore decided to include the entire chordate phylum in the present survey. We realise that this is a very ambitious enterprise. We shall therefore be obliged to restrict ourselves to early development, including only the initial interactions leading to the formation of the various organ anlagen and leaving organ morphogenesis out of the account. Moreover, we shall concentrate on the literature of the last two decades, referring to the monograph of Saxén & Toivonen (1962*) and to the two preceding monographs of Spemann (1962*, 1968*) for the older literature on primary induction. Considering the developmental period to be treated and the number of animal groups involved, it is evident that the entire literature is enormous, even when chiefly restricted to the last two decades. Therefore we feel obliged to refer wherever possible to review articles and monographs (marked with an asterisk) for detailed information and references, and to restrict ourselves to the main facts and interpretations. The reader should realise that an exhaustive discussion of the relevant literature would have led to at least a doubling of the present size of both text and reference list and would have made the book almost unreadable. The text of the book was written by the first author but was carefully checked by the other authors, so that the responsibility is shared equally by all three.

Acknowledgements

First of all we want to thank the editors of the Developmental and Cell Biology Series of Cambridge University Press for their interest in our attempt to approach the problem of embryonic induction from a more general biological point of view.

We want to express our deep gratitude to Dr J. Faber for the careful correction of the English text and for the many valuable suggestions for its improvement. We thank Mrs D. Parsons for the typing of the manuscript, Miss O. Kruijthof for bibliographical assistance and Mr L. Boom and Mrs C. L. Kroon (Hubrecht Laboratory) as well as Mr A. A. M. de Vries and coworkers (OMI, State University of Utrecht), for preparing the illustrations.

We extend our thanks to the Hubrecht Foundation for providing the means for the preparation of part of the illustrations.

Finally, we are indebted to several colleagues for allowing us to reproduce figures or for sending us original illustrative material.

1

Introduction

This volume starts with an outline of our own conceptions of embryonic induction in which we characterise certain aspects of 'inductive interactions' as distinct from 'non-inductive influences'. Particular properties of the action and reaction systems involved will be defined. These more theoretical considerations are meant to serve as a framework for the volume, and we hope they may also contribute to a better understanding of embryonic development. In the succeeding chapters we shall try to evaluate how far the experimental evidence available at present is in accordance with the various assumptions made. The notion that 'messages' released in inductive interaction may simply represent 'common products of cellular differention' (taking the latter term as broadly as possible) makes it desirable to place the discussion of their possible nature against the background of the general biochemical events occurring during early development. We shall therefore try to integrate morphological, biochemical and biophysical data in order to discern possible mechanisms involved in certain morphogenetic events.

The formation of the neural plate in the outer layer of an advanced amphibian gastrula under the influence of the underlying archenteron roof was, historically, the first typical case of an inductive interaction observed in early development; it was, therefore, called 'primary induction'. We now know that it is by no means the first inductive interaction. For instance, during the blastula stage, the meso-endoderm originates epigenetically in the totipotent animal 'ectodermal' moiety under an inductive influence emanating from the vegetal 'endodermal' yolk mass (see Chapter 10, p. 96). Moreover, it is very likely that intracellular inductive interactions are responsible for the establishment of dorso-ventral polarity in the uncleaved amphibian egg (see Chapter 7, p. 51), as well as for several events occurring during the preceding oogenesis, maturation and fertilisation of the egg. We must therefore abandon the blanket term 'primary induction' and distinguish specific events such as induction of dorso-ventral polarity, mesoderm induction, neural induction, etc. Indeed, we must be even more specific in our terminology since, for example, so-called mesodermal induction includes endoderm formation and comprises several successive phases (see Chapter 10, p. 101), while neural induction consists of at least two steps (see Chapter 12, p. 158).

There are a number of publications to which we shall frequently refer for more detailed information and references: the book by Saxén & Toivonen (1962*), which serves as starting point for this volume, and the subsequent reviews by Saxén & Kohonen (1968*) and Saxén *et al.* (1976*), which constitute a direct sequel to the 1962 book. Furthermore, we want to mention the excellent review written by Gerhart (1980*) entitled 'Mechanisms regulating pattern formation in the amphibian egg and embryo'. This actually was, in a sense, the outcome of a very fruitful cooperation between the US biochemists J. C. Gerhart and M. W. Kirschner and the Dutch morphogeneticists E. C. Boterenbrood, K. Hara, P. D. Nieuwkoop and G. A. Ubbels. It is essentially based upon the same general ideas about embryonic development as those developed in this volume. However, while Gerhart emphasises pattern formation, this volume focusses more explicitly on the nature of the inductive interactions which govern embryonic development.

We want to close this short introduction by informing the reader that we intend to treat development *chronologically*, following the increase in spatial multiplicity of the developing embryo as carefully as possible. An exception is made for the symmetrisation of meroblastic fish, reptilian and avian embryos and for that of the holoblastic mammalian embryo, which actually occurs during early development in the multicellular blastodisc or blastoderm. An exception is also made for the development of vertebrate asymmetry, the expression of which extends over a rather long period of development. Both topics will be discussed in Chapter 7 (on symmetrisation).

Since our present insight into inductive interactions during early development is largely based upon the analysis of amphibian development, in most chapters we shall first discuss the relevant amphibian literature and then add data from the analysis of other chordate groups, arranging them according to their significance for the inductive interaction in question.

2

Some general considerations on embryonic induction

The chordates are multicellular organisms composed of a large number of cells which belong to different cell types. Although a cell contains many organelles, each endowed with a certain structural and functional identity, these cannot really exist outside the cell, in which they form part of a highly complex and dynamic spatial and functional organisation. The cell must actually be considered as the smallest *autonomous* functional unit of the living organism. This autonomy finds its expression in the fact that many different cells can function more or less normally in an artificial tissue culture medium, where they maintain or reacquire the capacity for multiplication, so that cell clones can be formed. In higher organisms like cells are arranged in tissues forming functional units of the next higher order. Organ systems and organ anlagen composed of several tissues and endowed with particular functions form functional units of a still higher order, while organisms composed of various organs represent the highest functional order.

Such a complex, hierarchical organisation can only exist and be maintained when all parts of the organism interact more or less continuously (see P. Weiss, 1959* and Chandebois & Faber, 1983*). Interactions between different parts of the organism presuppose communication between these parts. Communication must occur at each of the different levels of organisation: between the various organelles in a cell, between the individual cells of a tissue, between the different tissues of an organ and between the different organs of an organism. The form of communication is likely to be different at each level of organisation. The most intimate contact may be that between individual cells of a tissue, e.g. through special membrane junctions. The communication between tissues, which are often separated by intercellular matrix, is less direct, while that between organs, which are usually relatively far apart, requires the intermediary of the nervous system or the vascular system.

We want to introduce two terms that will allow a better understanding of our concept of communication between different parts of the organism: the term 'message', defined as *any form of activity, chemical, physico-chemical or physical, which emanates from cells, tissues or organs* and which

3

may inform other cells, tissues or organs in a general way about the state of activity of the 'sender'; and the term 'signal', defined as *those messages released by the sender which are actually recognised by the receiving cells, tissues or organs as adequate information about the state of activity of the sender*. Signals may directly or indirectly affect the activity of the receiving cells, tissues or organs and may influence their ultimate direction of differentiation.

These considerations hold for the adult organism, for the preceding larval stage if present, and for the developing embryo, which is likewise a functional entity at every stage of its development. The individual cells of the early embryo must communicate with each other in order that the integrity of the embryo is maintained and its complexity or spatial multiplicity may increase. Communication may be temporally continuous or discontinuous; continuous communication will predominantly occur for functions which require a constant coordination, while discontinuous communication will be largely restricted to particular, possibly critical times or periods in development.

What are the basic requirements for communication inside a living organism? Each cell is surrounded by a cell membrane with particular structural and functional properties. Our present insight into membrane structure and function – although still highly fragmentary – implies much stress on the regulatory function of the cell membrane in the maintenance of a rather strict equilibrium between the cell interior and the surrounding medium. Electrical measurements have demonstrated that the permeability of the cell membrane is rather restricted (Ito, Sato & Loewenstein, 1974 *a*, *b*). Although passive transport of ions and small molecules by simple diffusion cannot be excluded, the often pronounced differences in ionic composition on either side of the cell membrane seem to be maintained by specific ion pumps incorporated into the cell membrane at particular sites. Although even large molecular complexes can be taken up by cells by means of pinocytosis – as e.g. the uptake of the large yolk proteins from the blood stream by the oocyte – it seems that many types of molecules do not pass the cell membrane barrier but react with particular receptor sites on the outer cell surface, leading to the production of so-called secondary messengers inside the cell, as e.g. in most (if not all) hormonal actions (see Branton, 1980*).

The most important aspect of the communication between cells, tissues or organs concerns the question whether a cell which receives certain messages from neighbouring cells or other parts of the organism is able to distinguish them from other messages, in other words, whether it can recognise certain messages as signals through its own external or internal receptor sites. These signals will be 'translated' into factors which can directly or indirectly influence the state of activity of the receiving cells. High-molecular weight messages, which cannot pass the cell membrane, may primarily be recognised as 'distinct signals' at the cell surface. Low-molecular weight messages, however, may pass the cell membrane and reach intracellular receptor sites (L. G. Barth & L. J. Barth, 1974). Nevertheless we may expect the cell membrane to play an important mediatory role in intercellular communication and interaction. Intracellular membranes may fulfil a similar function in intracellular interactions in the uncleaved egg (see below, p. 9).

No exchange of signals can be expected to occur between cells of the same cell type which are in the same physiological state. Although these cells will also release messages, these will not be recognised as signals by like cells, since they are identical to the messages these cells release themselves. We can therefore expect little or no effective signalling among the cells of a synchronously dividing monoculture. When cells of the same cell type are in different physiological states, e.g. in different phases of the cell cycle, effective signalling may occur, leading e.g. to synchronisation of cell division.

The notion of communication between the different parts of an organism implies that the greater the differences between adjoining cells, the stronger will be the effective signalling between the cells, since the messages released by the different cells will differ both quantitatively and qualitatively, so that their recognition as signals will be more effective. On the other hand, adjoining parts of an embryo may become so different from each other that they can no longer make effective contact, so that no interaction occurs, as between ecto- and endoderm cells of the amphibian neurula, which simply separate from each other when placed together experimentally (Townes & Holtfreter, 1955). However, it is significant that earlier in development, at the blastula stage, the presumptive ectoderm and endoderm cells make effective contact and interact strongly, leading to the formation of the meso-endoderm (see Chapter 10, p. 96).

We have suggested that cells of the same type which are in the same physiological state do not recognise each other's messages as signals. The same messages can however be recognised as signals by cells in a different physiological state, and even more effectively by cells of a different cell type. This assumption leads to the conclusion that the essential feature of cell communication lies in the *recognition* of the messages released as distinct signals. Therefore, in cell interaction the main emphasis is to be placed upon the *reaction system*.

The present notion of the ubiquitous nature of cell communication in both fully developed and developing organisms is clearly related to P. Weiss' (1969*) concept of 'cellular ecology' and to Chandebois' concept of 'cell sociology' (Chandebois, 1981; Chandebois & Faber, 1983*).

When applying our notion of cell communication to development, it must be concluded that no effective interaction can occur in a homogeneous system, e.g. in a synchronous monoculture of cells, even when the cells are pluripotent. In other words, a system must exhibit some degree of *heterogeneity* for interaction to occur. This conclusion is very important for the understanding of embryonic development, which is characterised by a steady increase in spatial multiplicity. An egg cell must possess a certain degree of spatial heterogeneity to be able to develop into a more complex organism. This heterogeneity usually manifests itself in polar differences, e.g. along the animal–vegetal polar axis. A most interesting, though exceptional, case is the *Fucus* egg, which initially exhibits spherical symmetry, showing no regional differences. Development only starts when the spherical symmetry is broken and a polar axis is formed, e.g. upon unilateral illumination (L. F. Jaffe, 1969*).

We now come to the concept of 'inductive interaction'. Not every form of interaction can be classified as inductive; an inductive interaction is a special type of interaction.

What are the proper criteria for qualifying cellular interactions as 'inductive', and what are 'non-inductive' interactions? We have already seen that in cellular interactions the recognition of released messages as distinct signals by the reaction system is essential. We want to relate the distinction between 'inductive' and 'non-inductive' interactions to a particular property of the reaction system: *its pluri-* or *unipotentiality*, respectively. A pluripotential system is a system which, under different circumstances, is able to realise more than one pathway of differentiation; representing a sequence of events which leads to the differentiation of a particular cell type, tissue, germ layer or organ system. A unipotential system, on the other hand, can only follow a single pathway of differentiation and therefore lacks the possibility of *choice*. Interactions which enable a developing system to make a choice from among different potential pathways of differentiation may therefore in general be classified as 'inductive', whereas interactions which only support the realisation of the single pathway of differentiation open to the developing system must be classified as 'non-inductive'.

However, this definition is at the same time slightly too broad and somewhat too narrow. It is too broad because many pluripotential systems seem to be able to proceed to a varying extent along one of the potential pathways of differentiation without inductive influences from outside; this represents the system's 'intrinsic developmental pathway' (cf. Chandebois & Faber, 1983*, 'autonomous progression'). For instance, the totipotent cells of the animal moiety of the amphibian blastula are able to proceed along the ectodermal pathway of differentiation when isolated from the vegetal, endodermal moiety. The definition is too narrow because interactions which lead to the expression of the toti- or pluripotentiality of the developing system usually do not include a choice between developmental pathways. They should nonetheless be considered as inductive; examples are hormonal action in egg maturation and sperm–egg interaction in fertilisation.

Therefore, 'inductive interactions' are *those interactions which enable a developing system to switch into a certain pathway of differentiation which differs from the pathway already partially in progress, as well as those which lead to the expression of its toti- or pluripotentiality.* This implies that interactions which only support the pathway of differentiation already in progress in the pluripotential system must be classified as non-inductive.

Any interaction involves a confrontation between two different components, called respectively the 'action system' and the 'reaction system'. Since either system may release messages which are recognised by the other system as distinct signals, in principle one always has to do with *reciprocity* or *mutual interaction.* Why then does development often clearly involve unilateral actions? In many instances the action seems to emanate from the system that is slightly more advanced in development, or slightly more firmly determined for a certain pathway of differentiation, while the reaction system is more responsive. Nevertheless, a reciprocal action can often be demonstrated, so that the reciprocity of interaction remains essentially valid.

Two fundamental questions arise when discussing inductive interactions: What is the nature of the messages released by the action system and how does the reaction system recognise these messages as distinct signals to which it can react by switching into a pathway of differentiation different from the one already in progress? The discussion of these questions constitutes the main subject of this volume.

Our understanding of these questions is still very limited. We can nevertheless make some suggestions about the answers on the basis of the notion of the general nature of cellular interaction as presented above. Although the possibility cannot be excluded that a particular action system produces highly specific messages for use by a certain reaction system with which it will come into contact in the course of development, it seems much more likely that the released messages are only '*transient, common products' of cellular differentiation* of the action system in question. The term 'common products' must however be taken to be as broad as possible, including not only all kinds of normal 'household' and 'luxury' molecules, but also particular components of the cell surface. Messages released by the action system during a certain stage of its development and differentiation are only recognised by the reaction system during a particular phase of its own development and differentiation. The release of messages by the action system and their subsequent recognition as distinct signals by the reaction system are primarily based upon *the confrontation between parts of the egg or embryo characterised by different pathways of differentiation,* parts which have come into contact with each other in a certain phase of development, often as a consequence of morphogenetic movements such as invagination, folding, evagination etc. In other words, *morphogenesis is the prerequisite for many inductive interactions. Inductive interactions are frequently also the prerequisite for subsequent morphogenetic processes*, e.g. mesoderm induction for gastrulation, neural induction for neurulation, etc., so that inductive and morphogenetic phases of development often alternate (Holtfreter, 1968*).

Since every cell or cell group *proceeds* along a given pathway of differentiation, the messages released by the cells may change with the progression of development, since they reflect the transient state of differentiation attained by the cells at a particular stage of development. In later development, messages may therefore be different from and more specialised than those released during early development.

The recognition of messages as distinct signals by the reacting cells will in particular be dependent upon the state of differentiation of the reaction system. In the older literature (see Saxén & Toivonen, 1962* and Saxén *et al.*, 1976*) one finds many examples of the time-dependency of both the inductive actions exerted by particular action systems and the responsiveness of the reaction systems concerned. One of the most striking features of inductive interactions is the rather strict time-dependency of the responsiveness of the reaction system. The reacting system passes through one or more rather sharply delimited periods of responsiveness to particular inductive stimuli, the so-called periods of 'competence'. During these periods the system is open to a choice between different pathways of differentiation. 'Competence' may be defined as *the transient ability of reacting cells to recognise certain messages released by neighbouring cells as distinct signals (signals which are instrumental in the choice of a particular new pathway of differentiation)*. Competence must always be related to the corresponding pathway of differentiation. Different competences usually refer to different periods of development. These periods may overlap temporally, however, so that several – but usually not more than two – alternative pathways may be available to the reaction system at a particular time. In general, the periods of competence seem to be shorter than the periods during which particular messages are released by the action systems, emphasising again the great importance of the reaction system in inductive interactions.

Since periods of competence represent periods of development during which the cells can be switched into one or the other pathway of differentiation, they represent '*labile*' periods during which development can still be *reversed* (see Wheldon & Kirk, 1973). Since periods of competence are of rather long duration – often many hours – they may encompass chains of individual steps characteristic for each of the alternative pathways of differentiation. It is therefore easily conceivable that different messages may be recognised by a reacting system as adequate signals for the same pathway of differentiation, affecting different steps of the process. A highly receptive reaction system may require only a more general stimulus, such as a change in intracellular pH, in ionic balance, in surface charge, etc., to be switched into a new developmental pathway. Messages that are quite different in nature, such as the various so-called heterogenous inductors, may guide the reaction system into the same new pathway of differentiation, as natural messages do.

Since different messages may be recognised by the reaction system as one and the same signal, the specificity of a given inductive interaction seems to be predominantly localised in the reaction system. The notion of the specificity of inductive interaction is closely associated with the terms 'instructive' and 'permissive' interaction. When the action system is largely responsible for the specificity of the interaction through the transfer of a specific message, to which the reaction system responds by entering into a particular pathway of differentiation, we speak of an 'instructive' action. When, on the other hand, the specificity of the reaction is largely due to the state of competence of the reaction system, so that even rather unspecific messages can serve as signals to open up a new developmental pathway, we speak of a 'permissive' action. The terms 'instructive' and 'permissive' were initially used by Saxén (1975). Saxén (1977*) no longer speaks of 'instructive' but of 'directive' actions as opposed to 'permissive' ones. We agree that this term is certainly more adequate than the older one but we consider the term 'permissive' also rather inadequate for influences which merely 'support' a pathway of differentiation already in progress. Moreover, the distinction between 'directive' and 'supportive' influences is painted too much in black and white. Although we cannot deny the existence of a certain directive element in the switching of a developing system into a *new* developmental pathway, the directiveness of the action highly depends upon the responsiveness of the reacting system, which may be so high that a very weak and possibly unspecific stimulus is sufficient to effectuate the switch.

On page 5 we referred to the possible occurrence of inductive inter-actions in the unicellular, uncleaved egg. This idea deviates from the generally accepted notion that inductive interactions represent interactions between cells or cell layers. Is such a restriction justified, however? There are at least three phenomena which plead against it: (1) an animal cell is a highly organised system which consists of several, if not many, separate compartments, which must interact during the normal functioning of the cell; (2) unicellular organisms, like the ciliates, show e.g. the formation of new rows of cilia and of mouth structures during cell division as a consequence of inductive interactions between particular parts of the cell (Sonneborn, 1970); (3) eggs of several amphibian species which are activated but not fertilised, so that they cannot undergo cell division and therefore remain unicellular, show cytoplasmic movements which strongly resemble the gastrulation movements of the normally developing embryo.

These pseudo-gastrulation phenomena occur at a given time after activation which more or less exactly corresponds to the timespan required to reach the gastrula stage in normal development. Since, in the multicellular embryo, gastrulation movements are closely associated with meso-endoderm induction (see Chapters 10 and 11, p. 101 and p. 138) this suggests that similar interactions can take place in the unicellular egg. However, the main argument for extending the concept of inductive interaction to the unicellular state is that important intracellular (particularly nucleocytoplasmic) interactions occur in the unicellular oocyte during oogenesis, maturation and fertilisation, which may contribute to our understanding of cellular and tissue interaction. It is even likely that intra- and intercellular interactions are at least partially based on the same or closely related mechanisms (see Gurdon, 1969*). It is for these reasons that we have included chapters on oogenesis, maturation and fertilisation in this book.

Summary

The constituent cells, tissues and organ systems of a living organism must interact continuously or at least periodically in order that the functional integrity of the organism be maintained. This holds for the adult organism as well as for the developing egg and embryo. However, in the hierarchically organised living organism the character of the interactions may differ at successive levels of organisation.

It seems likely that the cell membrane, being the barrier between the cell interior and the external environment, plays an important mediatory role in cellular communication and interaction.

In all interactions a distinction must be made between an 'action system' and a 'reaction system'. Interaction implies reciprocity, so that both interacting systems can act as action as well as reaction systems.

Interaction between different parts of an organism implies communication between the parts concerned. Communication only occurs between parts or systems which differ significantly, so that the 'messages' released by one system, the action system, can be recognised as 'distinct signals' by the other system, the reaction system, which may subsequently react to them.

In order to develop, an egg must have some degree of spatial heterogeneity, usually in the form of a bipolar structure. This heterogeneity is required for interactions to occur, which will lead to an increase in spatial multiplicity of the developing system.

Inductive interactions are a special form of interaction; they refer to pluripotential reaction systems, while non-inductive interactions refer to unipotential reaction systems. Since many pluripotential reaction systems have an intrinsic pathway of differentiation of their own, the term inductive interaction should be restricted to those interactions which switch the reaction system into a *new* pathway of differentiation, that differs from the one already in progress. However, those interactions which lead to the expression of toti- or pluripotentiality should also be included.

The 'messages' released by the action system are likely to be 'transient, common cell products' attendant upon the pathway of differentiation in progress.

It seems likely that the specificity of an inductive interaction resides chiefly in the responsiveness or 'competence' of the reaction system, rather than in the nature of the messages released by the action system. This qualifies inductive interactions more as 'supportive' than as 'directive'.

Embryonic development is primarily based upon the confrontation of parts of the egg or embryo engaged in different pathways of differentiation, which have come into contact during development as a consequence of morphogenetic movements. In other words, morphogenesis is the pre-requisitive for many inductive interactions, but the latter are often also the prerequisite for subsequent morphogenetic processes.

With the progression of development and differentiation, the messages released may change quantitatively as well as qualitatively.

The recognition of messages as distinct signals by the reaction system, and their effect upon the latter, are usually more strictly time-dependent than is the release of messages by the action system, emphasising the great importance of the reaction system in inductive interaction.

The term 'inductive interaction' should also be applied to intracellular interactions between different cellular components of the unicellular oocyte and egg.

3

Oogenesis and the origin of animal–vegetal polarity

The telolecithal amphibian egg

The origin of the animal–vegetal polarity of the amphibian egg leads us way back into oogenesis and possibly even into the preceding phase of primordial germ cell formation. However, an extensive treatment of these phases of development falls outside the scope of this book, so that their discussion will be restricted to the most relevant data. The reader is referred for oogenesis to Raven (1961*) and for primordial germ cell formation to Nieuwkoop & Sutasurya (1979*).

The primordial germ cells

In both anuran and urodele amphibians the primordial germ cells (PGCs) in the indifferent gonadal anlagen, as well as the primary oogonia and spermatogonia in the sexually differentiated gonad, are characterised by the so-called germ plasm, consisting of a conglomerate of mitochondria and germ-cell-specific germinal granules. The latter are not static structures but seem to go through alternating phases of activity and inactivity. The active phase is characterised by the fragmentation of the germinal granules and their association with ribosomes, and the inactive phase by their coalescence and dissociation from ribosomes.

Great significance has been attributed to the germ plasm, which in the anuran embryo at the end of the segregation of the PGCs from the endodermal blastomeres acquires a juxtanuclear position. This is considered to represent a phase of nucleocytoplasmic interaction important for germ cell determination (Mahowald & Hennen, 1971). However, in the urodeles, where germ cell formation seems to be purely epigenetic (Sutasurya & Nieuwkoop, 1974; Nieuwkoop & Sutasurya, 1979*), germ plasm in the form of nuclear emissions only appears in the PGCs at late tail bud stages, which is long after their cell-type-specific determination. Apparently germ plasm only represents a product of cellular differentiation (Ikenishi & Nieuwkoop, 1978) instead of playing the role of germ cell determinant as suggested by Bounoure (1939*), Blackler (1979*), Mahowald & Hennen

12

(1971), Kalt (1973) and L. D. Smith and Williams (1975*) and others for the anurans. Nieuwkoop & Sutasurya (1979*) suggest that the germ plasm may have the function of protecting the PGCs from entering somatic pathways of differentiation, thus maintaining the state of totipotency.

During the multiplication phase of the PGCs or during subsequent meiosis the germ plasm seems to disperse and ultimately to disappear during the early phases of vitellogenesis (Mahowald & Hennen, 1971; Kalt, 1973). According to L. D. Smith & Ecker (1971), Williams & L. D. Smith (1971) and L. D. Smith & Williams (1975*), nuage-like material reappears during oocyte maturation. However, Czołowska (1969) had already found precursor material of germ plasm in large ovarian oocytes. Germ plasm is evidently not present during the entire life cycle of the animal and therefore does not constitute a reliable criterion for the presence of a germ line in the sense of Weismann's 'Keimplasma Theorie' (1892; see further Nieuwkoop & Sutasurya, 1979*).

In the PGCs the germ plasm is situated on one side of the nucleus, possibly endowing the PGCs with an axial polarity. Since the germ plasm seems to be absent during the greater part of oogenesis, this axial polarity may be only a transient phenomenon.

Gametogenesis

In the amphibians sexual differentiation of the gonads normally occurs around metamorphosis, although neotenic species or individuals also show normal sexual differentiation. The PGCs may enter either the oogenetic or the spermatogenetic pathway (regardless of their own genetic constitution) in accordance with the sex of the somatic (mesodermal) tissues of the gonad (Blackler, 1970*). This rule seems also to hold for the higher vertebrates (see Witschi, 1967*; Chan & Wai-Sum, 1981*), but not in the same manner for the fishes, where in several families sex reversal is a normal phenomenon (Chan, 1977*).

In the ovarian anlage the primary oogonia divide, forming secondary oogonia which subsequently undergo four divisions, forming nests of 16 daughter cells connected by cytoplasmic bridges (incomplete cytokinesis; Coggins, 1973). In the testicular anlage the secondary spermatogonia, which only renew the stem cell population, go through complete cytokinesis, whereas the spermatogonia committed to progressive differentiation form intercellular bridges (Fawcett, 1972*). However, the number of 'incomplete' divisions varies from five to eight, so that nests of 32 and 256 cells are formed (Kalt, 1973, 1976).

The initiation of meiosis marks the transition of oogonia to oocytes and of spermatogonia to spermatocytes. Oocytes and spermatocytes go through the same successive stages of meiosis, viz. the leptotene, zygotene, pachytene and diplotene stages, although the duration of the various stages differs

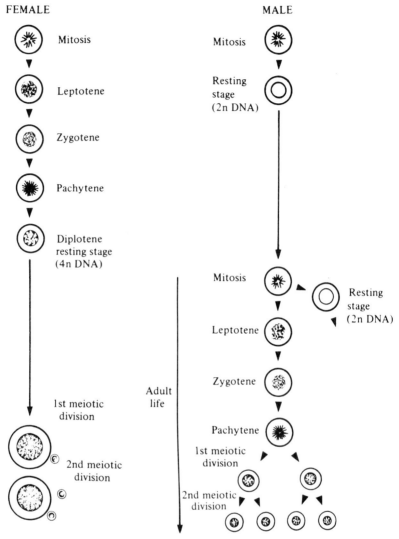

Fig. 1. Life cycles of female and male germ cells, dividing mitotically until or shortly after gonadal sex differentiation has set in. All female germ cells enter meiosis in foetal life, whereas male germ cells are maintained as a resting stem cell population, which can divide mitotically or from which meiotic cells emerge throughout adult life (after A. G. Byskov, 1982*).

in the two cell types (Fig. 1; see Byskov, 1982*). The oocytes arrest in the prophase of the first meiotic division. They require a special maturation process to proceed to the metaphase of the second maturation division (see Chapter 4, p. 26). Finally the completion of the meiotic divisions and the subsequent formation of the female pronucleus requires fertilisation. The male germ cells, on the other hand, complete meiosis before differentiation into spermatids and (subsequently) into spermatozoa sets in. Spermatozoa also go through a process of maturation, although this is of a different nature (see Chapter 5, p. 34).

Oogenesis

The reader is referred to the following reviews on oogenesis: Peters *et al.* (1972*) on mammals; Kunz & Schäfer (1978*) on amphibians, and Wallace & Selman (1981) on teleosts.

Oogenesis can be subdivided into a previtellogenic and a vitellogenic period.

The previtellogenic cytoplasmic growth phase

Previtellogenesis is characterised by extensive cytoplasmic growth, for instance in *Xenopus laevis* leading to a more than 200-fold increase in cell volume, i.e. from about 50 μm to about 300 μm oocyte diameter. This cytoplasmic growth involves a tremendous increase in 'common' cell organelles for metabolic and synthetic activities, such as mitochondria, endoplasmic reticulum (ER), Golgi complexes and ribosomes (see, among others, Webb & Smith, 1977 on mitochondrial DNA synthesis in *Xenopus laevis* oocytes). The cytoplasmic growth is accompanied by an enormous enlargement of the nucleus, which is then called the 'germinal vesicle'. According to Peterson (1971) the germinal vesicle of the previtellogenic oocyte may comprise about 7–13% of the cell volume.

During meiosis a large number of nucleoli are formed in the oocyte nucleus as a consequence of specialised amplification of DNA sequences coding for rRNA (Brown & Dawid, 1968). In the lampbrush chromosomes at the diplotene stage there is a *c*. 1000-fold increase in the number of DNA sequences coding for 28S and 18S RNA (for the two ribosomal subunits). There is also a high level of 5S RNA transcription but, in contrast to 28S and 18S RNA, the 5S RNA is transcribed from a large number of chromosomal genes which are not amplified. The separate formation of 28S and 18S RNA and that of 5S RNA is clearly demonstrated by the *o/o nu* mutant of *Xenopus laevis*, in which no 28S and 18S RNA synthesis occurs but which shows normal 5S RNA synthesis (see Ford, 1979*). 4S RNA (tRNA), which also accumulates during oogenesis, is likewise directly transcribed (see Gross, 1967*a*). The lampbrush chromosomes are also the site of a high level of transcription of a wide variety of unique DNA sequences.

The extra rDNA copies formed in the oocyte nucleus at the nucleolar organiser sites of each chromosome detach from them and form the initiation sites for the numerous nucleoli, which seem to float freely in the nuclear sap of the germinal vesicle (Gall, 1968; Brown & Dawid, 1968;

Evans & Birnstiel, 1968; see also Denis, 1974*). The nucleoli are the assembly sites of the ribosomes, for which (besides 28S and 18S RNA) 5S RNA and about 80 different ribosomal proteins are also required. The latter are synthesised on polysomes in the egg cytoplasm and transported to the nucleoli situated inside the germinal vesicle (Kumar & Warner, 1972; see also Ford, 1979*). The ribosomal subunits are finally transported from the nucleus to the cytoplasm through pores in the nuclear membrane.

E. H. Davidson *et al.* (1966) showed that 2% of the RNA formed on the lampbrush chromosomes is actually template RNA. This mRNA is not translated into protein but preserved throughout oogenesis. Most of the mRNA needed for early development of the embryo is indeed synthesised during the lampbrush chromosome stage (E. H. Davidson, 1976*).

Growing amphibian oocytes are active in protein synthesis (Bird & Birnstiel, 1971; Ficq, 1972; Wallace *et al.*, 1972), but synthesis comprises only a small fraction of the total protein accumulation during oogenesis; the great majority of proteins are synthesised in the liver and are transported to the developing oocyte (via the blood stream during vitellogenesis; see further below, p. 18.)

Mitochondrial multiplication, which is characteristic for the previtellogenic period, still continues during the initial phase of vitellogenesis until (for instance, in *Xenopus laevis*) the oocyte has reached a diameter of approximately 500 μm. Mitochondria are self-replicating cell organelles and contain a small amount of circular DNA, which has the same overall base composition as but no homology with nuclear DNA. Mitochondria can synthesise DNA, RNA and protein, essentially to support self-replication through growth and fission (see Denis, 1974*). Although the amount of DNA per mitochondrion is small, during oogenesis their number increases so much that the total amount of mitochondrial DNA ultimately comprises two-thirds of the extranuclear DNA (Dawid, 1966). The mitochondria supply the necessary energy to the egg and embryo by means of oxidative phosphorylation, using glycogen as a substrate (see also Patten & Villee, 1968*).

In many amphibian species the previtellogenic oocyte exhibits a distinct axial polarity, Brachet (1977*) mentioned that, at the pachytene stage, the rDNA copies accumulate locally inside the nucleus in the otherwise homogeneous oocytes. The diplotene previtellogenic oocyte contains a large cytoplasmic body, the so-called yolk nucleus of Balbiani, which adheres to one side of the germinal vesicle; this in its turn is situated slightly eccentrically in the oocyte. The yolk nucleus consists of a cloud of mitochondria and small vesicles and contains a pair of centrioles (Al-Mukhtar & Webb, 1971; Billett & Adam, 1976). Inside the germinal vesicle the nucleoli are concentrated on the side opposite the yolk nucleus, while a special area of nuclear emission is oriented towards the mitochondrial cloud (Balinsky & Devis, 1963; Billett & Adam, 1976; see also Fig. 2). This

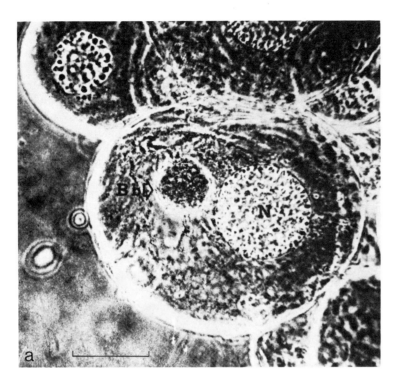

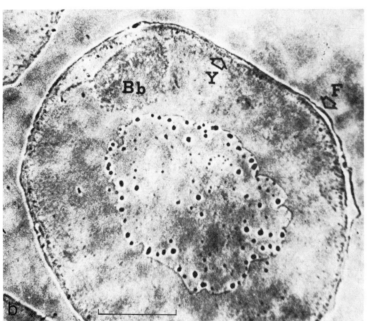

Fig. 2. (a) Phase-contrast photograph of living class-A oocyte from the ovary of a juvenile *Xenopus*, showing Balbiani body (Bb, the yolk nucleus of Balbiani) close to nucleus (N) with numerous nucleoli. Bar = 25 μm. (b) Unstained phase-contrast section of class-B oocyte with Balbiani body (Bb) and yolk granules (Y) in early stage of yolk formation. F, follicular membrane (courtesy B. I. Balinsky & R. J. Devis). Bar = 50 μm.

spatial configuration strongly suggests an interaction between the Balbiani body and the oocyte nucleus, but the nature and significance of this interaction are unknown. At the end of the previtellogenic period the mitochondrial cloud disperses and the mitochondria become uniformly distributed in the peripheral cytoplasm. The nucleoli are then also uniformly distributed along the inner surface of the germinal vesicle, so that the axial configuration of the previtellogenic oocyte once again is a transient phenomenon (Balinsky & Devis, 1963). These authors state that the yolk nucleus of the previtellogenic oocyte has nothing to do with yolk formation, but nevertheless they assume that the peripheral yolk is at least partially formed from transformed mitochondria.

As early as the leptotene stage, the oocyte becomes surrounded by follicle cells which form the future inner layer of the follicle. The follicle ultimately also comprises an intermediate, fibroblastic layer with blood capillaries and an outer layer with smooth muscle cells and connective tissue (Van Gansen & Weber, 1972; Coggins, 1973). As a result of an interaction between the oocyte and the follicle cells, the outer surface of the former develops many microvilli and the inner surface of the latter many macrovilli; micro- and macrovilli interdigitate. Both the oocyte and the follicle cells seem to be responsible for the deposition of the extracellular vitelline membrane, which tightly surrounds the growing oocyte.

The vitellogenic phase

Vitellogenesis starts uniformly along the periphery of the oocyte, the growing yolk platelets moving centripetally (Wittek, 1952). This primary yolk formation seems to depend exclusively on micropinocytosis by the microvilli of the oocyte membrane, which take up proteins from the intercellular space between the oocyte and the follicle cells (Wallace *et al.*, 1973). The small pinocytotic vesicles may either fuse into larger vesicles or may associate themselves with various cell organelles (such as mitochondria, multivesicular bodies, etc.). All these structures may be sites for yolk platelet formation (Wischnitzer, 1966*).

Yolk platelets contain three or four major proteins, which are formed from a single large precursor molecule called vitellogenin (Bergink & Wallace, 1974). Radioactive vitellogenin is actively removed from the blood by pinocytosis. Vitellogenin is synthesised in the liver and released into the blood stream under the influence of estrogen produced by the follicle cells. Estrogen formation and release are stimulated by follicle-stimulating hormone (FSH) from the pituitary. The vitellogenin is transferred to the oocyte by the surrounding follicle cells (Bergink & Wallace, 1974; see also Gurdon, 1968*; Denis, 1974*). The yolk proteins are laid down in crystalline form in the main body of the yolk platelet. This also has an amorphous peripheral layer which contains, among other things, a variety of enzymes as well as actin (which may play a role in cytoskeleton formation). The yolk platelets contain minor amounts of carotenoids, DNA and RNA in the outer granular layer, thus forming part of the nucleotide pool of the egg. The yolk platelets are surrounded by a unit membrane which may have been derived directly from the pinocytotic vesicles or from transformed mitochondria and multivesicular bodies (Balinsky & Devis, 1963).

According to Balinsky & Devis (1963) and Billett & Adam (1976) the vitellogenic oocyte passes through a phase of spherical symmetry, showing no axial polarity. In the nucleus the nucleoli are uniformly distributed along the nuclear membrane and the cytoplasm shows a uniform distribution of mitochondria, premelanosomes and cortical granules in the peripheral cytoplasm, and a concentric deposition of yolk platelets. The cortical granules, and possibly the premelanosomes, are derived from the Golgi complexes (Balinsky & Devis, 1963; Wischnitzer, 1966*).

Wittek (1952) has described a second phase of yolk formation, starting perinuclearly when primary yolk formation has reached the germinal vesicle. This secondary yolk formation would extend centrifugally. Since the yolk platelets gradually grow in size, a two-phasic yolk deposition would lead to a more or less concentric yolk arrangement, with peripheral and perinuclear regions of young, small-sized yolk platelets and an intermediate region of older, medium-to-large-sized platelets. According to Wittek, the subsequent transformation of the concentric configuration into a bipolar one finds its origin in the slightly eccentric location of the germinal vesicle inside the growing oocyte, causing an imbalance between the two modes of yolk formation, which extend in opposite directions. The growing vitellogenic oocyte would begin to show an animal–vegetal polarity when the secondary yolk reaches the egg surface on the future animal side and displaces the region of coarse yolk towards the opposite side, where it begins to form a coherent mass, the future vegetal yolk mass.

According to Gerhart (1980*) perinuclear yolk formation is difficult to reconcile with peripheral pinocytotic protein uptake. Moreover, Balinsky & Devis (1963) do not mention perinuclear yolk formation. In our opinion, secondary, perinuclear yolk formation is not required for the appearance of the animal–vegetal polarity in the growing oocyte. The slightly eccentric position of the germinal vesicle in itself may suffice to create an imbalance in the yolk distribution around the nucleus, a situation which may lead to the concentration of the more coherent heavy yolk on one side of the nucleus – the future vegetal side – for instance, as a result of contraction of cytoskeletal elements inside the more coherent yolk. Subsequently the yolk mass displaces the pigmented peripheral cytoplasm locally, giving rise to the appearance of an unpigmented (white) spot around the future vegetal pole (in *Xenopus laevis* when the oocyte is approximately 600 μm across). During further development the white spot gradually enlarges and finally forms the unpigmented vegetal moiety of the fully grown oocyte.

STAGE OF OOGENESIS						REFERENCES
pre-vitello-genesis	primary vitellogenesis		secondary vitellogenesis		mature oocyte	Wittek (1952)
I	II	III	IV	V	IV	Dumont (1972)
A and B	C		D	E		Balinsky & Devis (1963)
1 and 2	3	4	5	6		Davidson (1968)

Fig. 3. Diagrammatic representation of stages of amphibian oogenesis as distinguished by various authors, showing mutual relationships (after J. N. Dumont, 1972).

This differentiation of the animal and vegetal moieties occurs during stage IV of oogenesis (Dumont, 1972; see also Fig. 3).

Recently Capco & Jeffery (1981, 1982) and Capco (1982) have demonstrated that a poly(A)$^+$ RNA species, which is nearly uniformly distributed in the previtellogenic oocyte of *Xenopus laevis*, becomes distinctly localised in the vegetal subcortical region of the large vitellogenic and fully grown oocyte. A particular, though transient, localisation is also observed for newly synthesised rRNA, whereas the distribution of total RNA follows that of the egg cytoplasm. Jeffery & Capco (1978) found also differential distribution of maternal poly(A)$^+$ RNA in the egg of the ascidian *Styela* with accumulation in the animal hemisphere after fertilisation. These observations suggest that particular RNA species may accumulate locally by recognising specific cytoplasmic binding sites, thus enhancing the spatial segregation of the egg into regions destined for different pathways of differentiation. Such a mechanism was already suggested by L. D. Smith & Ecker (1970b*) and has been corroborated by C. R. Phillips (1982).

Brachet (1977*) suggests that the animal–vegetal polarity results from animal–vegetal differences in the properties of the cell membrane, thus causing regional differences in vitellogenesis. The animal–vegetal polarity of the oocyte has no relationship with the oocyte's orientation with respect to gravity, the oocytes being randomly oriented in the ovary. Neither is there a relationship between the polarity of the oocyte and the main direction of blood supply in the follicle (Van Gansen & Weber, 1972). Although there is no proof that the axial configuration of the primordial germ cell coincides with the transient polarity of the young previtellogenic oocyte, and later with the definitive animal–vegetal polarity of the egg, Nieuwkoop (1977*) has suggested that the animal–vegetal polarity of the egg may not be formed *de novo* during oogenesis but may be transferred from one generation to the next by means of cytoplasmic continuity through the germ cells. This notion would seem rather plausible for the anurans, where the germ plasm marks the vegetal pole region of the egg. In the urodeles, where germ cell formation is epigenetic (Sutasurya & Nieuwkoop, 1974) and the PGCs form germ plasm only late in development (Ikenishi & Nieuwkoop, 1978), the somatic ancestor cells of the germ cells must then preserve their cellular polarity.

The fully grown oocyte

The fully grown oocyte is an enormously enlarged cell. In *Xenopus laevis* it acquires some 20000 times or more the volume of the initial precursor cell, in other amphibian species this may even go up to 100000 times. This is primarily due to the extensive deposition of reserve material in the form of yolk platelets, lipid droplets and glycogen particles, but the oocyte also contains a large stock of 'common' cell organelles such as mitochondria and Golgi complexes, although the ER is only weakly developed. It also has a large amount of ribosomes formed by the numerous nucleoli of the germinal vesicle. The latter comprises about 3% of the egg volume in *Xenopus laevis*, but only about 1% of the egg volume in species with larger eggs. The full-grown oocyte has a large store of microtubular and actin-like proteins (Rutter, Pictet & Morris, 1973*). Burgess & Schroeder (1979*) state that the fully grown oocyte shows no indications of a microtubular cytoskeleton apart from the mitotic apparatus, but has a well-developed contractile microfilament system. One must nevertheless assume that the oocyte has a rigid cytostructure since its uneven yolk distribution is not affected by gravity prior to fertilisation. Actin-containing microfilaments are particularly abundant in the outer cortical layer. They are present in the microvilli and seem to be connected with a meshwork of microfilaments in the cortical layer (Gall, Picheral & Gounon, 1983).

In light microscope (LM) and transmission electron microscope (TEM) studies, Dollander (1962), Hebard & Herold (1967) and Ikushima & Maruyama (1971) noted that the cortical fibrillar network is thicker and more continuous on the animal and thinner and more discontinuous on the vegetal side. In a scanning electron microscope (SEM) study of *Xenopus laevis* eggs, Monroy & Baccetti (1975) observed differences in the organisation of the plasmalemma, in particular in the number and shape of the microvilli, which are more numerous and more slender at the vegetal than at the animal pole. Bluemink & Tertoolen (1978) found a significant difference between the two poles of the *Xenopus* egg in the range of intramembranous particles sizes in the E-face (external) of the plasma membrane. These studies show that the animal–vegetal polarity of the egg is expressed in the composition of the inner cytoplasm as well as in that of the cortical layer and the outer cell membrane. Dictus *et al.* (1984) recently found that in the unfertilised amphibian egg the egg membrane of the animal, pigmented hemisphere has a fluidity markedly different from that of the vegetal, unpigmented part. These differences become more pronounced after fertilisation. (See further Chapter 7, animal–vegetal polarity, p. 50).

The fully grown oocyte, which is metabolically inert, contains not only a complete machinery for energy supply and nucleic acid and protein synthesis and a large store of nutritional materials, but also transcribed regulatory and instructional information in the form of a store of stable proteins and 'masked' mRNA, which is apparently sufficient for subsequent development up to the early gastrula stage (see Gross, 1967a*; L. D. Smith & Ecker, 1970a*, b*).

The large oocyte shows a species-specific pigment pattern, the pigment being concentrated in the cortical layer of the animal moiety. Little is known about pigment formation in the growing oocyte. According to Balinsky & Devis (1963) the pigment granules derive from multivesicular bodies, which in turn are derived from transformed mitochondria. Pigment is often considered as a waste product of egg metabolism and does not seem to be essential for embryonic development, since it is not a general feature of animal eggs. Moreover, the periodic albinism mutant a^p of *Xenopus laevis*, which fails to form premelanosomes, shows normal development (Bluemink & Hoperskaya, 1975). The main function of the egg pigment seems to be the protection of the egg against harmful radiation.

The large oocyte shows a distinct animal–vegetal polarity in the distribution of cytoplasm and reserve material. Small-sized yolk platelets comprise only 10–20% of the animal egg moiety, the cytoplasm of which is particularly rich in mitochondria and ribosomes, whereas more than 50% of the vegetal moiety is occupied by large-sized yolk platelets. There is an intermediate region occupied by medium-sized yolk platelets, which is variable in size within a given species (Ubbels, 1977*; Malacinsky, 1984*). This region may actually be part of the animal cytoplasm, as suggested by Pasteels (1946). It seems to behave like animal cytoplasm in mesoderm formation (see Chapter 10, p. 96). This assumption is made more plausible by the recent observation by Dictus *et al.* (1984) that the plasma membrane of the fertilised egg consists of only two domains with markedly different membrane fluidity which coincide exactly with the pigmented animal and unpigmented vegetal moieties of the egg. It must therefore be concluded that the egg has minimal spatial heterogeneity: it consists of only two moieties, animal and vegetal. In view of its complex biochemical machinery the egg must be considered as a highly differentiated cell, with the unique feature that its differentiated state does not impair its totipotentiality (the reader is referred to Gerhart, 1980* and Brachet, 1980* for more detailed information).

The origin of animal–vegetal polarity in other chordate groups

Little is known of the origin of animal–vegetal polarity in the megalecithal eggs of teleosts, reptiles and birds. In the teleosts the animal–vegetal polarity is already expressed in an eccentric position of the nucleus in previtellogenic oocytes (Vakaet, 1955; Devillers, 1956*). The fully grown teleost oocyte has a thin, cytoplasm-rich superficial layer in the animal hemisphere, representing the future germ anlage, which contracts into the protruding germ disc upon fertilisation or activation (Devillers, 1956*). The flat, cytoplasm-rich germ anlagen of reptilian and avian eggs are likewise located at the animal pole of the very large, round egg cell. The very small, nearly alecithal mammalian egg seems to be completely apolar, however. Neither sperm entry nor gravity confer a polar axis on it (Hillman, Sherman & Graham, 1972). Polar heterogeneity is only acquired in the blastocyst, when the eccentric blastocyst cavity forms and the inner cell mass remains attached to one side to the outer trophoblastic cell layer (see Chapter 7, p. 66).

The small, oligolecithal eggs of many ascidians (urochordates) and of *Branchiostoma* (cephalochordates) do not show any heterogeneity before fertilisation. This is corroborated for the former by the observation that meridional, equatorial and oblique halves of unfertilised ascidian eggs can form normal mini-larvae upon fertilisation (Ortolani, 1958; Reverberi, 1961; Reverberi & Ortolani, 1962). It is not known whether the egg cytoplasm of the unfertilised *Branchiostoma* egg is likewise equipotential. In uro- and cephalochordate eggs both the animal–vegetal and the dorso-ventral polarity seem to be established immediately after fertilisation, the vegetal pole appearing near or at the sperm entrance point (Conklin, 1905*a*, *b* in *Cynthia*; Conklin, 1932 in *Branchiostoma*; see Chapter 6, p. 63).

Evaluation

In the anuran primordial germ cell an intracellular interaction seems to occur between the nucleus and the germ plasm, coincident with alternate periods of cellular activity and inactivity. A more continuous interaction takes place between the nucleus and the cytoplasm in the growing amphibian oocyte. This results in the establishment of an extensive machinery for nucleic acid and protein synthesis. In the previtellogenic oocyte, an interaction seems to occur between the Balbiani body and the oocyte nucleus. At the same time an intercellular interaction occurs between the oocyte and the surrounding follicle cells, resulting in the storage of an enormous amount of reserve material in the oocyte for use during the initial phase of embryonic development. The amphibian oocyte builds up a minimum of spatial heterogeneity; it consists of only two different moieties: animal and vegetal. This implies that the further development of the egg must be characterised by a maximum of epigenesis. The same seems to hold for the megalecithal oocytes of fishes, reptiles and birds. The very small, alecithal mammalian egg is apolar, acquiring polarity only in the blastocyst stage. The small oligolecithal eggs of uro- and cephalochordates apparently have no spatial heterogeneity before but acquire it immediately upon fertilisation.

4

Oocyte maturation

Definition

The full-grown, metabolically inert oocyte must be brought into a receptive state, before fertilisation by the sperm can occur. In different animal groups this state is reached either before, during or after the completion of the two meiotic maturation divisions. In the vertebrates the maturing oocyte passes through germinal vesicle breakdown, completion of the first and second meiotic divisions until metaphase II, and through the process of ovulation from the ovarian follicle. These processes are collectively called 'maturation' by L. D. Smith & Ecker (1970b*). Masui & Clarke (1979*) use a broader definition, extending maturation to the completion of the two meiotic divisions with extrusion of the two polar bodies and the formation of the haploid female pronucleus. However, for all vertebrates and a number of invertebrate groups, this definition would also include the fertilisation process, which (in our opinion) should be dealt with separately. We shall therefore follow Smith & Ecker's definition. Nonetheless it seems desirable to distinguish between the preparatory events leading up to the receptive state of the egg for fusion with the male gamete, and the nuclear events which may or may not take place (partially or entirely) during this cytoplasmic ripening process. In parthenogenetic development, the quiescent oocyte can be converted into a metabolically active cell without fertilisation and, often, even without meiotic divisions (Graham, 1974*).

A similar loose relationship seems to exist between ovulation and maturation. Although in many species ovulation and the beginning of maturation take place almost simultaneously, in other species ovulation occurs either before or after maturation. The two processes are not causally related and can proceed independently of each other under experimental conditions and, in some species, even under natural conditions. (See the extensive review on oocyte maturation by Masui & Clarke (1979*) and the general review by Gerhart (1980*). These reviews constitute the basis for the following subchapters.)

The successive morphological events

The rupture of the germinal vesicle is the first visible sign of oocyte maturation in amphibians. While breaking down, the germinal vesicle ascends towards the animal pole, where it causes the appearance of a white, unpigmented spot. Large basophilic bodies, rich in RNA and glycogen, accumulate in the cytoplasm of the maturing *Xenopus laevis* oocyte, below the nucleus. Feulgen-positive bodies, corresponding to the nucleolar organisers, appear on the inner side of the nucleus in its basal region, after which the basal nuclear membrane begins to undulate. After breakdown of the germinal vesicle, the Feulgen-positive bodies move towards the egg cortex, where they gradually disintegrate (Brachet, Hanocq & Van Gansen, 1970). Similar events were observed by Dettlaff & Skoblina (1969) in sturgeon oocytes. After germinal vesicle breakdown the chromosomes begin to condense, contract further, and are arranged in pairs on the equator of the first meiotic spindle. The paired homologous chromosomes subsequently separate, half of them forming the nucleus of the first polar body (which is then extruded), while the other half aligns again on the second metaphase plate; nuclear development then stops. During amphibian oocyte maturation the number and size of the surface microvilli are progressively reduced, and the attachment between the egg surface and the vitelline membrane is weakened (Schuetz, 1974, 1978). In mammalian follicles, on the other hand, where the oocyte ovulates together with the surrounding granulosa cells, the surface microvilli of the oocyte increase in number and are maintained during maturation (Zamboni, 1972*).

The hormone-dependency of the maturation process

Egg maturation is a hormone-dependent process controlled by the pituitary. In amphibians, reptiles, birds and mammals the pituitary gland secretes two different gonadotropic hormones, the luteinising hormone (LH) and the follicle-stimulating hormone (FSH). In fish only a single gonadotropic hormone (GTH) is produced which is, however, very similar to FSH in its action.

Rana pipiens oocytes from which the follicle cells have been completely removed, either with Ca^{2+}-free medium (Masui, 1967) or by pronase treatment (L. D. Smith, Ecker & Subtelny, 1968), do not react to gonadotropins but mature normally under the influence of progesterone. The pituitary hormones apparently act on the follicle or granulosa cells surrounding the oocyte, which in turn produce a progesterone-like hormone which affects the oocyte itself. Maturation can also be induced, though less effectively, in follicle-enclosed oocytes by other steroid hormones such as deoxycorticosterone and testosterone, but not by oestradiol (Masui, 1967;

Smith *et al.*, 1968). This conclusion was corroborated by Dettlaff & Skoblina (1969) for the sturgeon and by Jacobelli *et al.* (1974) for the trout.

Progesterone acts at the oocyte surface membrane since injection of the hormone fails to release the maturation response (L. D. Smith & Ecker, 1971). Steroid action on the oocyte membrane probably involves a change in the conformation of surface protein molecules, including either formation or dissociation of S—S bonds (Brachet, 1974*, 1978*; Brachet *et al.*, 1975): proteolytic enzymes and SH reagents can trigger maturation. Steroid action in oocyte maturation resembles the action of polypeptide hormones and steroids during larval development and adult life. However, these involve cAMP as a second messenger, while cAMP does not mimic progesterone effects on the oocyte (L. D. Smith & Ecker, 1971 and L. D. Smith, 1975*), so that there is no convincing evidence for a role of cAMP in the hormonal induction of maturation.

In mammals the situation is more complex. Although LH and FSH can induce maturation in follicle-enclosed oocytes *in vitro* (Tsafriri *et al.*, 1972), oocytes removed from their follicular environment always undergo spontaneous maturation. The gonadotropic hormones seem to affect the condition of the follicle by causing a degeneration of the cumulus cells, a process that in turn may stimulate the initiation of maturation. LH stimulates prostaglandin (PG) production in the follicle cells, and both LH and PG stimulate adenylate cyclase, increasing the cAMP level in the follicle cells (Lamprecht *et al.*, 1973). Extrafollicular cAMP application has no effect on maturation, but its injection into the follicle induces it. Follicular steroids may also play a significant role in the maturation of mammalian oocytes, although gonadotropins alone may sufficiently alter the mammalian follicle to initiate maturation. Growing oocytes usually do not respond to gonadotropins. Oocytes removed from their follicles can mature under progesterone treatment when they have attained a certain size. This holds for amphibian (Smith *et al.*, 1968) as well as for mammalian oocytes (Tsafriri & Channing, 1975).

Biochemical and biophysical aspects of maturation in amphibian oocytes

Oocyte maturation is an energy-consuming process. Germinal vesicle breakdown is inhibited by anaerobiosis and by respiratory inhibitors such as potassium cyanide (Brachet, Pays-de Schutter & Hubert, 1975). The energy is required for protein synthesis (L. D. Smith & Ecker, 1970*a*), as demonstrated by the strong inhibitory effect on maturation of protein synthesis inhibitors such as cycloheximide and puromycin (Brachet, 1967*c*; L. D. Smith & Ecker, 1969).

The stimulation of protein synthesis following the initiation of maturation is maximal around the first meiotic metaphase and is maintained at that level throughout maturation (L. D. Smith, 1975*). It does not require the presence of the germinal vesicle (L. D. Smith & Ecker, 1969) and is therefore independent of the transcriptional activity of the nucleus. This is in agreement with the observation that, although the maturation of follicle-enclosed oocytes following gonadotropin treatment is inhibited by actinomycin D (Dettlaff, 1966; Brachet, 1967 *c*; Schuetz, 1967) and by α-amanitin, follicle-free oocytes induced to mature by progesterone treatment are not inhibited by these drugs (Wasserman & Masui, 1974). This strongly suggests that RNA synthesis is required for progesterone production in the follicle cells, but not for oocyte maturation.

In the maturing oocyte, protein synthesis occurs on stored mRNA; this demonstrates that maternal RNA is used many hours before fertilisation (Smith & Ecker, 1970 *a**, *b**). Progesterone-induced maturation in *Xenopus laevis* oocytes leads to a two-fold increase in protein synthesis, due to recruitment of mRNA rather than to a change in translational efficiency (Richter, Wasserman & Smith, 1982; Wasserman, 1982; Wasserman, Richter & Smith, 1982).

Bellé, Boyer & Ozon (1982) showed that carbon dioxide does not inhibit the initial steps of progesterone-induced maturation but prevents germinal vesicle breakdown by inhibiting the formation or the amplification (or both), of a maturation-promoting factor (see below). There is no detectable nuclear DNA synthesis during progesterone-induced maturation; only some mitochondrial DNA is synthesised (Hanocq *et al.*, 1974; Brachet *et al.*, 1974). There is a sharp decrease in RNA synthesis at the time of germinal vesicle breakdown, leaving only some residual mRNA synthesis (Webb, La Marca & Smith, 1975).

The induction of oocyte maturation is dependent on the presence of divalent cations, since no maturation following progesterone treatment occurs in Ca^{2+}- and Mg^{2+}-free medium (Merriam, 1971 *a*, *b*). Removal of internal Ca^{2+} ions inhibits maturation. Merriam suggests that germinal vesicle breakdown induces changes in the cytoplasmic distribution of cations, including a compartmentalisation of intracellular Ca^{2+} and Mg^{2+}. R. N. Johnston & Paul (1977) observed a rapid but transient release of Ca^{2+} into the external medium, followed by a Ca^{2+} influx from the medium, which reaches a maximum shortly before germinal vesicle breakdown. Ionophore A 23187, which facilitates transport of divalent ions across membranes, acts as a maturation-inducing agent (Reed & Lardy, 1972). The site of action of Ca^{2+} is near the cell surface of the oocyte, since superficially injected Ca^{2+} causes maturation, whereas more deeply injected Ca^{2+} does not (Moreau, Dorée & Guerrier, 1976). Progesterone-induced maturation is accompanied by a depolarisation of the oocyte membrane, resulting in a decrease of internal K^+ and an increase of internal Na^+ (Morrill & Watson, 1966), the depolarisation reaching a low level just before germinal vesicle breakdown (Ziegler & Morrill, 1977). Depolarisation can also be achieved by deprivation of external Ca^{2+}, indicating that the major portion of the Na^+–K^+ transport is regulated by Ca^{2+}. There is an increase in membrane resistance followed by a decrease just before germinal vesicle breakdown. The latter phenomenon is accompanied by a marked increase in ATPase activity (Morrill, Kostellow & Murphy, 1971, 1974; Moreau, Guerrier & Dorée, 1976). The ATPase activity is Ca^{2+}-activated and ouabaine-insensitive.

The maturation-promoting cytoplasmic factor (MPF)

L. D. Smith & Ecker (1971) demonstrated that cytoplasm of progesterone-stimulated *Rana pipiens* oocytes prior to germinal vesicle breakdown can induce maturation when injected into untreated oocytes or into oocytes from which the germinal vesicle has been removed (Masui & Markert, 1971). The steroid apparently causes the production of a second effector, which then induces actual maturation. They called this cytoplasmic factor, which is independent of the nucleus, the 'maturation-promoting factor' (MPF). These observations were confirmed by Dettlaff, Felgengauer & Chulitskaia (1977) for the sturgeon. Reynhout & Smith (1974) showed that MPF is not species-specific. Masui & Markert (1971) found that MPF appears shortly before germinal vesicle breakdown, its activity remaining high in the fully matured oocyte and decreasing only after egg activation. MPF is still periodically detectable in cleaving blastomeres, so that it may be acting both in meiosis and in mitosis (Wasserman & Smith, 1978*). The relation between MPF and germinal vesicle breakdown is still unknown, however.

MPF formation requires protein synthesis; its action is not dependent on Ca^{2+} ions and is resistant to proteolytic enzymes, so that it involves neither synthesis nor degradation of proteins. Serial transfer of cytoplasm (5–10 times) does not lead to a weakening of the effect, which points to autocatalytic amplification of MPF (Masui *et al.*, 1977).

Protein phosphorylation is known to be dependent on the cAMP level regulating the activity of protein phosphokinases. An initial decrease in cAMP level seems to be important for the initiation of maturation, since a high cAMP level prevents maturation. The initial decrease in cAMP may be caused by an increase in phosphodiesterase activity, while Ca^{2+} ions may play an important role in the activation of membrane-bound phosphodiesterase (Masui & Clarke, 1979*). MPF of *Xenopus laevis* oocytes can be stabilised with adenosine $5'O$-(3-thiotriphosphate) (ATP-y-S). It is neither inhibited by Ca^{2+} ions nor by calmodulin (Hermann *et al.*, 1983). MPF seems to be a normal agent in meiosis and mitosis and may act as an autophosphorylating kinase. Mitotic factors derived from mammalian cells cause breakdown of the nuclear membrane and chromosome condensation in amphibian oocytes (Sunkara, Wright & Rao, 1979).

The cytostatic cytoplasmic factor (CF)

The arrest of meiotic division at metaphase II was interpreted by Monroy (1965*) as being due to the building-up of a self-inhibitory factor during maturation; activation of the egg following fertilisation would entail the removal of this block, leading to the completion of meiosis and the initiation of mitosis. Cytoplasm of unfertilised eggs of *Rana pipiens* injected into blastomeres at the 2-cell stage leads to arrest of development at metaphase (Masui & Markert, 1971), while no arrest occurs when cytoplasm of activated eggs is used. Masui & Markert call this cytoplasmic factor, which is again independent of the presence of the nucleus, the 'cytostatic factor' (CF). Its activity develops during maturation, shortly after germinal vesicle breakdown, and remains high until egg activation. This factor is not species-specific either, but different species may show quantitative differences in activity (Meyerhof & Masui, 1977, 1979*a*). The factor is destroyed upon fertilisation or upon artificial activation by pricking. Chulitskaia & Felgengauer (1977) found a cytostatic effect of injected cytoplasm of mature non-activated oocytes of *Rana temporaria* and *Acipenser stellatus* on cleaving blastomere nuclei only when it was treated with ethylene glycol bis-β-aminoethylene-N,N,N',N',-tetraacetic acid (EGTA). This was confirmed by Ryabova (1980). CF arrests the cell cycle at metaphase, an arrest which in turn induces chromosome condensation (Meyerhof & Masui, 1979*b*; see below). It seems likely that a similar factor is responsible for the initial arrest of the oocyte nucleus at first meiotic prophase during the previtellogenic growth phase (see p. 14).

Chromosome condensation

Gurdon (1967*b**, 1968*a*) showed that nuclei from blastulae or from adult brain, when exposed to oocyte cytoplasm, undergo morphological and biochemical changes similar to those of the oocyte nucleus. Ziegler & Masui (1973) and Leonard, Hoffner & Diberardino (1982) noted that nuclei introduced into oocytes undergoing progesterone-induced maturation show condensation of chromosomes shortly after germinal vesicle breakdown. Ziegler & Masui (1976) demonstrated the necessity of RNA and protein synthesis for chromosome condensation. Chromosome condensation requires substances released from the germinal vesicle, since no effect occurs when nuclei are injected into enucleated oocytes induced to mature with progesterone. The proteins from the germinal vesicle which associate with the condensing chromosomes are predominantly histones. This is in accordance with the observed synthesis of histones in *Xenopus* oocytes undergoing maturation (Adamson & Woodland, 1977) and likewise in mouse oocytes (Wasserman & Letourneau, 1976*b*). The histones may protect the chromosomal DNA from disintegration by cytoplasmic enzymes

(Wyllie, Gurdon & Price, 1977). The reader is further referred to the reviews by L. D. Smith & Ecker, 1968*, 1970*a**, *b**; L. D. Smith, 1975*; Wasserman & Smith, 1978*; Tsafriri, 1978*; Masui & Clarke, 1979*; Gerhart, 1980 and Brachet, 1980*.

Evaluation

The maturation process clearly demonstrates that the oocyte is a fully differentiated cell capable of reacting to a hormonal stimulus by performing a chemical chain reaction very similar to that of specialised cells in the adult animal. This reaction starts at receptor sites on the cell membrane, where cAMP may play a role in transferring the stimulus to the egg cytoplasm; here a cytoplasmic factor is formed which acts upon the nucleus, unblocking its meiotic divisions. When the latter have reached metaphase II another cytoplasmic factor is formed, leading to a second arrest of the nucleus. The maturation process also leads to a loosening of the egg microvilli from those of the follicle cells, and to the subsequent ovulation of the egg from the ovary.

In the process of maturation, we meet several forms of inductive interaction; first a long-range interaction between the pituitary and the egg follicle, which leads to another, in this case shorter-range interaction between follicle cells and oocyte. These hormonal actions release nucleo-cytoplasmic interactions which bring the egg to a highly receptive state for the process of fertilisation, leading to the expression of its totipotentiality. The mature, ovulated oocyte, which becomes surrounded by jelly layers during its passage through the oviduct, again enters a metabolically inert state for temporary storage in the oviduct until the time of fertilisation.

The gonadotropic hormonal action of the pituitary, which causes the synthesis and release of progesterone from the follicle cells and the subsequent action of the hormone on the cell membrane of the oocyte, must be classified as rather specific and undoubtedly directive in nature. It strongly resembles hormonal action in the adult organism. The subsequent intracellular nucleocytoplasmic interactions, in which the cytoplasmic factors MPF and CF play an important role, bring the oocyte to the mature, receptive state. These interactions seem to be elements of normal cellular activity, which probably take part in both meiosis and mitosis. The process of egg maturation may therefore yield more insight into cellular interaction during embryonic development.

5

The development and maturation of the male gamete

An extensive discussion of spermatogenesis falls outside the scope of this volume. We shall restrict ourselves to the most relevant data. For more detailed information the reader is referred to Roosen-Runge (1977*), Kunz & Schäfer (1978*, pp. 73–95) and to other reviews cited below.

Spermatogenesis

The male gonad or testis consists of a large number of seminiferous tubules surrounded by connective tissue (see Fig. 4a). The tubules are lined with large Sertoli cells held together by tight junctions, thus forming the major part of the blood–testis barrier that isolates the developing germ cells from the surrounding tissue fluid. The tight junctions separate a basal compartment, formed by the Sertoli cell layer and the connective tissue capsule, from an adluminal compartment, to which also the lumen of the seminiferous tubule belongs. It is an interesting observation that the diploid spermatogonia are located exclusively in the basal compartment, while the meiotic spermatocytes and haploid spermatids are only found in the adluminal compartment, suggesting that the transfer of the germ cells from one compartment to the other may control the process of meiosis. This transfer requires a local and temporal breakage of tight junctions (see Fig. 4b). Whereas the spermatogonia are not connected with the Sertoli cells, all other stages of sperm cell formation are intimately associated with them. It is unknown whether this association is the cause or consequence of the differentiation process (see Hogarth, 1978*; Byskov, 1982*; and Setchell, 1982*).

Like other stem cells in the body, spermatogonial stem cells multiply only very slowly. A minor proportion (type A) form new stem cells (c. 10%), while the majority (c. 90%) form spermatogonia of the intermediate and B types respectively, representing the next stages of spermatogenesis. The type B spermatogonia form first primary and then secondary spermatocytes while going through the meiotic divisions leading to the formation of the haploid spermatids. The transformation of the spherical spermatid into the strongly elongated spermatozoon, a process called

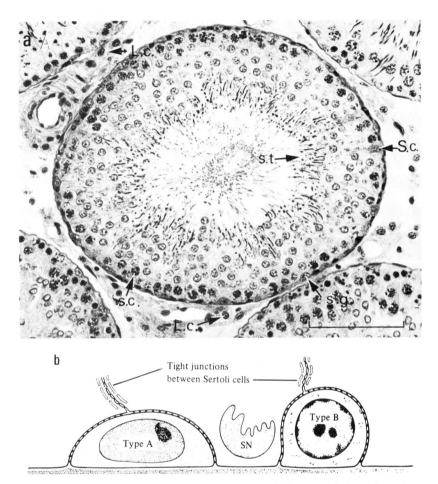

Fig. 4. (a) Transverse section through mouse seminiferous tubule with various stages of spermatogenesis: s.g., spermatogonia; s.c., spermatocytes; s.t., spermatids; L.c., Leydig cell; S.c., Sertoli cell. Bar = 100 μm. (b) Diagram of the Sertoli cell layer with tight junctions which form a barrier between the basal compartment, containing type A and type B spermatogonia, and the adluminal compartment, containing all the stages of spermiation. SN, Sertoli cell nucleus (courtesy B. P. Setchell).

spermiation, occurs in intimate association with the Sertoli cells. One of the two centrosomes of the spermatid extends into the flagellum of the spermatozoon, while the Golgi complex (which is situated on the opposite side of the nucleus) gives rise to the acrosomal vesicle. The nucleus becomes highly condensed and often spindle-shaped, while the mitochondria accumulate in the midpiece. The full-grown spermatozoa are finally released from the Sertoli cells which retain, however, much of the sperm cytoplasm. The spermatozoon moves with the help of the flagellum; the energy for the movement is supplied by the small number of mitochondria associated with the base of the flagellum in the midpiece. Apparently no further DNA, RNA or protein synthesis is required, so that mRNA and ribosomes are absent and the haploid nucleus is fully inert. The spermatozoa are retained in the nutrient medium of the seminiferous tubules and genital ducts, so that little storage of metabolites is required (see Hogarth, 1978*; Bedford & Cooper, 1978* and Fig. 1 on p. 14).

In the highly polarised spermatozoon, various regions can be distinguished, viz. the acrosomal region with the acrosomal vesicle, the post-acrosomal region with the condensed nucleus, the equatorial zone with a high actin content, the tail midpiece with its mitochondria, and finally the flagellar tail region. There seem to be different plasmalemma domains with different lectin-binding capacities, surface charges and antigenic properties in the acrosomal, post-acrosomal, midpiece and tail regions. This is demonstrated, among other things, by the different head-to-head and tail-to-tail agglutination of sperm (see Bedford & Cooper, 1978*). The spermatozoon has a very complex microtubular cytoskeleton, while the microfilament system plays an active role in the acrosome reaction at fertilisation (Burgess & Schroeder, 1979*).

Sperm maturation and capacitation in fishes and amphibians

Typical sperm maturation and capacitation seem to be restricted to meta- and eutherian mammals. Related phenomena may however play a role in other chordates. In fish and in anuran amphibians, which have external fertilisation, sperm taken from the testis seem to be fully capable of penetrating the egg investments. It must however be realised that sperm has to be stored in the male genital ducts for some length of time without losing its fertilising capacity. It is therefore likely that in one way or another the sperm is inactivated during passage and storage in the male genital ducts. Inactive sperm must then be activated upon release by the male. In several invertebrate groups, e.g. the sea urchins, substances released by the egg or its investments act upon the sperm and stimulate its activity (Tyler & Tyler, 1966*). Similar substances seem to play a role in urochordates, where they in addition have a chemotactic function (Sabbadin & Zaniolo, 1979). In the vertebrates, diffusible egg substances are unknown. The release of the sperm into the external environment in fish and anuran amphibians may serve as the activating stimulus, simply as a result of dilution of inhibitory factors, a process which in certain respects may be compared to capacitation of the mammalian sperm in the female reproductive tract (see below).

Sperm maturation in meta- and eutherian mammals

In mammals the immobile spermatozoa from the rete testis undergo a real maturation process during their passage through the epididymis. Sperm is carried through the caput and corpus epididymidis by peristaltic contraction and is stored in the cauda epididymidis, where it stays viable for weeks and even months. During this maturation process the last residue of cytoplasm, the so-called cytoplasmic droplet, is shed. Moreover, the appearance of the acrosome may be altered and the nucleus becomes further stabilised by extensive S—S crosslinkage, while the other parts of the spermatozoon are mechanically strengthened. While the testicular sperm depends chiefly on oxidative glucose metabolism, the sperm in the cauda epididymidis predominantly shows glycolytic assimilation of glucose. The epididymis withdraws much water from the seminal fluid, which is probably secreted by the Sertoli cells, and secretes glycoprotein and other substances (see Gwatkin, 1976*).

The surface membrane of the spermatozoon undergoes modifications during passage through the epididymis, showing a decreasing activity of Na^+/K^+-dependent ATPase, increasingly negative surface charges (G. W. Cooper & Bedford, 1971), changes in the distribution of receptors for fluorescein- and ferritin-labelled lectins (Nicholson, 1974*), and the acquisition of special membrane glycoproteins (Olson & Hamilton, 1978). The main function of the epididymis, however, may be the provision of a congenial environment for the development of full motility. Sperm motility develops gradually during passage of the sperm through the epididymis. Sperm which is artificially retained in the rete testis also gradually achieves motility but still lacks fertilising capacity. Sperm motility is required for the penetration of the egg investments, but not for transport through the female tract (Gwatkin, 1976*). Shapiro & Eddy (1980*) consider the main function of the male reproductive tract to be the lowering of the metabolic rate of the sperm in order to preserve its fertilising capacity (see further Bedford, Cooper & Calvin, 1972* and Bedford & Cooper, 1978*).

Sperm capacitation in the mammalian female genital tract

Gwatkin (1976*) believes that maturation of the spermatozoon in the epididymis involves some kind of coating of the surface membrane with specific antigens, as e.g. Concanavalin A (Con A) binding sites (Gordon, Dandekar & Bartoszewicz, 1974; Yanagimachi, 1981*), which are subsequently removed during capacitation within the female genital tract. Capacitation of mammalian sperm is not species-specific. Capacitation enables the spermatozoon to undergo the acrosome reaction, which is essential for the penetration of the egg investments. (See Chapter 6, p. 38.)

During capacitation, sperm motility increases. Since cyclic nucleotides and phosphodiesterase inhibitors can enhance sperm motility, Hoskins & Casillas (1975*) suggest that cyclic nucleotides activate a cAMP-dependent protein kinase, which in turn activates a Ca^{2+}-controlled motility-regulating protein. Cyclic nucleotides also regulate the metabolic activity of spermatozoa (see Gwatkin, 1976*). During capacitation, alterations also occur in the distribution of intramembranous particles (Fawcett, 1975*), and in the number and kind of receptor sites in the plasma membrane overlying the acrosomal vesicles; these changes enable the spermatozoon to undergo the acrosome reaction when meeting the egg investments at fertilisation.

Evaluation

In the development of the male gamete the process of meiosis may be controlled by the transfer of the spermatogonia from the basal to the adluminal compartment of the seminiferous tubule. In the latter compartment, spermiation occurs in intimate association with the Sertoli cells. The nature of the cell interactions during transfer of the germ cells from one compartment to the other and their subsequent association with the Sertoli cells is not known, however.

In chordates with external fertilisation sperm maturation may involve some form of inactivation in order to preserve fertilising capacity, while the release of the sperm into the external environment may activate it by dilution or active removal of inhibitory factors. In meta- and eutherian mammals the spermatozoa undergo a typical maturation process during their passage through the epididymis, where sperm motility gradually develops and special surface membrane properties are acquired. During the subsequent capacitation of the sperm in the female genital tract, sperm motility is further enhanced and some final cell surface changes take place which are essential for sperm penetration of the egg investments at fertilisation. All these changes form part of the cellular differentiation process of the male gamete.

6

'Fertilisation' as the interaction of sperm and egg

Some introductory remarks

Sperm–egg fusion was described for the first time by Van Beneden in 1875 in the rabbit and by Hertwig in 1876–8 in the sea urchin. The process of sperm–egg fusion has been most extensively studied in the sea urchins and some other invertebrate groups, groups in which rather large quantities of sperm and eggs can easily be obtained for biochemical analysis. Fertilisation has also been thoroughly investigated in mammals, particularly in domestic species. Little is known about fertilisation in other vertebrates, however. We must therefore chiefly rely on sea urchins and mammals as our sources of information. (See the more general reviews by Løvtrup, 1965*; Monroy, 1965*; Gwatkin, 1976*; Austin, 1978*; Bedford & Cooper, 1978*; Epel, 1978*; Epel & Vacquier, 1978*; Shapiro & Eddy, 1980*; Shapiro, 1981*; Yanagimachi, 1978*, 1981*; Burgess & Schroeder, 1979*; Brachet, 1980*; Shapiro, Schackmann & Gabel, 1981*; and Vacquier, 1981*).

'Fertilisation' is a complex process characterised by, among other things, a number of successive membrane fusion phenomena resulting from cell–cell interaction (Bedford & Cooper, 1978*; Epel & Vacquier, 1978*). The first membrane fusion reaction is the 'acrosome reaction'; the second reaction occurs between the newly formed fusible membrane area of the spermatozoon and the egg plasmalemma in the actual sperm–egg interaction process; it leads to sperm–egg fusion and may be called 'insemination'; the third reaction takes place between the membranes of the individual cortical granules and the egg plasmalemma and is called the 'cortical reaction'; the fourth reaction, which is not always a typical membrane fusion (see below, p. 47), concerns the fusion of the male and female pronuclei ('syngamy'), which may be called the actual 'fertilisation process'.

The acrosome reaction represents the preparatory phase of the fertilisation process. Sperm–egg fusion leads to a rapid depolarisation of the egg plasma membrane constituting a rapid, but incomplete, block to polyspermy; this is followed by the much slower cortical granule exocytosis

37

with concomitant alterations of the egg surface and egg investments, constituting a slow but complete block to polyspermy. Finally, the metabolic activation of the egg occurs with the onset of mitosis.

The acrosome reaction

The acrosome reaction is a very common phenomenon in the animal kingdom, with the sole exception of some invertebrate groups where fertilisation of the 'naked' egg occurs by sperm lacking an acrosome. All animals with eggs surrounded by egg investments have spermatozoa with acrosomes, which contain enzymes capable of digesting a passage through these investments (Austin, 1978*).

The acrosome reaction serves at least two purposes: it exposes or prepares a specialised fusible surface on the head of the spermatozoon, and it leads to the release of the contents of the acrosomal vesicle. The latter contains several lysins instrumental in the penetration of the egg investments. In mammals it houses lysins that prepare the sperm plasma membrane. Moreover, it contains binding moieties for the proper attachment of the sperm to the egg investments and subsequently to the egg membrane itself.

The acrosome reaction usually comprises a fusion of the plasma membrane of the sperm head with the outer membrane area of the acrosomal vesicle and their subsequent breakdown, leading to the extrusion of the content of the acrosomal vesicle and the exposure of the inner acrosomal membrane area. In the mammalian spermatozoon a vesiculation of the fusing membranes occurs, so that the acrosomal content can escape through the pores thus formed without complete destruction of the sperm plasma membrane. A fusible surface is subsequently formed posterior to the acrosome.

The acrosome reaction is released when the spermatozoon approaches or actually makes contact with the outer egg investments, i.e. the jelly layers in the great majority of invertebrate and lower vertebrate eggs, or the granulosa cells, with their gelatinous matrix, of the cumulus oophorus of the mammalian egg. Artificial removal of the cumulus shows that the surface of the zona pellucida can also release the acrosome reaction (see Yanagimachi, 1981*). Lopo & Vacquier (1981*) have recently isolated a 84 K glycoprotein from sea urchin sperm plasma membrane preparations which may play a role in the induction of the acrosome reaction. There is apparently only a limited species-specificity in the triggering of the acrosome reaction (Metz, 1978*).

The first detectable event in the acrosome reaction is an alteration of the Ca^{2+} permeability of the plasma membrane of the sperm head, leading to an influx of Ca^{2+} ions and an H^+ efflux (Collins, 1976; Collins & Epel, 1977; Epel, 1978*). In the presence of Ca^{2+} in the outer medium, the acrosome reaction can also be induced by the ionophore A 23187, which facilitates bivalent ion transport across biomembranes. A high Ca^{2+} concentration in the outer medium can do the same, as does an elevation of the pH of the external medium, which in the presence of extracellular Ca^{2+} leads to an H^+ efflux. The Ca^{2+} influx as the primary trigger may secondarily induce a cytoplasmic pH change, which results in metabolic activation of the spermatozoon, increasing its motility and respiration (see Epel & Vacquier, 1978*; Shapiro & Eddy, 1980*). The acrosome reaction is accompanied by a net K^+ efflux. External Na^+ is an absolute requirement for Na^+ uptake and H^+ efflux (see Shapiro & Eddy, 1980*; compare egg activation, p. 45).

It is interesting to note that membrane fusion during exocytosis of secretory granules in somatic cells also requires a Ca^{2+} influx. Here membrane fusion can also be induced by ionophores (see Poste & Allison, 1973*).

The acrosome reaction releases the content of the acrosomal vesicle or granule, which contains several lysins and binding moieties. The lysins comprise, among other things, hyaluronidase and a trypsin-like protease called 'acrosin'. In the mammals the former reacts with the hyaluronic acid of the matrix of the cumulus oophorus, while the latter seems to play a role in the perforation of the zona pellucida (see Yanagimachi, 1981*).

Motility and respiration of the spermatozoon are markedly increased as a consequence of the acrosome reaction. This results in a marked increase in both Ca^{2+} and cAMP level, which may both act as intracellular messengers (see Shapiro & Eddy, 1980*). According to Yanagimachi (1981*) the spermatozoon may actually function as a localised 'catalytic drill' by means of its vigorous flagellar movements. As soon as the sperm head makes contact with the egg plasma membrane, sperm motility ceases rapidly (Gwatkin, 1976*; Epel, Cross & Epel, 1977). Sperm motility is apparently not required for the engulfment of the sperm into the egg cytoplasm (see below under sperm–egg fusion, p. 40).

Since cells are usually non-fusible, sperm–egg fusion requires the formation of specialised, fusible surfaces on both gametes. As we have seen above, the acrosome reaction leads to the formation of such a specialised fusible surface on the sperm head. In sperm lacking an acrosome, as found in some invertebrate groups, a fusible surface may be formed by the interaction of a particular area of the sperm plasma membrane with diffusible substances emanating from the egg (see Austin, 1978*). The mature oocyte seems to contain large fusible surface areas (see further under sperm–egg fusion, p. 40).

Sperm–egg attachment

Sperm–egg interaction shows a high species-specificity which is mediated by species–specific receptors on the complementary gametes. In some invertebrate groups, among others in certain urochordates, receptor molecules are released by the egg (e.g. fertilisin), which act on the sperm at a distance as a chemotactic and acrosome-reaction-initiating agent (J. C. Dan, 1967; Miller, 1983*). For both actions, Ca^{2+} is an absolute requirement. In vertebrates specific receptors bind the sperm to the egg investments and to the egg plasma membrane (see Metz, 1978*). In the sea urchin, Aketa (1973) found species-specific sperm-binding receptors on the vitelline membrane, which could be blocked by the lectin Con A or could be removed by trypsin treatment, indicating their glycoprotein nature. Denuded sea urchin eggs can also be fertilised, though less readily, which indicates that there are also sperm-binding receptors at the level of the egg plasmalemma. These are however less specific, since cross fertilisation can easily occur under these conditions (see Metz, 1978*). Vacquier & Moy (1980) isolated a species-specific protein from sea urchin sperm, called 'bindin', which mediates in specific sperm–egg adhesion, while Glabe & Vacquier (1977) could isolate a species-specific bindin receptor glycoprotein from sea urchin eggs (see also Lopo & Vacquier, 1981*). In mammals the species specificity of the sperm–egg interaction is predominantly localised in the zona pellucida, the outer surface of which contains specific sperm receptor sites (see Yanagimachi, 1978*). It is the penetration of the zona pellucida rather than the adherence to the zona that seems to be species-specific (see Bedford & Cooper, 1978*). According to Yanagimachi a 'bindin'-like substance with high species-specificity may also be involved in the attachment of the spermatozoon to the zona pellucida of the mammalian egg.

In mammalian spermatozoa which form an acrosomal filament as a result of the acrosome reaction, a polymerisation of actin occurs. Actin is present in inactive form in the equatorial region of the sperm head (see Poste & Allison, 1973*). Unlike other microfilament systems this polymerisation is not inhibited by cytochalasin B (Sanger & Sanger, 1979*).

Sperm–egg fusion

Sperm–egg fusion is a rapid process which is followed by the much slower engulfment of the sperm into the egg (see Epel & Vacquier, 1978*).

In 'naked' eggs of some invertebrates, fusion of sperm and egg can occur over the entire egg surface. The same seems to hold for eutherian mammals, where the spermatozoon may fuse with any region of the egg except that overlying the mitotic spindle, an area which is devoid of microvilli. In the majority of amphibian eggs, which have penetrable egg capsules, sperm entrance takes place almost exclusively in the animal hemisphere (Elinson, 1975; Charbonneau & Picheral, 1983). In the anuran *Discoglossus*, sperm penetration is restricted to the animal pole (see Elinson, 1980*). In eggs with an impenetrable chorion, as in many fish, the spermatozoon can only reach the egg through the micropyle. Fish eggs without chorion, however, can be fertilised anywhere (T. Yamamoto, 1961*), so that the micropyle only directs the sperm's approach. In the teleost *Plecoglossus altivalis* a fertilisation cone is formed at the sperm entry point (Kudo, 1983). In the tunicates and in *Branchiostoma* the sperm penetrates at the vegetal pole of the egg (Conklin, 1905*a*, *b* and Conklin, 1932; Monroy *et al.*, 1973, respectively).

Austin (1968*) suggested that the microvilli play a role in sperm–egg fusion. Epel & Vacquier (1978*) believe that a part of the egg surface, that has a specific microvillous pattern, may function as the site for sperm–egg fusion. Bedford & Cooper (1978*), however, think that in mammalian eggs the intervillous regions of the oolemma possess a special capacity to fuse with the spermatozoon, while the microvilli subsequently enfold the spermatozoon, leading to its incorporation into the egg cytoplasm. In amphibians the sperm entrance point is initially marked by a local absence of microvilli but later by long microvilli (Elinson & Manes, 1978).

Y. M. Takahashi & Sugiyama (1973) found that, in contrast to the acrosomal reaction, external Ca^{2+} is not required for sperm–egg fusion, possibly because the released intracellular stores of Ca^{2+} in egg and sperm are sufficient for the fusion process. Mg^{2+} plays a specific role in binding and fusion of egg and sperm (see Epel, 1978*; Shapiro & Eddy, 1980*).

The uptake of the spermatozoon into the egg involves the mobilisation of actin in the cortical layer of the egg as well as in the microvilli. Sperm incorporation is prevented by cytochalasin B (Schatten & Mazia, 1977*). The actin of the spermatozoon may also play a role in pulling the sperm inwards by means of bundles of actin filaments. A cluster of microtubules forms around the area of fusion, possibly guiding the sperm inside.

Egg membrane depolarisation

In the sea urchin the fusion of sperm and egg results in a rapid electrical depolarisation of the egg plasma membrane, the potential changing from -30 mV to $+20$ mV. This depolarisation lasts about 1 min. A depolarisation to at least $+5$ mV is required to prevent polyspermy (L. A. Jaffe, 1976). There is also a dramatic change in membrane resistance (L. A. Jaffe & Robinson, 1978). Membrane depolarisation acts as a fast block to polyspermy, which is either due to a change in electrically coupled receptors or to the formation of an electrostatic barrier (see also Austin, 1978*; Epel & Vacquier, 1978*).

A similar depolarisation occurs in anuran amphibian eggs, from the resting potential of the unfertilised egg of -19 mV in *Xenopus laevis* and -28 mV in *Rana pipiens* to $+8$ mV, the so-called fertilisation potential (Cross & Elinson, 1980; Grey *et al.*, 1982). The potential remains positive for about 15 min (Grey *et al.*, 1982). It results from a net Cl^- efflux, since it is reduced by raising the Cl^- concentration in the external medium (Ito, 1972; Cross & Elinson, 1978). A reduced fertilisation potential causes polyspermy. In marine animals the action potential is dependent on an increased Na^+ permeability (Na^+ influx) (Shapiro, 1981*). The ionophore A 23187 also produces membrane depolarisation, but the reaction is slower (see Shapiro & Eddy, 1980*). Pricking of immature *Rana pipiens* oocytes produces a slow depolarisation of smaller amplitude (Schlichter & Elinson, 1981). Membrane depolarisation furnishes a block to polyspermy that is rapid but incomplete, since it can be overruled by a high sperm/egg ratio (see Epel & Vacquier, 1978*).

Cortical granule exocytosis

Sperm entry releases cortical granule exocytosis in many invertebrate and vertebrate eggs. However, this does not seem to be a universal phenomenon, since it does not occur e.g. in the polyspermic urodele egg (Løvtrup, 1965*a*; see further on p. 44). In anuran eggs the so-called cortical reaction starts at the sperm entrance point and spreads as a wave over the egg surface towards the opposite side (Picheral & Charbonneau, 1982). It is accompanied by a local stiffening of the cortical layer of the egg (Hiramoto, 1974). Hara & Tydeman (1979) describe a contraction wave, called 'activation wave', which travels over the surface of the *Xenopus* egg in 2–3 min at a speed of about 10 μm s^{-1}. In the larger *Rana pipiens* egg the speed is 20–30 μm s^{-1} (see Elinson, 1980*). In immature anuran oocytes the capacity for cortical granule exocytosis in response to sperm penetration does not appear until the initiation of the first meiotic division (L. D. Smith & Ecker, 1969; Katagiri & Moriya, 1976). In the zebra fish, *Brachydanio rerio*, two types of cortical granule are found, arranged in irregular rows beneath the egg plasmalemma. Exocytosis starts 30 s after sperm addition and is completed about 5 min later (Hart & Yu, 1980).

Cortical granule exocytosis leads to a release of intracellular Ca^{2+} and an exchange with the environment as a result of an increase in membrane permeability. In sea urchins, tunicates, amphibians and mammals, the ionophore A 23187 can also trigger the cortical reaction independent of the presence of extracellular Ca^{2+}, so that intracellular Ca^{2+} release must play an important role (see Shapiro & Eddy, 1980*). This can be clearly demonstrated with the aid of the Ca^{2+}-sensitive luminescent phosphoprotein 'aequorin'. In the medaka fish a 'firing-off' of cortical granules spreads progressively from the micropyle, where the sperm has entered, to the opposite side of the egg. Injected aequorin shows that a wave of Ca^{2+} release precedes the cortical reaction. The wave is followed by a rapid removal of free Ca^{2+} to compartments inaccessible to aequorin (Ridgway, Gilkey & Jaffe, 1977). In the sea urchin, where similar phenomena occur, evidence is found for the presence of a Ca^{2+}-binding modulator protein called 'calmodulin' (Head, Mader & Kaminer, 1979). Shapiro & Eddy (1980*) believe that a calmodulin–Ca^{2+} complex may be involved both in triggering the cortical reaction and in its propagation over the egg surface after the release of Ca^{2+} from the ooplasmic reticulum.

In anuran amphibians the cortical reaction, which is normally released by the sperm, can also be evoked by many different artificial means and agents such as pricking, hypotonic shock, electrical shock, detergents, ionophores, etc., demonstrating the non-specificity of the trigger. Under these conditions the cortical reaction starts simultaneously at several sites (Charbonneau & Picheral, 1983). Since injected Ca^{2+} triggers the cortical reaction in unfertilised amphibian eggs (Hollinger & Schuetz, 1976) the sperm may actually act as a 'Ca^{2+} bomb' discharging the cortical granules in the immediate vicinity of the sperm entrance point (Shapiro & Eddy, 1980*). The exocytosis of the cortical granules is probably a self-propagating reaction, the release of Ca^{2+} by the discharging cortical granules initiating the reaction in adjacent ones. Campanella & Andreuccetti (1977) found in *Discoglossus pictus* an extensive system of membranous cisternae and ER in close association with the cortical granules, which may play a role in conducting the stimulus from one cortical granule to the next. In invertebrates the cortical reaction can be prevented by procaine, which occupies Ca^{2+} binding sites on biomembranes, and by La^{3+}, which has a high affinity for Ca^{2+} binding (see Epel & Vacquier, 1978*).

The cortical reaction leads to the release of the content of the cortical granules into the virtual space between the egg plasmalemma and the vitelline membrane. In the sea urchins the released cortical granules contain several proteases; one which cleaves bonds between the egg plasmalemma and the vitelline layer and another one which destroys the 'bindin' receptor sites on the outside of the vitelline layer (Carroll & Epel, 1975), thus preventing other spermatozoa from adhering to the latter.

The cortical reaction in sea urchins also involves a hardening or 'tanning' of the vitelline layer, changing it into the impenetrable fertilisation membrane (see Giudice, 1973*). According to Markman (1958), Ca^{2+} is required for the formation of the fertilisation membrane, while a peroxidase may play a role in the hardening process (see below, p. 45).

In the anuran amphibia, where the cortical reaction is completed in *c.* 8–9 min after sperm addition, a lectin released by the cortical granules passes through the vitelline membrane and forms the fertilisation envelope at the boundary of the vitelline membrane and innermost jelly layer

(Wyrick, Nishihara & Hedrick, 1974; Grey, Wolf & Hedrick, 1974). The formation of the fertilisation envelope constitutes the slow but complete block to polyspermy (Wolf et al., 1976).

Ginzburg (1971) found that in the urodele, *Ambystoma mexicanum*, cortical granule exocytosis does take place, but already during the passage of the egg through the oviduct, resulting in a very ineffective block to polyspermy. This nevertheless indicates that sperm–egg interaction in monospermic anuran and polyspermic urodele eggs may be essentially similar in nature. Polyploidy is prevented in the polyspermic urodele egg by the fact that the female pronucleus exerts a stimulating influence upon the nearest sperm nucleus. The two nuclei moreover show mutual attraction. As soon as the pronuclei have fused and a zygote nucleus is formed, the latter exerts an inhibitory influence on accessory sperm nuclei, affecting first the nearest and later the more distant ones (Fankhauser, 1925, 1932, 1948*).

In the majority of mammalian eggs a trypsin-like protease is released by the cortical granules, which alters or removes sperm receptor sites at the periphery of the zona pellucida. Another factor may be responsible for the hardening of the zona pellucida, making it impenetrable to other spermatozoa. This process is called the 'zona reaction' (see Gwatkin, 1976*; Wolf, 1981*). It usually takes 15 min or less, but the complete plasma membrane block to polyspermy requires 2–3.5 h (see Gwatkin, 1976*). Polyspermy is however normally prevented, since only a few spermatozoa establish contact with the zona pellucida *in vivo*. In fertilisation experiments *in vitro*, with a much higher sperm/egg ratio, a significant degree of polyspermy is observed (Austin, 1969*). The egg plasma membrane apparently also undergoes changes which interfere with further sperm penetration since, in the rabbit (where no zona reaction occurs upon cortical granule discharge), the plasma membrane nevertheless loses its capacity to bind spermatozoa.

Although mammalian eggs are usually monospermic, the pig egg is often polyspermic, showing an incomplete block to polyspermy. Here polyploidy is possibly prevented by a suppressing mechanism similar to that described by Fankhauser for the urodele egg (Hunter, 1976). In polyspermic mouse eggs, obtained by fertilisation of zona-free eggs with capacitated sperm in modified Krebs–Ringer solution, the majority of supernumerary sperm nuclei are eliminated by abstriction into cytoplasmic blebs during or slightly after second polar body formation, as well as by an unknown suppression mechanism (Yu & Wolf, 1981).

In the sea urchin cortical granule exocytosis leads to a two-fold increase in membrane surface area, and must be accompanied by a temporary but profound disorganisation of the egg cortex. Some of this extra surface membrane, derived from the cortical granules, is taken up in the elongation of the microvilli (Eddy & Shapiro, 1976; Charbonneau & Picheral, 1983).

In *Xenopus* hardly any change in the microvillous pattern is to be seen after fertilisation, so that here the extra membrane must be rapidly internalised (J. G. Bluemink, personal communication).

In the zebra fish, *Brachydanio rerio*, the excess membrane derived from the cortical vacuoles also seems to be rapidly removed by endocytosis (Hart & Yu, 1980).

The cortical reaction causes a separation of the vitelline layer from the egg plasmalemma and the formation of a subvitelline space, which enables the egg to rotate inside its investing membranes under the influence of gravity (see Chapter 7, p. 58).

Egg activation

Fertilisation or artificial (parthenogenetic) activation of the egg switches on metabolic processes required for the reinitiation of meiosis and the extrusion of the second polar body, as well as for subsequent mitosis. This process is called 'egg activation' (Masui & Markert, 1971). Amphibian oocytes can be precociously activated by the divalent ionophore A 23187, the onset of activation being related to the time of germinal vesicle breakdown (Belanger & Schuetz, 1975).

After both fertilisation and parthenogenetic activation of sea urchin eggs a stimulation of respiration occurs (Ohnishi & Sugiyama, 1963), demonstrating that the latter is a consequence of activation (see above, p. 43). However, two-thirds of the oxygen uptake during the first 15 min after fertilisation seems to be due to the synthesis of hydrogen peroxide. The latter substance is not only toxic to sperm, but also plays a role in the hardening of the fertilisation membrane and therefore acts as an additional block to polyspermy (Shapiro, 1981*).

Egg activation causes new DNA synthesis. The amphibian egg contains a great excess of DNA nucleotides, equivalent to the total DNA content of 5000–10 000 cells. The greatest part of this by far represents mitochondrial DNA, amounting to two-thirds of the extranuclear DNA (Dawid, 1966). A certain amount of DNA or DNA precursors is moreover present in the yolk platelets, but this only becomes available during neurulation (Denis, 1974*), so that the actual pool of DNA precursors available for rapid DNA replication is much more restricted than was originally thought (Brachet, 1965*). Egg activation is characterised by the onset of a nearly complete preponderance of DNA replication over other nucleic acid biosyntheses (Legros & Brachet, 1965).

Activation of the *Pleurodeles* egg leads to a dramatic increase in protein synthesis. This is clearly demonstrated by the use of puromycin or cycloheximide, which strongly inhibit protein synthesis and block cleavage. Actinomycin D, which blocks RNA transcription, does not affect cleavage, demonstrating that the egg contains the necessary mRNAs for protein synthesis in the form of 'masked' mRNA stored in the oocyte during oogenesis (Legros & Brachet, 1965).

Egg activation gives rise to a marked decrease in macroviscosity rendering the cytoplasm of the amphibian egg more susceptible to gravity (see further Chapter 7, p. 58). Fertilisation is also accompanied by changes in water permeability (Løvtrup, 1962).

In sea urchins the exchange of Ca^{2+} with the external medium following upon the cortical reaction leads to the activation of a Na^+/H^+ counter transport, with a resulting increase in intracellular pH. The Na^+ requirement for egg activation (Chambers, 1975, 1976) and the Na^+/H^+ exchange reaction at fertilisation (J. D. Johnson, Epel & Paul, 1976) strongly suggest that the H^+ efflux is coupled with Na^+ uptake and with an increase in intracellular pH. The latter seems to be a prerequisite for the activation of various synthetic processes and transport changes, including DNA replication. Treatment of the egg with ammonia also causes activation of the egg metabolism by increasing the intracellular pH (Steinhardt & Mazia, 1973; Epel et al., 1974; Epel, 1978*). Contrary to Epel (1978*), Shapiro & Eddy (1980*) believe that in sea urchins the increase in intracellular pH is the necessary and sufficient prerequisite for the metabolic activation of many fertilisation-related events.

The causal link between the intracellular pH change and the stimulation of protein and DNA synthesis is still unclear. Increased intracellular pH may affect protein phosphorylation and may thus influence cellular metabolism, but other explanations are possible (Shapiro, 1981*). Benbow & Ford (1975) found that cleavage stages of *Xenopus* eggs contain an initiation factor for DNA synthesis. It seems to be a protein, which increases in amount during maturation and tends to disappear during later stages of development. Løvtrup (1965 a*) and Grant & Youngdahl (1974) suggested that the grey crescent region of the fertilised amphibian egg may contain a factor which stimulates DNA synthesis. This suggestion should be further substantiated, however.

Gilkey (1981), studying fertilisation in fish, found that sperm–egg interaction leads to a transient increase in free Ca^{2+} in the egg cytoplasm. Since there is no detectable change in intracellular pH in the fish egg, Gilkey believes that the increase in free Ca^{2+} ions must be responsible for removing the egg's developmental block and turning on its synthetic activity.

Masui & Markert (1971) noted that maturing *Rana pipiens* oocytes and unfertilised eggs contain a powerful metaphase-arresting agent, which they called 'cytostatic factor' (see Chapter 4, p. 30). Re-establishment of meiosis requires the destruction of this factor. According to Wasserman & Smith (1978*) the factor disappears within 15 min of fertilisation and reappears shortly before first cleavage. It is inactivated by Ca^{2+} addition, thus clearing the way for the re-establishment of the mitotic interphase (Masui & Clarke, 1979*).

Chromatin decondensation and the fusion of the pronuclei

Sperm penetration into the egg cytoplasm leads to a rapid hydration and swelling of the condensed chromatin of the sperm nucleus, preceded by the breakdown of the nuclear envelope.

In mammals a disruption of S—S bonds in the chromatin seems to be involved in this decondensation process (Bedford, Cooper & Calvin, 1972). A cytoplasmic factor responsible for the decondensation of the sperm chromatin begins to appear in the mammalian oocyte about the time of germinal vesicle breakdown, increases with the progression of maturation, and only diminishes after fertilisation (see Yanagimachi, 1981*). It reappears during cleavage, probably representing a cyclic phenomenon associated with mitosis. Activation of egg metabolism does not seem to be a prerequisite for decondensation of the sperm nucleus but is certainly required for the transformation of the decondensed nucleus into the actual sperm pronucleus (Yanagimachi, 1981*).

In the rabbit, 4–5 h after sperm penetration the male and female pronuclei become visible as a result of the coalescence of perinuclear vesicles (Longo, 1973*). Activation of the egg is required for the rearrangement of the egg cytoskeleton, while the expanding sperm aster guides male and female pronuclei to their site of union (see Yanagimachi, 1981*). In the majority of vertebrate species the zygote nucleus is formed by the actual fusion of the two pronuclear envelopes following the disappearance of the apposed membrane areas. In mammals the nuclear envelopes of the two pronuclei first undergo vesiculation and complete breakdown after which their chromatin contents fuse (Longo, 1973*; see further Bedford & Cooper, 1978*).

Evaluation

Fertilisation represents an interaction between two different cells, which on the one hand have differentiated in quite separate directions – the highly mobile spermatozoon and the nearly immobile, particularly receptive mature oocyte – yet on the other hand must have important features in common, particularly in their membrane composition, since the two cells do not only recognise each other but are able to fuse into a single cell.

The interaction between the fully active spermatozoon and the egg investments and egg surface membrane, or both, releases the acrosome reaction in the sperm head, as a result of which a specialised fusible surface is formed and enzymes are set free that are needed for the penetration of the egg investments. The subsequent, direct interaction between the sperm and the egg surface, which ultimately leads to the rather exceptional phenomenon of both cellular and nuclear fusion, evokes a series of reactions in the egg. These comprise a rapid electrical depolarisation of the egg membrane and the subsequent exocytosis of the cortical granules, which constitute a rapid but incomplete, and a slow but complete block to polyspermy, respectively. These reactions, which are peculiar to sperm–egg interaction, guarantee that only a single sperm nucleus will be able to fuse with the female pronucleus. The latter is formed after the resumption of meiosis as the consequence of another feature of sperm–egg interaction, the activation of the metabolically inert egg through a deblocking mechanism.

Sperm–egg fusion brings the sperm nucleus under the direct influence of the egg cytoplasm. This causes a rapid decondensation of the highly condensed chromatin of the sperm nucleus. The accompanying centrosome gives rise to the sperm aster, which then plays an important role in bringing together the two pronuclei and in their ultimate fusion into the zygote nucleus. This constitutes the last step in the formation of the diploid egg, which will now develop autonomously into a new individual.

Sperm–egg interaction has one very specific feature, its high species specificity. The latter resides in specific, either identical or complementary receptor sites on the fusible membrane of the sperm head and on the egg investments and egg plasmalemma, or both. This specificity is essential to prevent hybridisation which usually leads to abnormal or abortive development.

Shapiro & Eddy (1980*) state that gamete interaction mechanisms, although particular with regard to the ultimate fusion of the cells concerned, strongly resemble mechanisms common to all cells. The acrosome reaction of the sperm and the cortical granule exocytosis of the fertilised egg, which show great similarity (see Austin, 1978* and Epel & Vacquier, 1978*), also have much in common with normal exocytosis of cellular products, one of the most general features of cellular activity. Sperm–egg fusion resembles the fusion of myoblasts into multinuclear myofibrils, except that in the latter neither nuclear fusion nor DNA replication occur.

The process of meiosis which, for the greater part, takes place at the beginning of oogenesis (but which is completed after sperm–egg interaction) has many features in common with the process of mitosis – the second most common process of cellular activity. Meiosis and mitosis seem to be regulated by the same or related cytoplasmic factors. The decondensed sperm nucleus matches the female pronucleus; the egg cytoplasm may be responsible for this.

We may therefore suggest that the special nature of the fertilisation process lies more in the particular combination of common cellular processes than in the specific nature of the processes involved. Fertilisation may thus turn out to be one of the most illuminating phenomena for our understanding of cellular interaction during embryonic development. Epel (1978*) indeed believes that the activation of development at fertilisation may provide clues to the triggering of specific developmental programmes in later development, particularly in induction and determination. All the various steps in the complicated process of fertilisation seem to be based on common cellular mechanisms.

7

Egg symmetrisation or the origin of dorso-ventral polarity

Symmetrisation in the telolecithal amphibian egg

Where does the animal–vegetal polarity reside?

Immediately after fertilisation the amphibian egg shows radial symmetry about the animal–vegetal polar axis. The egg consists essentially of two different moieties; a pigmented animal one with fine yolk granules and relatively abundant cytoplasm, and a nearly unpigmented vegetal one which is very rich in coarse yolk granules and rather poor in cytoplasm. The animal–vegetal polarity is also expressed in differences in membrane configuration and membrane fluidity. Monroy & Baccetti (1975) observed characteristic differences in surface protrusions between the animal and vegetal hemisphere of the unfertilised egg of *Xenopus laevis*. Bluemink & Tertoolen (1978) reported an animal–vegetal difference in the distribution of intramembranous particles in the egg plasmalemma, while Dictus *et al.* (1984) found pronounced differences in membrane fluidity of the two egg moieties, with a sharp boundary between the pigmented and unpigmented domains.

Pasteels (1938, 1939) showed that complete inversion of *Rana* eggs under certain conditions led to complete reversal of the animal–vegetal polarity as a result of rearrangement of the internal cytoplasm with respect to the pigmented cortex. Recent reversal experiments by Malacinski & Chung (1981) and H.-M. Chung & Malacinski (1983) confirmed Pasteels' observations. In their most successful cases there was a complete reversal of the pigment pattern, the cleavage pattern and the site and direction of invagination of the dorsal blastoporal lip. *Xenopus laevis* eggs which were rotated in two steps 90° + 90° and kept in this position at a low temperature showed nearly complete reversal of the cytoplasmic content with respect to the pigmented and unpigmented cortex domains. In these eggs, which could develop normally, the animal–vegetal polarity was nearly reversed (Cleine, Boorman & Dixon, 1982; Cleine, 1983). These experiments demonstrate that the spatial configuration of the internal cytoplasm rather than the structure of the cortex is responsible for the establishment of animal–vegetal polarity. In inverted eggs the membrane properties of the two egg moieties seem to be secondarily altered under the influence of the displaced internal cytoplasms (W. J. A. G. Dictus & J. G. Bluemink, personal communication). The primary role of the cytoplasm in animal–vegetal polarity is also supported by the recent observation of Capco & Jeffery (1981) that, after injection of regionally specific (vegetal) poly(A)$^+$ RNA into fertilised *Xenopus* eggs, this RNA accumulates strongly in the vegetal hemisphere of the egg in the period between fertilisation and first cleavage. Although the yolk may play a role in animal–vegetal axis formation, the properties of the different cytoplasms of the two egg moieties primarily determine this polarity (see also Gerhart *et al.*, 1983*).

The externally visible symmetrisation and its significance

Some hours after fertilisation an area of intermediate pigmentation, called the 'grey crescent', is formed on one side of the anuran egg at the boundary of the pigmented and unpigmented moieties, as the first clear manifestation of dorso-ventral polarity. In the more lightly pigmented urodele egg, an equivalent light crescent can be distinguished. The strict topographical relationship between the grey crescent and the site of the future dorsal blastoporal lip, from which the dorsal axial organs of the embryo will develop, is the most significant aspect of grey crescent formation. This strict relationship has led many investigators to attribute special significance to the grey crescent in dorso-ventral polarisation and axis formation (see Dalcq & Pasteels, 1937, 1938; Ancel & Vintemberger, 1948*; Pasteels, 1964*; Løvtrup, 1965*a*, *b**; Brachet, 1977*; Gerhart, 1980*; Vacquier, 1981*; and Gerhart *et al.*, 1983*).

Sperm entrance point and symmetrisation

Grey crescent formation has been studied particularly in the rather darkly pigmented eggs of anurans, which have the great advantage of being monospermic, so that the relationship between sperm entrance point and grey crescent formation can be analysed. Urodele eggs, which are usually less heavily pigmented and are moreover polyspermic, are obviously less suitable for such an analysis. In many anuran species a distinct correlation has been found between the location of the sperm entrance point in the animal hemisphere of the egg and the appearance of the grey crescent on the opposite side of the egg (Ancel & Vintemberger, 1948*; Elinson, 1975 and many others). The correlation is good but not perfect (Kirschner *et al.*, 1980), a fact to which we shall return later (see p. 56) (see also Pasteels, 1964*; Brachet, 1977*; and Elinson, 1980*).

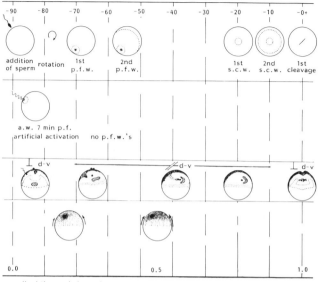

Fig. 5. Diagrammatic representation of surface phenomena in anuran eggs, after K. Hara *et al.* (1977) and K. Hara & P. Tydeman (1979) (first and second row); of cytoplasmic segregation, after G. A. Ubbels & R. T. M. Hengst (1973) (third row); and of grey crescent formation, after J. Paleček, G. A. Ubbels & K. Rzehak (1978) (last row). Upper scale: time before first cleavage in minutes; lower scale: normalised time scale in tenths of fertilisation/first cleavage interval (slightly modified after G. A. Ubbels, 1977), p.f.w., post-fertilisation waves; s.c.w., surface contraction waves; a.w., activation wave; $\perp$d–v, perpendicular to dorso/ventral axis ('transverse' section); //d–v, parallel to dorso/ventral axis ('sagittal' section).

Insemination and the activation wave

Elinson (1975) showed by means of localised insemination that in *Rana pipiens* the sperm may penetrate anywhere in the animal hemisphere, but preferentially within 60° of the animal pole. Sperm can also enter the vegetal hemisphere down to 30° below the equator, but it cannot pass the yolk, so that only activation and no fertilisation ensues. In *Xenopus laevis* a similar situation seems to exist (Hara & Tydeman, 1979). As we have already discussed on p. 42, sperm entrance releases the cortical reaction, consisting of a wave-like propagation of cortical granule exocytosis accompanied by a local cortical contraction, the so-called 'activation wave' (Hara & Tydeman, 1979; see also Fig. 5). The two phenomena start at the sperm entrance point, travel around the egg towards the opposite side in about 2 min at a speed of 600 μm min^{-1} (Elinson, 1975; Páleček, Ubbels & Rzehak, 1978; and Hara & Tydeman, 1979) and seem to be Ca^{2+}-dependent (Gingell, 1970). Local application of polycations and ionic detergents to the *Xenopus* egg leads to a local contraction of the cortex (Gingell & Palmer, 1968). This is accompanied by a change in electrostatic surface potential. Artificial activation by pricking with subsequent cortical contraction also requires Ca^{2+} (Wolf, 1974*a*, *b*, *c*). Merriam & Sauterer (1983) and Merriam, Sauterer & Christensen (1983) recently observed the presence of contractile granular material immediately below the surface of the *Xenopus* egg. This so-called subcortical contractile matrix, consisting of two layers, is much thicker in the animal than in the vegetal hemisphere. In contrast to the deeper pigment-containing layer of the matrix, the thin, superficial pigment-free layer is cytochalasin-B-insensitive, so that its contractility is not related to actin microfilaments. Fertilisation triggers a pronounced decrease in cytoplasmic macroviscosity (Merriam, 1971*a*, *b*), making the egg more susceptible to the influence of gravity. There are also changes in the permeability of the membrane for water and for divalent ions (Løvtrup, 1962; Morrill, Kostellow & Murphy, 1971).

Post-fertilisation waves

Both in anuran and urodele eggs Hara, Tydeman & Hengst (1977) observed the propagation of two 'post-fertilisation waves', which in *Xenopus laevis* start at 20 and 25 min post-fertilisation at the sperm entrance point, travel at a speed of 60–70 μm and propagate around the egg in 15–20 min (see Fig. 5). T. Kubota (1967) found that fertilisation leads to a local increase in cortical rigidity spreading from the sperm entrance point concomitant with the growth of the sperm aster. In eggs activated by pricking, an overall change in rigidity occurs with no regional pattern. Kubota suggested that the sperm aster may play a role in dorso-ventral polarisation. Elinson & Manes (1978), who observed a concomitant shortening of the microvilli, believe that the cytoplasm rather than the egg cortex becomes more rigid. Ubbels *et al.* (1983) found that the growth of the sperm aster is spatially correlated with the progression of the post-fertilisation waves, the aster rays reaching the egg surface at the time the waves are seen externally. They consider the sperm centriole as a microtubule-organising centre that structures the animal egg cytoplasm.

The post-fertilisation waves, which represent local surface contractions in which the microfilament system seems to be involved, can also be inhibited by the anti-microtubule drugs vinblastine and colchicine (Kirschner *et al.*, 1980*). This suggests that a local interaction occurs between the microtubular system of the sperm aster and the microfilament meshwork of the egg cortex. The experiments of Manes & Barbieri (1976, 1977), who injected a homogenate of sperm into *Bufo arenarum* eggs and observed grey crescent formation opposite the point of injection, while the injection of Ringer solution had no such effect, point in the same direction.

The authors assume that the active factor in the sperm homogenate is the centriole; this has been confirmed by Maller *et al.* (1976) and by Heidemann & Kirschner (1978) in *Xenopus laevis*.

The yolk-free cytoplasm

As already mentioned, the expansion of the sperm aster may play an important role in the dorso-ventral polarisation of the amphibian egg (T. Kubota, 1967). When the sperm aster enlarges it carries the sperm pronucleus and the centriole towards the centre of the egg, where the former meets the descending female pronucleus at about time 0.5 (see legend of Fig. 5, p. 52). Klag & Ubbels (1975), Ubbels (1977*). Páleček *et al.* (1978) and Ubbels & Hengst (1978) observed that the extension of the sperm aster causes a dorsad displacement of an area of yolk-free cytoplasm which probably originates in the centre of the egg at the time of germinal vesicle breakdown (see Fig. 5, p. 52; see also Brachet, 1977*). Sperm aster expansion is apparently also responsible for the dorsad displacement of the male and female pronuclei after they have come together in the centre of the egg. These displacements take place before the appearance of the grey crescent. Although the dorsal yolk-free cytoplasm shows morphological symptoms of high metabolic activity in the form of abundant ribosomes and small vesicles (Herkovits & Ubbels, 1979) it does not seem to play an important role in dorso-ventral polarisation, since in eggs showing reversal of the dorso-ventral axis upon rotation it is not displaced towards the future dorsal side (Ubbels *et al.*, 1979).

Grey crescent formation

Dalcq & Dollander (1948) noted that the cortical layer of the symmetrised egg is markedly thinner dorsally than ventrally. Balinsky (1966) observed dorso-ventral differences in the microvillous pattern. These observations were confirmed by Hebard & Herold (1967), Ikushima & Maruyama (1971) and Sawai & Yoneda (1974), both in anuran and in urodele eggs. Dollander & Melnotte (1952) and Dollander (1957) found differences in cortical permeability for vital dyes, with a maximum in the thinner grey crescent region, whereas O'Dell *et al.* (1974) did not find any obvious differential staining of the grey crescent region with radioactive lectins, such as Con A and soybean agglutinin.

Grey crescent formation also occurs after artificial activation by puncturing the egg with a needle, by electric or osmotic shock treatment, or by exposure to divalent ionophores. This shows that the release as such of the symmetrisation reaction is rather unspecific. There is no correlation between the insertion point of the needle and the site of the grey crescent, so that here grey crescent formation must be governed by factors other than sperm aster expansion. Cytochalasin B can also induce grey-crescent-like pigment displacements in non-activated eggs (Manes, Elinson & Barbieri, 1978). These observations suggest the presence of a weak, intrinsic asymmetry in the unfertilised egg, which is easily overruled by other factors (see further on p. 56).

Grey crescent formation is prevented by colchicine treatment, which affects the microtubular system. Its inhibiting action is not due to a destruction of the sperm aster, but it may affect the connection between the aster rays and the egg cortex. Cytochalasin B treatment does not prevent grey crescent formation (Manes *et al.*, 1978), but this may be due to the low permeability of the egg plasma membrane for this drug. Two different mechanisms have been proposed for grey crescent formation.

The first was proposed by Ancel & Vintemberger (1948*); they concluded from their observations that in *Rana temporaria* (=*fusca*) the entire pigmented animal cortex shifts about 30° with respect to the egg interior, its future dorsal side moving up and being stretched, while its future ventral side moves down and is compressed. This seems also to hold for *Discoglossus* (Klag & Ubbels, 1975).

The second, forwarded by Løvtrup, (1965a*), Rzehak (1972), Elinson (1975), Ubbels (1977*), Elinson & Manes (1978) and Páleček *et al.* (1978), assumes that in *Xenopus laevis* and in *Rana pipiens* symmetrisation of the egg occurs as a result of an eccentric contraction of the entire pigmented cap around the sperm entrance point, followed by partial relaxation. This results in a local stretching of the cortical layer on the side opposite the sperm entrance point, which is then accentuated during actual grey crescent formation (see Fig. 5, p. 52). Elinson (1975), Elinson & Manes (1978) and Páleček *et al.* (1978) believe that the contraction carries along cytoplasmic material deep to the pigmented cortical layer and may therefore be responsible for the formation of the so-called dorsal vitelline wall (see below). According to Ubbels *et al.* (1983), the spatial interaction between the sperm aster and the cortex defines the asymmetrical contraction of the pigmented cortex, which is then assumed to evoke asymmetry in the vegetal hemisphere through the formation of the dorsal vitelline wall.

These alternative mechanisms seem to operate in different amphibian species. They have been extensively discussed by J. Clavert (1962*), Pasteels (1964*), Løvtrup, (1965a*), Brachet (1977*), Gerhart (1980*) and Malacinski (1984*). The forces involved in the contraction of the pigmented animal cap may, in our opinion, be provided by the contractile microfilament meshwork underlying the plasmalemma, the asymmetrical character of the contraction being due to its initiation at the eccentric sperm entrance point. A shift of the entire pigmented animal cortex with respect to the egg interior, notably the yolk mass, is mechanically far less easy to understand (see also Nieuwkoop, 1977* and Elinson, 1980*). According to the contraction hypothesis, the extent of the displacement of the dorsal boundary of the contracting pigmented cap should be proportional to the eccentricity of the point of sperm entry. However, grey crescent formation does not seem to be affected by this: it is similar regardless of whether insemination occurs near the equator or near the animal pole, a grey crescent even forms after artificial activation.

Precocious symmetrisation and possible presymmetrisation during oogenesis

Grey crescent formation normally starts about halfway between fertilisation and first cleavage, defined as time 0.5 (see legend to Fig. 5, p. 52), but can be evoked precociously by heat-shock treatment. The grey crescent appears immediately after fertilised axolotl eggs have been brought to 37 °C for 10 min, if this is done less than 1.5 h after fertilisation (Beetschen, 1979 a, b). Precocious grey crescent formation can also be induced in non-activated axolotl eggs by cycloheximide (Grinfeld & Beetschen, 1982) and by diphtheria toxin, both being protein synthesis inhibitors (Gautier & Beetschen, 1983). These authors conclude that inhibition of protein synthesis elicits grey crescent formation. Benford & Namenwirth (1974) showed that the precociously induced grey crescent in axolotl eggs corresponds to the site of the future dorsal blastoporal lip and thus constitutes a reliable criterion for dorso-ventral polarisation. These authors concluded that there is some form of weak symmetrisation in unfertilised eggs, which manifests itself after heat treatment.

Although in the majority of amphibian species the oocytes are radially symmetrical, Pasteels (1937c) observed an eccentric position of the pigment cap with respect to the yolk mass in eggs of *Rana esculenta* and *Discoglossus pictus*. After fertilisation the dorso-ventral axis of the embryo always corresponded to the plane of symmetry through pigment cap and yolk mass and showed no relation to the sperm entrance point. Wittek (1952) noted a bilaterally symmetrical configuration of the germinal vesicle, yolk mass and pigment cap in ripe oocytes of several *Rana* and *Triturus* species. Wolff (1969*) also assumed that there is a pre-existing potential bilateral symmetry in the egg, which can be easily overruled by other factors such as sperm entry, gravity, etc. In our opinion this pre-symmetrisation may be due to the long-term action of gravity on ovarian oocytes, particularly on those with less viscous cytoplasm. The long-term influence of gravity may also be responsible for the imperfect relationship between sperm entrance point and grey crescent site (see p. 51).

Suppression of sperm-induced bilateral symmetry

Grey crescent formation, which usually occurs opposite the sperm entrance point, can be affected or overruled by a number of different agents.

(1) It has been known for more than 40 years that ultraviolet (UV) irradiation can affect embryonic development (see Brachet, 1977*). Manes & Elinson (1980) were able to prevent grey crescent formation by UV irradiation from the vegetal side. The effect was dose-dependent. They suggested that UV inhibits the cortical movements involved in grey crescent formation. H.-M. Chung & Malacinski (1980, 1981) and Malacinski & Woo Youn (1981 b) have recently shown that UV irradiation can alter the dorso-ventral polarisation of the egg. Whereas the grey crescent and, later, the dorsal blastoporal lip normally form on the side opposite the sperm entrance point, in UV-irradiated eggs they often develop on the same side. By means of nuclear transplantation Grant & Wacaster (1972) were able to show that UV irradiation damages the cortex but does not affect the nucleus. Using a UV microbeam Grant & Wacaster (1972) could demonstrate that irradiation of the marginal zone of the egg causes stronger morphogenetic disturbances than vegetal pole irradiation. Malacinski, Benford & Chung (1975) found that the grey crescent region is most sensitive to UV treatment, the time of maximal sensitivity being prior to grey crescent formation. Beal & Dixon (1975) and Züst & Dixon (1975), studying the effect of UV on germ cell migration in *Xenopus laevis*, found inhibition of cytokinesis in the irradiated area, probably due to damage to the microfilament system.

(2) Glade, Burrill & Falk (1967) observed that a horizontal temperature gradient of approximately 2 °C, acting upon fertilised *Rana pipiens* eggs which were oriented randomly with respect to the dorso-ventral axis, led to a significant increase in dorsal axis formation in the quadrant that was warmest. This was the case when the heat gradient was applied immediately after fertilisation and a few hours before the beginning of gastrulation. The authors suggest that during the former period of heat sensitivity a new plane of symmetry arises, while a change in metabolism occurs during the latter period, when bilateral symmetry is expressed. The mechanism of this biphasic heat sensitivity is not well understood, however.

(3) Landström & Løvtrup (1975) observed differences in energy consumption along the dorso-ventral axis. They studied the influence of an oxygen gradient on dorsal blastopore formation by rearing eggs in close-fitting glass tubes closed at one end, so that the oxygen supply became restricted on the closed side. Irrespective of the original dorsal-ventral orientation of the egg, dorsal blastopore formation was always directed towards the open end. They concluded that anaerobiosis can lead to a reversal of the dorso-ventral polarity by suppressing the original polarity and inducing a new one. Their experiments, as well as Glade's, showed

that dorso-ventral polarity is not irreversible and is probably based on a metabolic gradient (see Løvtrup, 1965*a**).

(4) Pasteels (1946) and Ancel & Vintemberger (1948*) found that fertilised or activated eggs can be symmetrised by lateral compression, the plane of symmetry always being parallel to the compressing plates and the orientation of the dorso-ventral axis depending upon the slant of the animal-vegetal axis at the time of compression (see further under paragraph 5, below).

(5) Ancel & Vintemberger (1948*) demonstrated that in *Rana fusca* the polarising influence of the sperm can be overruled by the effect of gravity, i.e. by a rotation of the egg through more than 135°. When two rotations were made successively in opposite directions, only the second one was effective. The so-called orientation rotation can act as long as the grey crescent has not yet formed. The grey crescent always appears on the side along which the yolk mass has descended under the influence of gravity. Ancel & Vintemberger (1948*) and Pasteels (1964*) considered the symmetrisation of the egg to be a two-step process, with first labile and then definitive fixation of the dorso-ventral axis.

In eggs surrounded by their jelly layers a rotated position can be maintained for only a short period, i.e. until the perivitelline space has been formed. Kirschner & Hara (1980) were able to keep eggs in a rotated position for any length of time by placing them in a 5% Ficoll solution, which counteracts the normal water uptake into the perivitelline space and thus prevents reorientation of the egg inside its capsules. A 30° rotation for 30 min or a 90° rotation for 5 min turned out to be sufficient to determine dorso-ventral polarity. H.-M. Chung & Malacinski (1980, 1981) and Malacinski & Chung (1981) recently confirmed the predominant influence of gravity in anuran as well as urodele eggs.

Kirschner *et al.* (1980*) and Gerhart *et al.* (1981) showed that, in eggs rotated in 5% Ficoll, the dorsal blastopore develops in whatever region of the egg equator that is turned upward. Complete reversal of the dorso-ventral axis determined by the sperm entrance point occurs after rotation at early stages, from time 0.0 to 0.5, called the pre-grey crescent period. At time 0.5 and later, called the post-grey crescent period, eggs become more and more refractory to the effect of gravity, but application of higher gravity forces can still cause axis reversal at later stages; e.g. at time 0.9 after application of $26 \times g$ for 4 min, and even at time 0.95 after application of $50 \times g$ for 4 min. At time 0.5–0.7 and with application of $30 \times g$ for 4 min a large fraction of the eggs developed into duplicitas embryos. Double axis formation can be explained by assuming that the high gravity force is strong enough to create a new 'dorsalising centre' by the formation of a new vitelline wall (see next section), but not strong enough to obliterate the old dorsal side determined by the sperm entrance point.

In all these polarity changes the alterations in egg orientation are accompanied by a rearrangement of the yolk distribution (H.-M. Chung & Malacinski, 1983). The latter authors attribute the determinative role in dorso-ventral axis formation to the spatial configuration of the internal cytoplasms, particularly the animal cytoplasm and the yolk mass, and not to that of the egg cortex, the grey crescent being only a transient marker of the location of the future blastopore (Malacinski, 1984*; Neff *et al.*, 1984).

Vitelline wall formation

Grey crescent formation seems to be accompanied by an upward movement of coarse yolk on the dorsal side of the egg, described by Pasteels (1948, 1964*) as the establishment of the 'mur vitellin' or vitelline wall. The latter is situated directly underneath the dorsal egg cortex. Opinions differ as to the way this vitelline wall is formed, i.e. whether this happens by a pulling-out of the dorsal margin of the yolk mass by the upward displacement of the dorsal cortex, as suggested by Elinson (1975), Elinson & Manes (1978) and Páleček *et al.* (1978), or by the yolk margin being pushed up by the expanding sperm aster, as proposed by Heidemann & Kirschner (1978) and by P. Hausen (personal communication). It is evident that in rotation experiments, where the yolk mass slides down under the influence of gravity, a vitelline wall is formed by a thinning-out of the dorsal margin of the yolk mass, which is rather firmly attached to the cortical layer of the egg. This represents the formation of the so-called Born's crescent (G. Born, 1885; see Pasteels, 1964*). Nieuwkoop (1977*), Kirschner *et al.* (1980*) and Gerhart (1980*) consider the vitelline wall region as the 'dorsalising centre': a site of more intense interaction between the animal cytoplasm and the vegetal yolk mass leading to a dominance of the dorsal equatorial region of the egg. Gerhart (1980*) suggests that the absence or scarcity of medium-sized yolk platelets in the area of the dorsal vitelline wall may be responsible for an intensified interaction between the two moieties of the egg. Malacinski (1984*) comes to a similar conclusion. Gerhart *et al.* (1981) consider eccentric animal cap contraction as the primary event in the symmetrisation process, which then leads to vitelline wall formation, while grey crescent formation is only an epiphenomenon. Recently Vincent & Gerhart (in preparation) applied a grid-like vital stain pattern to the vegetal hemisphere of the anuran egg and found that, concomitant with dorsal grey crescent formation, the vegetal yolk mass shifts about 30° ventrally relative to the egg surface. In our opinion this may explain the formation of the dorsal vitelline wall by the adhesion of the yolk to the dorsal cortex during the displacement of the yolk relative to the egg surface.

Egg rotation and UV damage

A relocation of the grey crescent and subsequently of the dorsal blastopore as a consequence of egg rotation suppresses the deleterious effect of UV treatment (Scharf & Gerhart, 1980; H.-M. Chung & Malacinski, 1980, 1981; Kirschner *et al.*, 1981). Duncan (1979) suggests that UV irradiation may raise the intracellular Ca^{2+} concentration, so that its morphogenetic effect would be due to changes in the Ca^{2+} distribution at critical stages in development.

Curtis' cortical grafting experiments

We now come to a discussion of the much-cited cortical grafting experiments by Curtis (1960, 1962 *a*), which seem to constitute important experimental evidence in favour of Dalcq & Pasteels' (1937, 1938) developmental theory. This theory is based on a presumed interaction between an internal animal–vegetal gradient in yolk distribution and a superficial cortical morphogenetic field centred around the grey crescent region. It was supported, among other things, by Schultze's (1894) and Penners & Schleip's (1928 *a*, *b*) inversion experiments and by G. Born's (1885) rotation experiments, extensively restudied by Pasteels (1938, 1939, 1940 *a*, *b*, 1941 *a*, *b*; see also Pasteels, 1946, 1948 and 1964*).

In our opinion, the theory has some important shortcomings. First, the amphibian egg clearly shows a *discontinuously* rather than continuously graded yolk distribution along the animal–vegetal axis. Second, Dalcq & Pasteels only focus attention on altered interactions between the yolk mass and the dorsal egg cortex in rotated eggs but ignore the fact that a displacement of the heavy yolk or parts of it must necessarily affect the localisation of the animal cytoplasm, thus creating new possibilities for interaction between the two main moieties of the egg, the animal cytoplasm and the yolk mass.

Curtis (1960) concluded that the dorsal cortex plays a morphogenetic role after grafting small pieces of dorsal cortex of the *Xenopus* egg undergoing first cleavage either to the same (dorsal) side or to the opposite (ventral) side of host eggs, and obtaining normal development in the former and double embryo formation in the latter case. In 1962 *a*, he reported that excision of dorsal cortex of uncleaved *Xenopus* eggs leads to an arrest of development at the blastula stage. Grafting of grey crescent cortex of 8-cell embryos into the ventral side of uncleaved eggs induced secondary axis formation, but grafting of grey crescent cortex of uncleaved eggs or of 8-cell embryos into the ventral region of 8-cell embryos had no such effect.

Gerhart (1980*) and Gerhart *et al.* (1981) have reinterpreted Curtis' grafting experiments on the basis of the outcome of their previously reported rotation experiments (see p. 58). Their main criticism concerns the absence of some of the necessary controls in Curtis' experiments, i.e. (notably) the absence of ventral cortex grafts into the ventral side of host embryos. Curtis' grafting technique unavoidably involved a partial rotation of the egg devoid of its investments. The absence of the latter must have prevented reorientation of the egg and must also have led to strong deformation. Cortical grafting to the dorsal side requires a turning-up of the dorsal side of the egg, while grafting to the ventral side implies a turning-up of the ventral side, the side of sperm entrance. In the former situation gravity will enhance the existing dorsal vitelline wall, whereas in the latter case sliding-down of the yolk may lead to a secondary vitelline wall forming on the ventral side. When the operation is performed at the 2-cell stage, the primary vitelline wall of the deformed egg can no longer be adequately 'suppressed' and double embryo formation may result (as is actually the case). At the 8-cell stage the cell membranes formed by the third cleavage plane may prevent any displacement of the yolk mass with respect to the egg cortex, so that rotation of the egg no longer has any effect on the pre-existing dorso-ventral polarisation.

The prevention of embryonic axis formation by excision of dorsal cortex at the uncleaved egg stage (Curtis, 1962*a*) was attributed by Brachet & Hubert (1972) to mitotic abnormalities due to wounding, leading to an arrest of development at a late blastula stage (see also Brachet, 1972*). Tompkins & Rodman's (1971) finding that in *Xenopus laevis* dorsal egg cortex can induce a secondary axis when introduced into the blastocoelic cavity of an early gastrula, whereas animal cortex rarely does, could not be reproduced by Malacinski, Chung & Asashima (1980), nor could it be verified by Asashima (1980) in sandwich experiments. In the experiments of Tompkins & Rodman, a critical point may have been that subcortical cytoplasm of the vitelline wall was perhaps included in the grafts, since yolk platelets can have inductive capacity (Faulhaber, 1972; Faulhaber & Lyra, 1974; Asashima, 1975; Wall & Faulhaber, 1976). The suggestion made by Curtis (1962*b*) that damage to the dorsal cortex could lead to lethality in succeeding generations due to 'cortical inheritance' was seriously put in question by Brachet & Hubert's (1972) observation of chromosomal aberrations caused by injury to the dorsal egg cortex. Summarising, it must be concluded that *no particular morphogenetic significance can be attributed to the dorsal cortex of the egg, so that grey crescent formation must be considered as no more than an epiphenomenon of dorsalisation* (see Gerhart, 1980* and Gerhart *et al.*, 1981).

Some biochemical data relevant to egg symmetrisation

Little is known about the biochemical events during symmetrisation. Brachet (1965*) states that RNA synthesis is more active dorsally than ventrally in the egg. He assumes that dorso-ventral polarisation is due to a gradient in ribosome distribution. According to Duspiva (1962*) respiration and RNA, carbohydrate and glycogen distribution, as well as the distribution of various enzymes, seem to show a steep animal–vegetal and a flatter dorso-ventral gradient. However, this may be only the consequence of differential yolk distribution, since the respiratory gradient can no longer be demonstrated when calculated on the basis of cytoplasmic nitrogen content. Tompkins & Rodman (1971) analysed different regions of the egg cortex by means of gel electrophoresis and found that its composition differs regionally.

Fertilisation is accompanied by changes in water permeability. This reaches a maximum during the symmetrisation contraction and remains rather high after grey crescent formation (Løvtrup, 1962). Capco & Jeffery (1981) observed an uneven distribution of maternal poly(A)$^+$ RNA in the oocyte cytoplasm and in the developing embryos of *Xenopus laevis*. Injected vegetal pole poly(A)$^+$ RNA redistributes itself in a vegetal–animal gradient. Examining animal/vegetal, dorso/ventral and right/left portions of *Xenopus* embryos, C. R. Phillips (1982) found approximately four times more total RNA in animal than vegetal regions and a 1.5-fold higher RNA concentration in the most dorsal region than in the other five regions. Poly(A)$^+$ RNA distribution shows temporally and regionally specific changes during development, with a higher concentration in the most animal regions.

Using the monoclonal antibody technique, Hausen and coworkers recently found that after germinal vesicle breakdown, during egg maturation and subsequent fertilisation, the majority of proteins present in the germinal vesicle are distributed more or less evenly over the animal cytoplasm of the egg. However, a particular protein accumulates predominantly or exclusively in a subcortical region of the vegetal moiety of the egg, including the dorsal vitelline wall region (Dreyer *et al.*, 1983; P. Hausen, personal communication). Hausen suggests that an interaction between the animal cytoplasm and the vegetal subcortical cytoplasm of the vitelline wall may play an important role in the ultimate symmetrisation of the egg (see further Chapter 9, p. 81, and Chapter 10, p. 99).

Symmetrisation in other chordate groups

J. Clavert (1962*) distinguished two different types of vertebrate eggs on the basis of the time of dorso-ventral polarisation: eggs with total cleavage, as those of amphibians and chondrosteans, where symmetrisation is accomplished before cleavage, and eggs with discoidal cleavage, as those of reptiles, birds, selachians and teleosts, where symmetrisation occurs when the blastoderm already comprises a large number of cells. The

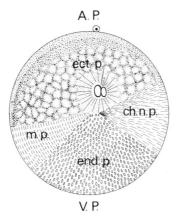

Fig. 6. Cytoplasmic segregation in the fertilised, uncleaved egg of *Branchiostoma lanceolatum*. Ch.n.p., chordo-neuroplasm; ect.p., ectoplasm; end.p., endoplasm; m.p., mesoplasm; A.P., animal pole; V.P., vegetal pole (after E. C. Conklin, 1932).

mammalian egg actually constitutes a third type, with total cleavage but late determination of bilateral symmetry.

Origin of dorso-ventral polarity in uro- and cephalochordates and in holoblastic fishes

In the uro- and cephalochordates animal–vegetal and dorso-ventral polarisation of the egg seem to occur simultaneously. After fertilisation a complex cytoplasmic segregation process takes place in the initially homogeneous egg, leading not only to animal–vegetal but also to dorso-ventral polarisation. In this segregation process an unpigmented yolk-poor animal plasm and a slightly darker yolk-rich vegetal plasm are formed, as well as a dorsal grey, crescent-shaped chordo- and neuroplasm and a ventral yellow, crescent-shaped mesoplasm (see Conklin, 1905*a*, *b* for *Cynthia* and Conklin, 1932 for *Branchiostoma*; see also Fig. 6). Little is known about the forces which act in this spatial segregation process. The fact that in the polarised egg the sperm entrance point always coincides with the vegetal pole points to a directing influence of the sperm in animal–vegetal polarisation, but gravity may also play a role in the spatial segregation process.

   In the chondrosteans, symmetrisation was studied in three *Acipenser* species by Dettlaff & Ginzburg (1954*). Fertilisation occurs through the micropyle at the animal pole, so that a directing influence of the sperm in symmetrisation can be excluded. Since the unfertilised egg is slightly oval in shape it usually lies with the animal–vegetal axis parallel to the substrate.

After fertilisation the egg rounds up and an orientation rotation takes place similar to that in amphibians, leading to the appearance of a clear crescent on the side where the vegetal pole has descended. Dettlaff & Ginzburg (1954*) showed that a 90° rotation is sufficient for the fixation of the position of the clear crescent (see also Dettlaff, Ginzburg & Shmalgauzen, 1981*). A similar determination of the plane of bilateral symmetry may occur in the eggs of *Petromyzon* and *Protopterus* (J. Clavert, 1962*).

Symmetrisation in meroblastic vertebrate embryos

Although the symmetrisation of meroblastic eggs of fishes, reptiles and birds occurs at much later stages of development (multicellular blastoderm) than that of the holoblastic amphibian eggs, where it takes place before cleavage, it is discussed in this chapter since the processes involved are very similar. The same holds for the symmetrisation of the holoblastic mammalian egg.

Meroblastic fish eggs

In the selachian, *Scylliorhinus canicula*, the blastoporal groove marking the dorsal side of the embryo always appears at the upper edge of the blastodisc of the slightly tilted egg. The plane of bilateral symmetry, which is probably formed under the influence of gravity, seems to be firmly determined as soon as a subgerminal cavity develops (Wintrebert, 1922).

The young oocyte of the teleost *Lebistes reticulatus* shows bilateral symmetry due to the eccentric position of the germinal vesicle, but this is no longer visible in the fully grown oocyte (J. Clavert & Filogamo, 1957). The still totipotent blastodisc of the teleost embryo segregates from the liquid yolk by the differentiation of an interjacent yolk syncytial layer, the periblast. Devillers (1951, 1956*) found that symmetrisation of the trout embryo occurs through a streaming of syncytial periplasm to the future dorsal side. Symmetrisation is first clearly expressed in the eccentric position of the subgerminal cavity. Devillers was not able to affect symmetrisation of the trout blastodisc by rotation, but J. Clavert & Filogamo (1957, 1959) could show that in *Lebistes reticulatus* the position of the blastodisc in space determines its dorso-ventral polarity, the germ anlage appearing at the lower edge of the blastodisc. J. D. Dasgupta & Singh (1981) suggested from the study of toluidine and methylene-blue-stained embryos that the yolk syncytial layer plays a role in the spatial differentiation of the blastodisc (see Chapter 9, p. 93).

Meroblastic reptilian eggs

The multicellular blastoderm of the lizard, *Lacerta vivipara*, segregates into a compact epiblastic layer with cylindrical cells and a thin hypoblastic layer with flat cells. Symmetrisation is revealed by the formation of a blastoporal plug in the future posterior half of the epiblast (Hubert, 1962). Pasteels (1954*b*) and J. Clavert & Zahnd (1955) demonstrated that the rule of von Baer, which was originally deduced for the avian egg (see below), can also be applied to the reptilian egg, in which group it shows about the same variation as in birds. In both groups the rotation of the egg inside the oviduct and uterus, when gravity is acting upon the slightly tilted blastoderm, determines the orientation of the embryo. The time of axis determination in reptiles is unknown, however.

Meroblastic avian eggs
It seems very likely that during oogenesis gravity already plays a role in the establishment of the animal–vegetal polarity of the very 'bottom-heavy' avian oocyte.

The avian egg acquires its surrounding albumen layers, shell membrane and calcareous shell during its passage through the oviduct, while it is rotating on its long axis in a clockwise direction. Under the influence of egg rotation in the oviduct, the chalazae (which connect the yolky egg with the inner shell membrane at the blunt and pointed poles of the egg) begin to coil, one clockwise and the other counterclockwise, because the heavy yolk counteracts egg rotation (see J. Clavert, 1962*). The cleaving blastodisc first consists of a single layer of blastomeres but is gradually converted into a multilayered (though still homogeneous) blastoderm. During this period glycogen accumulates, which seems to be required for further development. During area pellucida formation yolk-rich cells detach from the blastoderm and accumulate in the subgerminal cavity. Subsequently a primary hypoblast layer segregates from the epiblast by poly-invagination. Cell detachment and hypoblast formation start at the future posterior side of the blastoderm and progress towards the future anterior side (Eyal-Giladi & Kochav, 1976; Fabian & Eyal-Giladi, 1981).

There is a critical period for the symmetrisation of the chick blastoderm 6–8 h before laying (Vintemberger & Clavert, 1959; J. Clavert, 1961). At that time the multicellular germ anlage detaches from the underlying yolk through the formation of the subgerminal cavity. Simultaneously, the germ anlage becomes thinner and begins to spread over the yolk. The area pellucida appears off-centre as the first visible manifestation of symmetrisation (Eyal-Giladi & Kochav, 1976; Kochav, Ginsburg & Eyal-Giladi, 1980). Kochav & Eyal-Giladi (1971), using eggs removed from the uterus 8 h before laying, when the blastoderm is still single-layered, demonstrated that suspending the egg in a beaker by one of the chalazae determines the embryonic axis, which is always directed upward. In these suspended eggs, the area pellucida begins to appear at the upper side of the blastoderm (Eyal-Giladi & Fabian, 1980). Older suspended eggs, however, showed the normal orientation perpendicular to the long axis.

In the chick it is usually the blunt end of the egg that leads during the passage through the oviduct, less frequently the pointed end. Seen sideways, the embryonic axis usually points away and sometimes towards the observer (Rule of von Baer). In accordance with J. Clavert's observations, this rule can be explained by the effect of gravity upon the rotating egg having its blastoderm tilted slightly sideways (Eyal-Giladi & Fabian, 1980). In different bird species the long axis of the egg during its passage through the oviduct may be either nearly horizontal or more or less tilted, influencing the orientation of the embryonic axis, so that it is clear that in birds the egg's position in space determines the plane of bilateral symmetry. In the single-layered blastoderm, polarity is still labile and can be altered experimentally, e.g. by transverse folding (Eyal-Giladi, 1969, 1970 a, b).

Lutz (1962, 1964) and Lutz *et al.* (1963) found that longitudinal transection of early quail and duck blastoderms always leads to formation of parallel double axes in normal orientation. Transverse transection at different antero-posterior levels results in a posterior embryo of normal orientation and an anterior embryo of varying orientation – normal, perpendicular or reversed. According to Lutz (1965), these experiments indicate that embryonic axis formation depends on a strong postero-anterior and a weak antero-posterior gradient. In our opinion a single gradient would suffice.

Symmetrisation in the holoblastic mammalian embryo

The very small, nearly alecithal mammalian egg seems to be completely apolar. Neither sperm entry nor gravity influences its initial spatial development (Hillman, Sherman & Graham, 1972); 4- and 8-cell mouse embryos consist of wholly equipotential cells (Tarkowsky & Wróblewska, 1967). According to Seidel (1969*) isolated $\frac{1}{2}$ or $\frac{1}{4}$ blastomeres of rabbit embryos may either develop into a normal embryo or form a trophoblastic vesicle without inner cell mass. Seidel explained these different outcomes by postulating the existence of a formative centre in the mammalian egg but, in our opinion, it may simply be due to variations in the process of regulation after the separation of the blastomeres. During cleavage a compact morula is formed, consisting of many peripheral and some internally situated cells. They develop respectively into early differentiating outer trophoblast cells and still totipotent inner cell mass (ICM) cells (Gardner, 1970*, 1972). This development seems to be governed by purely epigenetic factors, i.e. by the differential physico-chemical effect of peripheral v. internal localisation of the blastomeres (Mulnard, 1967; Tarkowsky & Wróblewska, 1967; Graham, 1971*; Hillman *et al.*, 1972; Wilson, Bolton, & Cuttler, 1972; see Fig. 7).

Mammalian blastocyst formation in many respects resembles blastocoel formation in the amphibian embryo but is of much greater magnitude. It

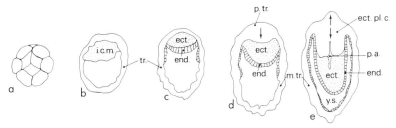

Fig. 7. Early mammalian development: a, morula; b, trophoblast (tr.) and inner cell mass (i.c.m.); c, segregation of endoderm (end.) from embryonic ectoderm (ect.) and outgrowth of endoderm along inner surface of trophectoderm; d, segregation of polar trophectoderm (p.tr.) from mural trophectoderm (m.tr.); and e, separation of polar trophectoderm (ectoplacental cone) (ect.pl.c.) from embryonic ectoderm (ect.) by formation of proamnion (p.a.), y.s., yolk sac (after R. L. Gardner, 1978*).

is based upon very active solute transport into the interior, in which membrane-bound Na^+/K^+ ATPase plays a crucial role (Benos & Biggers, 1981*). This active transport system requires energy from oxidative processes, as shown by the inhibitory action of ouabaine and potassium cyanide. During early blastocyst formation the initially 'leaky' epithelium changes into a 'tight' transporting system.

Fernandez & Izquierdo (1980) counted the numbers of cells at the morula–blastocyst transition in half, normal and double mouse embryos and found that neither the total number of cells nor the number of cell cycles correlates with the time of appearance of the blastocyst cavity; rather, this seems to depend upon the length of time which had elapsed since fertilisation.

Although gravity may play a role in the separation of the ICM from the vesicular trophoblast, Graham (1971*) assumes that the site where the ICM is attached to the trophoblast is the result of a purely random separation of the two components. The ICM controls trophoblast proliferation by inhibiting trophoblast differentiation near the attachment site, which leads to the formation of the ectoplacental cone (Gardner, Papaioannou & Barton, 1973). Animal–vegetal polarisation of the inner cell mass may be due to the development of the ectoplacental cone (Gardner, 1970*, 1972). It is subsequently expressed in the spatial segregation of the endoderm from the still totipotent ectoderm (Gardner & Papaioannou, 1975*). Under experimental conditions such a segregation can still occur after 5 days of gestation (Pedersen, Spindle & Wiley, 1975*). The differentiation of giant trophoblast cells is characterised by replication of (probably) the entire genome up to several hundred times the haploid amount of DNA, without nuclear or cell division (endopolyploidy) (M. J. Sherman, McLaren & Walker, 1972). This differentiation is independent of the ICM (M. J. Sherman, 1975*).

Although no experimental evidence is available, it seems rather likely that in mammals, like in birds, gravity plays a role in the subsequent symmetrisation of the embryonic anlage. L. J. Smith (1980) noted that the blastocyst's ICM–adembryonic polar axis is always oriented obliquely, the abembryonic pole being fixed either to the right or to the left uterine wall. This orientation at least allows the assumption of a possible influence of gravity. While the primary endoderm spreads over the inner surface of the trophoblast, thus forming the visceral endoderm of the yolk sac, the double-layered embryo (ss), suspended between the amniotic cavity and the yolk sac, may become polarised by gravity and as a consequence acquire bilateral symmetry (see further Gardner, 1978* and Chapter 9, p. 93).

Searle & Jenkinson (1978) could characterise the trophectoderm and its derivatives immunochemically, while Stinnakre *et al.* (1981), using antisera, could identify the primary endoderm as a derivative of the ICM. Dziadek (1979), Gardner (1982) and Papaioannou (1982) recently verified the origin of both the parietal and visceral endoderm from the ECM. Their differentiation depends on contact with different non-endodermal cells (Hogan & Tilly, 1981; Gardner, 1982).

Asymmetry in the lay-out of the vertebrate embryo

Although asymmetry is only expressed during early larval development, a short discussion of the problem of asymmetry nevertheless seems appropriate in this chapter.

In the amphibian larva asymmetry is chiefly expressed in the *situs viscerum*, the heart, and the *corpora habenularia* of the diencephalon. The asymmetry of the *situs viscerum* finds its expression in the left-sided stomach, the right-sided liver, and the intestinal curvature (urodeles) or coiling (anurans). The asymmetry of the heart finds its origin in the fusion of the right and left primary heart anlagen, the left anlage being ahead in development. When fusion is prevented two heart anlagen develop, of which the left one starts beating earlier (Zwirner & Kuhlo, 1964). The two *corpora habenularia* of the dorsal diencephalon are of unequal size, the left one being larger than the right (see von Kraft, 1971*b**). There is a pronounced (though not absolute) correspondence between heart, intestinal and *habenular* asymmetry in *Triturus alpestris* (see von Kraft, 1971*b**).

In nature a certain, though low incidence of *situs inversus* is found, which must have a genetic basis, since it appears particularly in certain populations. An increased rate of *situs inversus* can be evoked experimentally by centrifugation (T. Yamamoto, 1971*b**; Nishimura, 1967), X-irradiation (Wehrmaker, 1969), lithium-treatment (Wehrmaker, 1964, 1967), application of an electric current (von Kraft, 1968*a*), UV irradiation (von Kraft, 1968*b, c,* 1969*a, b*), unilateral heating (Zwanzig, 1938), experimental left or right translocations (Dalcq, 1947), and left or right defects (von Woellwarth, 1969, 1970; von Kraft, 1971*a*; see von Kraft, 1971*b**).

Spemann & Falkenberg (1919) and Fankhauser (1930) observed normal *situs viscerum* in larvae developing from left halves, and *situs inversus* in 50–75% of those developing from right halves of medially constricted amphibian eggs. Spemann attributed the origin of the *situs viscerum* to the body curvature of the larva showing left-sided dominance. Defect experiments disproved this hypothesis, however. von Kraft (1968*b*, *c*, 1969*a*) observed a stage-dependent susceptibility to UV irradiation of *Triturus alpestris* embryos. Late blastulae to mid-gastrulae (stages 9–13 Harrison) show a higher incidence of *situs inversus* as a result of right-side UV irradiation, older gastrulae (stages 12–13 Harrison) showed equal left and right irradiation effects, whereas there is a dominance for left-side irradiation effects by the end of gastrulation/incipient neurulation (von Kraft, 1968*c*). Mirror image asymmetry of the *corpora habenularia* developed in parabiotic young neurulae, a variable, transitional situation was obtained in parabiotic mid-neurulae, while no mirror image asymmetry was seen in parabiotic post-neurulae (von Kraft, 1971*c*). From all these observations von Kraft (1971*b**) concluded that the determination of bilateral asymmetry occurs during the period from the early gastrula to the mid-neurula, when UV irradiation or other experimental intervention leads to a disturbance of normal morphogenesis.

What may be the primary *origin* of the bilateral asymmetry of the vertebrate embryo? Boterenbrood & Nieuwkoop (1973) observed an asymmetry in the mesoderm-inducing capacity of left and right portions of the vegetal endodermal yolk mass, the capacity diminishing earlier on the left than on the right. The asymmetry in the induction of the mesoderm may subsequently affect both the endoderm and the neurectoderm (see Chapter 10, p. 102). Recently Boterenbrood, Narraway & Hara (1983) found a still earlier asymmetry, i.e. one in the direction of propagation of the cleavage waves, which from the 10th cleavage cycle onwards run predominantly from ventral/left to dorsal/right in the animal hemisphere, leading to a lengthening of the cleavage cycles that is more pronounced dorsally and on the right than ventrally and on the left side (see Chapter 9, p. 85). How the latter asymmetry relates to the asymmetry in the induction of the mesoderm is not yet clear.

Evaluation

In unfertilised eggs of several amphibian species a weak intrinsic asymmetry has been demonstrated, which might be due to gravity acting during the long period of oogenesis upon the very viscous cytoplasm of oocytes whose vegetal pole is fortuitously turned upwards.

The dorso-ventral polarisation of the monospermic anuran egg is primarily determined by the site of sperm entry, which ultimately leads to the appearance of a grey crescent at the boundary between the pigmented and unpigmented moieties of the egg on the side opposite the sperm entry point. The same seems to hold for light crescent formation in the polyspermic urodele egg. The extension of the sperm aster is correlated with the propagation of the so-called post-fertilisation wave(s), probably due to an interaction of the sperm aster rays with the microfilament system of the egg cortex. The sperm aster seems also to be responsible for bringing together the two pronuclei in the centre of the egg and for their subsequent displacement towards the dorsal side. It is not yet clear whether the expansion of the sperm aster is responsible for the formation of the so-called vitelline wall underlying the dorsal grey crescent region of the egg. Alternatively, the vitelline wall may be formed either by an asymmetrical contraction of the pigment cap or by a shift of the egg surface with respect to the inner cytoplasm (particularly the yolk mass) leading through local adhesion to a stretching of the dorsal cortex. It is very probable that the grey crescent itself is only an epiphenomenon of the dorsalisation process and does not play a morphogenetic role in development. This has become evident from rotation experiments in which gravity completely overrules the sperm-dependent symmetrisation as a result of, among other things, a thinning-out of the dorsal margin of the yolk mass when it slides down inside the egg under the influence of gravity. The outcome of the cortical grafting experiments of Curtis (1960, 1962a), which seemed to confirm Dalcq & Pasteels' (1937, 1938) developmental theory, has been re-interpreted in the light of the necessary rotation and deformation of the 'naked' egg during preparation for cortical grafting (Gerhart et al., 1981). Dorsal vitelline wall formation is now considered to be one of the principal events in the symmetrisation process, establishing a locally intensified interaction between the two primary moieties of the amphibian egg, the animal and the vegetal cytoplasms.

The successive intracellular interactions occurring in the egg represent nucleocytoplasmic and intercytoplasmic interactions between its different moieties, in which 'normal' nuclear and cytoplasmic factors are acting likewise as in any other cell.

In the uro- and cephalochordates animal–vegetal and dorso-ventral polarisation seem to occur simultaneously directly after fertilisation, a process in which the sperm entry point and gravity may play a role. The nature of the spatial segregation process involved is still little understood, however.

In the holoblastic sturgeon and *Protopterus* eggs, symmetrisation is caused by a reorientation of the egg inside its investments under the influence of gravity, a process which is very reminiscent of the gravity-guided orientation rotation in the amphibians.

In the meroblastic fish, reptilian and avian eggs, symmetrisation occurs in the multicellular blastodisc or blastoderm, likewise under the influence of gravity. Although nothing is known about symmetrisation of the holoblastic mammalian embryo, the oblique position of the blastocyst's ICM–adembryonic pole axis with respect to the uterine wall (L. J. Smith, 1980) at least allows the assumption of a possible influence of gravity.

The bilateral asymmetry of the vertebrate embryo can be traced back to left/right differences in mesoderm induction (see Chapter 10, p. 102) and possibly already to an asymmetry during cleavage (see Chapter 9, p. 85).

8

The process of cleavage

There is a very extensive literature on the process of cleavage and cell division; here we can only mention the more significant data. For further details the reader is referred to the general reviews by Wolpert (1960*); Roberts (1961*); Dettlaff (1964*); Zotin (1964*); Løvtrup (1965a*); Fautrez-Firlefyn & Fautrez (1967*); Rappaport (1971*); Gerhart (1980*) and Vacquier (1981*).

Cleavage represents the process of fragmentation of the fertilised egg into cells with a nucleocytoplasmic ratio more similar to that of differentiated cells, by means of cell division without intercalated cell growth. Fautrez-Firlefyn & Fautrez (1967*) state that cleavage is only to be understood [as a normal cellular activity] when one considers that it is preceded by the extensive growth of the oocyte, during which developmental phase mitotic activity is inhibited.

Surface contraction waves

Timelapse-cinematographic analysis of cleaving eggs of *Ambystoma mexicanum* has brought to light that 'surface contraction waves', probably two, pass over the egg surface prior to the appearance of the cleavage furrow. They originate in the future initiation point of the cleavage furrow near the animal pole of the egg and spread in animal–vegetal direction (Hara, 1971; see also Fig. 5 on p. 52). These waves coincide with a local increase in cortical stiffness (Sawai & Yoneda, 1974). Surface contraction waves are insensitive to the antimitotic drugs colchicine and vinblastine and occur also in artificially activated eggs lacking a centriole (Hara, Tydeman & Hengst, 1977). These cyclic changes in the cortical and subcortical layers of the egg during cleavage take place with nearly the same timing in enucleated egg fragments (Sawai, 1979; Hara, Tydeman & Kirschner, 1980; Sakai & Kubota, 1981). They must therefore be brought about by some cytoplasmic or cortical clock mechanism. According to Gerhart (1980*) these changes resemble the cyclic changes in sulfhydryl/disulphide ratios, in thymidine kinase activity and in Ca^{2+}-dependent ATPase activity as observed in sea urchin eggs (P. Harris, 1978). However, it is not clear whether the cyclic nuclear changes predominate over the cytoplasmic ones or vice versa. Normal cleavage undoubtedly requires the proper coordination of nuclear and cytoplasmic cyclic events. Shinagawa (1983) recently observed that the cytoplasmic clock in non-nucleated egg fragments has a longer cycle length than in normal *Xenopus* eggs, and concludes that this basic cytoplasmic cycle length is modulated by the mitotic apparatus.

Cleavage furrow formation

The first cleavage division starts 1–3 h after fertilisation, depending on the species and on the temperature. In the anuran, *Xenopus laevis*, the primary constriction phase accounts for 15% of the reduction of the egg diameter in the plane of cleavage, followed by a secondary phase of membrane ingrowth (Bluemink, 1971 *b*). Similar data hold for the urodele *Ambystoma mexicanum*. First, a stripe of more intense pigmentation appears at the animal pole of the egg. Some minutes later this splits into two and the area in between deepens and widens, while on both sides stress marks appear (Bluemink, 1970; Sawai, 1976 *b*). The surface area in the region of the single or double stripe, which corresponds to the future site of new membrane insertion, is greatly enlarged by surface protrusions (Bluemink, 1970; Denis-Domini, Baccetti & Monroy, 1976). Inside the egg cortex previously non-oriented actin-like microfilaments, which bind heavy meromyosin, become polarised and aggregate, forming a thick contractile ring in the direction of the furrow (Bluemink, 1970; Selman & Perry, 1970; Perry, John & Thomas, 1971; Perry, 1975; Bluemink & de Laat, 1977*). Zotin & Poglazov (1962) already suggested that cleavage furrow formation is due to a local contraction in the zone of the cleavage furrow caused by factors originating from the diastema at anaphase (see below, p. 76). Light vesicles and osmiophilic stacked lamellae are exclusively found in that region of the primary furrow where new membrane will be inserted (Bluemink, 1970; Singal & Sanders, 1974*a, b*; Singal, 1975*a, b*). At the end of the constriction phase the layer of filaments beneath the furrow bottom is split into two by local ingrowth of new membrane, after which the filaments take up lateral positions. Furrow growth proceeds by bilateral insertion of new membrane (Bluemink, 1970; see also Fig. 8).

The cleavage furrow, which starts at the animal pole of the egg, extends rapidly over the circumference of the egg in opposite directions until the two ends meet at the vegetal pole. While in the animal hemisphere extensive shrinkage of surface area occurs in the furrow region and the adjacent area, in the vegetal hemisphere contraction is restricted to the furrow itself (Sawai, 1976 *b*). The terminal stage of the cleavage process involves a change in the diastema as it is split into two over its entire length by vacuolisation along the midline (see Fig. 8). This vacuolisation is preceded by the migration of mucopolysaccharide granules from the cytoplasm

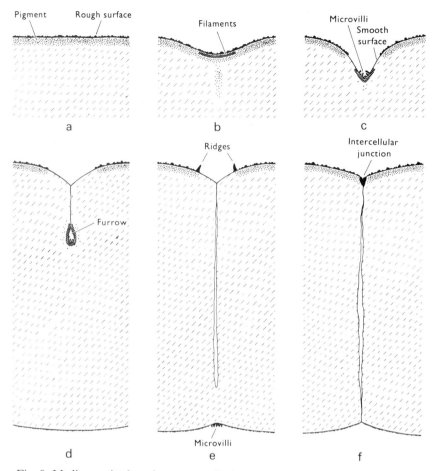

Fig. 8. Median sagittal sections perpendicular to cleavage groove and furrow of stages of first cleavage of the newt egg. a, Before cleavage; b, onset of cleavage, with formation of band of microfilaments; c, early cleavage, with insertion of new smooth and unpigmented surface in the groove and formation of microvilli in its deepest part; d, mid-cleavage, with furrow extending below the lowest part of the groove and microvilli and microfilaments at its leading edge; e, late cleavage, with furrow approaching the vegetal pole and formation of microvilli in vegetal groove; and f, post-cleavage, with formation of intercellular junctions (after G. G. Selman & M. M. Perry, 1970).

surrounding the diastema towards the site of new membrane insertion (Motomura, 1960; Zotin, 1964*). The distribution of pyrophosphatase (TPPase) activity in the outermost cisternae of the Golgi complexes and in vesicles and membrane cisternae near the furrow strongly suggests that new membrane and surface material is added to the developing furrow (Singal & Sanders, 1974a, b; Singal, 1975a, b; Sanders & Singal, 1973,

1975). At the end of cleavage the new membrane is 'sealed off' from the external environment by the formation of junctional complexes (Bluemink, 1970, 1971 b; Schroeder, 1973). Terminal junctional complex formation is preceded by the formation of a row of long microvilli (Singal & Sanders, 1974 a, b) and of cytoplasmic ridges (Selman & Perry, 1970; Okamoto & Eguchi, 1975; see also Fig. 8).

Cytochalasin B (CCB), which is known to disorganise the microfilament system and the anchorage sites of the microfilaments to the cell membrane, also affects the cell membrane itself, particularly interblastomeric junction formation (Bluemink, 1971 a) and leads to regression of the furrow and flattening out of the newly formed membrane (see also Estensen, Rosenberg & Sheridan, 1971). Externally applied CCB is only effective when new membrane insertion has started, i.e. approximately 7 min after the onset of furrow formation in *Xenopus laevis* (de Laat, 1972; de Laat, Luchtel & Bluemink, 1973; Selman, Jacob & Perry, 1976). CCB can apparently only penetrate the new membrane. Furrow regression can be advanced by injection of CCB underneath the egg membrane in the furrow region, leading to disruption of the microfilament system (de Laat *et al.*, 1973; Luchtel, Bluemink & de Laat, 1976; see also Bluemink, 1978*).

The pre-existing and the new membrane differ markedly in composition; the pre-existing membrane possesses a dense population of intramembranous particles on the external face (EF) of the membrane and only few particles on the cytoplasmic (PF) face, whereas the new membrane is relatively poor in particles (Bluemink *et al.*, 1976; Sanders & Dicaprio, 1976 a, b).

De Laat *et al.* (1975), using flame spectrophotometry and ion-selective electrodes, showed that during the first cleavage of the *Xenopus* egg a water influx occurs at a constant rate without intracellular compartmentation, while the Na^+, K^+ and Cl^- concentrations remain constant. Whereas all the Cl^- ions are free, 30% of the K^+ ions and 60% of the Na^+ ions are bound. Since the newly formed membrane has a much higher permeability than the pre-existing membrane, particularly for K^+ ions, the membrane potential begins to drop 7 min after the initiation of furrow formation and only reapproaches the original value when the new membrane becomes 'sealed off' by the formation of junctional complexes (Morrill & Watson, 1966; Woodward, 1968; M. Takahashi & Ito, 1968; S. Ito, 1972; de Laat, 1972; Bluemink & de Laat, 1973; de Laat *et al.*, 1973; de Laat & Bluemink, 1974; de Laat, 1975; see also Selman *et al.*, 1976).

In Chapter 7, p. 53, we have already discussed the relationship between Ca^{2+} ions and the contraction of the cortical microfilament system. Schroeder & Strickland (1974) found that ionophore A 23187 elicits a rapid cortical contraction in *Rana pipiens* eggs, both in the presence and absence of extracellular Ca^{2+}. Intracellular ethylenediaminetetraacetic acid (EDTA) or EGTA inhibits contractility, indicating the importance of intracellular Ca^{2+} for the contractile response. The action of ionophore A 23187 was confirmed by Osborn, Duncan & Smith (1979) in the 16-cell *Xenopus* embryo. During its action, Ca^{2+} seems not to be lost from the embryo but only to be redistributed as a result of release from yolk platelets and pigment granules, or both. A rise in intracellular Ca^{2+} interferes with normal cleavage, but a minimal Ca^{2+} concentration is required for cleavage.

Using fluorescent Con A, O'Dell *et al.* (1974) found that in *Xenopus laevis* eggs, binding sites for Con A are only present near the cleavage furrow, particularly at the boundary of pre-existing and nascent membrane. Tencer (1978 *a, b*) and Geuskens & Tencer (1979 *a, b*) could inhibit furrow formation by wheat germ and soybean agglutinin. These lectins interfere with intramembrane protein movement and affect the relationship between surface receptors and the alignment of microfilaments of the cytoskeleton. In dechorionated eggs of the ascidian, *Phallusia mammillata*, cleavage can be prevented by Con A, but ooplasmic segregation, which occurs before first cleavage, is normal, so that a multinuclear syncytium is formed with regionally segregated cytoplasmic areas. Con A is thought to act by immobilising surface structures that are connected to microfilaments (Zalokar, 1979).

In *Rana* eggs, the rate of oxygen uptake is minimal shortly before the beginning of each cleavage and maximal halfway between two successive minima, with the rate of carbon dioxide production varying inversely (Zeuthen & Hamburger, 1972).

Furrow induction by the diastema of the mitotic apparatus

Rotation of the mitotic apparatus in amphibian and sturgeon eggs, achieved by compression of the egg between glass plates, led Zotin & Pagnaeva (1963) and Zotin (1964*) to conclude that the time of determination of furrow formation coincides with late anaphase/early telophase, when the diastema approaches the egg surface in the animal pole region. T. Kubota (1966) reported that autonomous furrow formation was established in the cortex between metaphase and late anaphase. According to Rappaport (1971*) it is likely that in eggs and early blastomeres the asters represent the components of the mitotic apparatus active in inducing furrow formation in the egg cortex. In laterally compressed sea urchin eggs and newt kidney cells Rappaport & Rappaport (1974) found that, in both cleavage and ordinary cell division, furrow formation starts where under experimental conditions equatorial surface is pushed inwards and makes contact with the mitotic spindle. These experiments demonstrate that the diastema induces the furrow-forming process. This begins with the formation of a thickened layer of subcortical cytoplasm, which induces excitability and subsequent contraction of the surface layer. These observations were recently confirmed by Selman (1982), who found that furrow determination is completed 0.46 Dettlaff units before the onset of furrowing (a Dettlaff unit being the time interval between the onset of the first and second cleavage).

T. Kubota (1969) showed that, in *Rana nigromaculata*, transplanted subcortical cytoplasm of the median zone has the capacity to induce a furrow in cortex which has no autonomous capacity for furrow formation. From experiments in which the surface layer was affected by cutting and rubbing, he concluded that the reactivity of the cortical layer appears 45 min before first cleavage, which is immediately after the apposition of the pronuclei, so that determination of the first cleavage plane must occur during the period between syngamy and centrosome migration. Injection of taurodeoxycholate induces the formation of a furrow-like depression, a reaction which is inhibited by lowering the external Ca^{2+} concentration (T. Kubota, 1979b). Heidemann & Kirschner (1978) reported that in *Xenopus laevis* oocytes the ability to form cleavage furrows arises 6–8 h after germinal vesicle breakdown, probably through certain maturational changes in the egg cortex. Asters stimulate furrow formation, but abortive furrows can also be formed when the eggs are artificially activated.

Raff, Brothers & Raff (1976) studied the mutation *nc* of *Ambystoma mexicanum*, which conditions failure to cleave. Homozygous *nc* eggs form no mitotic apparatus or cytasters, although there is a normal pool of soluble tubulin (Raff, 1977). Raff suggests that the mutant has a defect in the microtubule assembly mechanism.

Cleavage-initiating factors (CIFs)

Sentein (1961) analysed the nuclear events during the very complex process of cell division. The polar regions of the cleavage spindle are the seat of several activities. They direct fibrillogenesis, lead to an accumulation of non-fibrillar material, and divide and separate the centrioles (Sentein, 1968). The antimitotic agent chloral hydrate destroys the spindle fibres, whereas selenium dioxide affects the chromosomes (Sentein, 1967).

Gerhart (1980*) emphasises that cleavage requires the presence of a centriole furnished by the sperm. This acts as the primary cleavage-initiating factor (CIF) and allows the formation of an astral spindle, the spatial position of which at anaphase determines the position of the cleavage furrow. Artificially activated eggs contain only a small aster-like yolk-free region near the nucleus, but no full-size aster or bipolar spindle (Manes & Barbieri, 1977). Enucleated *Rana pipiens* eggs injected locally with homogenates of sperm centrioles fail to form a deep cleavage furrow, whereas nucleated controls do, so that some nuclear component is also required (Briggs & King, 1953). Enucleated eggs fertilised with heavily irradiated sperm were almost devoid of chromosomal material, as shown by Feulgen staining, but cleaved normally (Briggs, Green & King, 1951). This rules out the chromosomes as the component in question, leaving the centromere or the nuclear envelope as possible candidates.

Huff & Preston (1965) found a CIF in *Rana pipiens* embryos around the beginning of gastrulation. The factor was also produced in artificially activated eggs, which do not undergo normal cleavage. Masui, Forer & Zimmerman (1978) showed that sea urchin mitotic apparatuses (MA), isolated in glycerol–dimethylsulphoxide, induced cleavage in unfertilised *Rana pipiens* eggs. A cleavage-inducing agent (CIA) isolated from frog brain induced cleavage in 40–60% of the cases when injected into normal frog eggs, but not when injected into enucleated eggs, so that the CIA must act through or in association with the nucleus. Injection of MA plus CIA led to a significantly higher frequency of cleavage (70%).

Nuclei from a synchronised population of blastula cells in a late phase of their cleavage cycle frequently prematurely induced an additional cleavage furrow when injected into *Bombina orientalis* eggs. Donor nuclei do not need to have initiated or completed the S phase, nor do they have to be synchronised with the recipient egg cytoplasm to induce such an additional furrow (Ellinger, 1978).

The propagation of the cleavage furrow

K. Dan & Kojima (1963) and Sawai, Kubota & Kojima (1969) could demonstrate that cortical and subcortical cytoplasm are 'prepared' for furrow formation in front of the advancing tip of the furrow. Incisions made close to the tip do not affect the furrow progression but incisions made far enough ahead of the tip block it. Insertion of impermeable cellophane 0.5–0.7 mm ahead of the furrow tip also stops furrow progression, whereas insertion of permeable millipore filter does not (Kojima, 1972). This suggests that a diffusible factor is involved. In *Triturus pyrrhogaster* eggs the linear extension of the 'prepared' region is longest in the animal hemisphere (about 1.0 mm) and decreases towards the vegetal pole. In the direction perpendicular to the furrow the capacity for furrow induction extends for only 0.1 mm (Sawai, 1972). Sawai distinguishes between the inductive capacity of the subcortical cytoplasm and the reactivity of the overlying cortex. The furrow-inducing capacity is not species-specific; it also acts between *Triturus* and *Xenopus* eggs. The reactivity of the cortex propagates meridionally over the surface of the egg from the animal towards the vegetal pole. After furrow completion the reactivity begins to be lost, first in the animal region but eventually from the entire surface. It reappears in the animal region with the onset of the next cleavage. Grafting of non-reactive cortex in front of the advancing furrow does not interfere with the passage of the furrow. This demonstrates that in the graft the furrow is actually formed under the inductive influence of the subcortical cytoplasm of the host egg (Sawai, 1974, 1976*a*, 1983; see also T. Kubota, 1979*a*).

Wound healing

A few words must be said about cortical wound healing in the egg and early embryo. Holtfreter (1943*a*) studied the properties of the egg surface layer in wound healing, attributing these properties to the contractility of a purportedly extracellular 'surface coat'. However, the existence of an extracellular surface coat could not be confirmed by TEM studies (Perry, 1975), which showed that the contractile properties of the egg surface are localised in a meshwork of microfilaments which directly underlies the plasmalemma.

When a wound is made in the surface of an unfertilised or fertilised egg the wound edges first retract by relaxation of the existing tangential tension in the cortical layer; this is followed by a circular constriction of the wound edges (Holtfreter, 1943*a*). Wound healing also involves new membrane formation inside the exovate that protrudes from the wound (Luckenbill, 1971; Bluemink, 1972). In hypotonic media wound healing usually does not occur. In weakly hypotonic saline medium no wound healing occurs in unfertilised eggs, but between fertilisation and first cleavage wound healing does take place in the animal hemisphere, extending gradually to the vegetal hemisphere at later stages (Dollander & Vivier, 1954). Løvtrup (1965*a**) mentioned that Ca^{2+} ions influence the mechanical properties of the cortical layer. Gingell (1970), studying surface contraction by local application of polycations and ionic detergents, pointed out the striking similarity between cleavage and wound closure. He suggests that the same contractile elements are operative in both cases. Stanisstreet & Panayi (1980) showed that colchicine treatment does not prevent wound healing, whereas cytochalasine B does. This was recently confirmed by Merriam & Christensen (1983). Microtubules are apparently not involved in wound healing but wounding, by allowing an influx of Ca^{2+}, results in a Ca^{2+}-activated contraction of the microfilament system.

Wound healing in the multicellular early embryo is very similar to that in the fertilised, uncleaved egg. This is due to the fact that the contractile surface layer of each cell is intimately connected with that of adjacent cells by means of junctional complexes, so that the contractile system acts on a supracellular level (Dollander, 1961*).

Wound healing in the ectoderm of *Xenopus laevis* early neurulae and chick primitive streak stages likewise goes through an initial phase of wound 'gaping' under the influence of tangential tension, followed by a contraction of the marginal cells of the wound area. In the chick embryo cell proliferation also plays a role in wound closure (Stanisstreet, Wakely & England, 1980).

Perry, Selman & Jacob (1976), using tritiated cytochalasin B, observed the highest contractile activity in the apices of dissociated *Triturus* prospective ectodermal cells, where the pigment and the fibrillar meshwork are located. Tencer (1978 *a*, *b*) and Geuskens & Tencer (1979 *a*, *b*) showed that in *Xenopus laevis* lectins affect wound healing by blocking the contraction of the microfilament system surrounding the wound. Wound closure in the ectoderm of *Xenopus* neurulae seems to be initiated by a local influx of Ca^{2+} ions (Stanisstreet, 1982).

Evaluation

Cleavage represents the fragmentation of the egg into an (at least initially) logarithmically increasing number of cells of increasing nucleocytoplasmic ratio, as a compensation for the earlier enormous growth of the oocyte in the absence of mitotic activity.

Cleavage is a very complex process in which nucleocytoplasmic interactions play a major role. Cleavage furrow formation, which is preceded by surface contraction waves, is initiated by the diastema of the mitotic spindle; this activates the subcortical cytoplasm to exert an inducing action upon the egg cortex, which then leads to the assembly and orientation of microfilaments. Furrow progression represents a self-propagating interaction between the subcortical cytoplasm and the egg cortex. Several cleavage-initiating factors have been claimed to exist by various authors, of which the centriole seems to represent the primary one.

Furrow formation in cleaving amphibian eggs accounts for 15–30% of the reduction in cell diameter, so that 70–85% of the cleavage process is due to the ingrowth of new membrane from the tip of the cleavage furrow. The old egg surface membrane, resealed after each division, has special properties, such as a very low ion permeability, screening the egg interior from the external environment. The newly formed intercellular membrane has quite different properties, allowing for intercellular communication. During each division it becomes 'sealed off' from the external environment by junctional complexes. Cleavage represents a special form of mitosis, i.e. cell division without intercalated growth, and is based on intracellular interactions common to all cells.

Wound healing shows much similarity with cortical contraction during cleavage, both being based on the contractile properties of the cortical microfilament system.

9

Early development up to gastrulation

In the anamnia the period of early development is characterised by the process of cleavage and the formation of the blastocoelic cavity. In the amniotes early development extends until the segregation of the embryonic anlage into an essentially double-layered embryo. In both subphyla the period of early development constitutes the preparation for mesoderm induction and for the very important morphogenetic process of gastrulation.

The period of cleavage in holoblastic anamnia

Brachet (1965*) states that cleavage is essential to obtain relatively small cells which have sufficient plasticity to undergo the morphogenetic processes of gastrulation and neurulation. The logarithmic increase in cell number moreover leads to an increase in potential nuclear activity, which is indispensable for development. We want also to emphasise the morphogenetic significance of blastocoelic cavity formation. The blastocoel separates the two primary moieties of the egg, which have developed from the animal and vegetal ooplasms, respectively. It provides the necessary space for the invagination of the mesoderm and endoderm and constitutes the internal medium of the developing embryo. Dollander (1961*) stressed the point that during cleavage the outer layer of the egg maintains its properties as a continuous outer coat as a result of the formation of interblastomeric junctional complexes, allowing it to function in a global as well as a regional manner, thus safeguarding the polarities of the developing system (see also R. D. Campbell, 1967 on desmosome formation).

Hausen and coworkers recently demonstrated that nuclear proteins, which are distributed throughout the cytoplasm after germinal vesicle breakdown, are taken up stepwise into the cleavage nuclei as their number increases. In the beginning the uptake seems to be more or less uniform, but at later stages of development certain proteins are differentially taken up by particular cell nuclei. This emphasises the significance of cell division for nucleocytoplasmic interactions (Dreyer et al., 1983).

81

The cleavage pattern

In 60–70% of the eggs of *Ambystoma* and *Xenopus* the first cleavage plane coincides with the plane of bilateral symmetry (Ancel & Vintemberger, 1948*; Ubbels *et al.*, 1979). In our opinion this may be due to the position of the zygote nucleus, which is slightly eccentric towards the dorsal half of the egg, where it lies close to the dorsal vitelline wall underlying the grey crescent. This position leads to an orientation of the first cleavage spindle perpendicular to the plane of bilateral symmetry, offering the largest free space for the spindle. However, it is not a strict correlation, since 30–40% of the eggs show more or less pronounced deviations from this rule. Similar spatial relationships play a role in the orientation of the cleavage spindles in the succeeding cleavages. The second cleavage is oriented vertically, but perpendicular to the first. The third cleavage plane is horizontal. It intersects the animal/vegetal axis at about one third from the animal pole. The fourth cleavage is again vertical and the fifth horizontal. Sirakami, Gejo & Hirose (1962) call attention to the rather wide variation in cleavage plane directions in the 8–16-cell embryo of *Bufo vulgaris*, showing that in the amphibian embryo the accuracy of cleavage is not important for normal development. In the anurans the great majority of cleavage planes intersect the outer surface but some are oriented parallel to the surface (see further under anuran gastrulation, p. 126). In urodeles there are no tangential cleavages.

The cleavage pattern of the axolotl egg, analysed by timelapse-cinematography, shows synchronous divisions up to and including the tenth cleavage, with a roughly constant cycle length of about 90 min (Hara, 1977). Cycle length in the synchronous period varies considerably between species: it is 90 min in *Pleurodeles*, 60 min in *Bufo* and only 30 min in *Xenopus*. Each species possesses a species-specific cleavage timing mechanism in its cytoplasm, since cytoplasm of one species injected into the uncleaved egg of another species affects the cleavage timing of the latter. A cleavage-timing factor is already present in the cytoplasm of maturing oocytes; it is not identical to the maturation promoting factor (MPF) (Aimar, Delarue & Vilain, 1981).

After the tenth cleavage, synchrony is lost owing to a variable lengthening of the cleavage cycles in individual blastomeres. According to Chulitskaia (1967) in *Rana temporaria* embryos asynchrony starts around the tenth division in not more than 5% of the nuclei, while the next division already exhibits asynchrony of a significant proportion of the nuclei. This leads to a sharp drop in the mitotic index, which gradually decreases further until the beginning of gastrulation, when it reaches the low level of 5%. Signoret & Lefresne (1971, 1973) and Signoret (1977 *a, b*) came to similar conclusions. In cleaving axolotl eggs the rather abrupt onset of asynchrony is expressed in a variation of only 8 min in the length of the 10th cleavage (which, on average, lasts 90 min), and a variation of at most 130 min in the 11th cleavage (which, on average, lasts 127 min). Graham & Morgan (1966) and Flickinger, Freedman & Stambrook (1967a) concluded from continuous ³H-thymidine labelling of tissue squashes that the lengthening of the cleavage cycles is not due to longer S and M phases but to the appearance of G_1 and G_2 phases, which are previously undetectable. Boterenbrood *et al.* (1983) call attention to the fact that in *Xenopus laevis* the lengthening of the cleavage cycles may start at the 10th, 11th or 12th cleavage.

Dettlaff (1964*) was the first to suggest that a shortage of cytoplasmic materials originally provided by the germinal vesicle and needed for rapid cleavage may be a possible cause of asynchrony. Benbow *et al.* (1977*) assumed that the exhaustion of an initiator of DNA synthesis during cleavage may lead to the appearance of G_1 and G_2 phases. The transition from the initial period of synchrony to the period of asynchronous cleavage, during which the cells acquire motile and transcriptional activities, is called the mid-blastula transition (MBT) (Newport & Kirschner, 1982a). This transition, as studied in *Xenopus laevis* embryos, does not depend on the number of cell divisions, on the time since fertilisation or on a counting mechanism involving sequential modification of DNA, but solely on the embryo reaching a critical nucleocytoplasmic ratio. The authors suggest that the MBT is the consequence of the titrating-out of a factor originally present in the egg by the exponentially increasing amount of nuclear material. When the substance is exhausted a new developmental programme is started, leading to the acquisition of new cell properties. In *Cynops pyrrhogaster* desynchronisation in nucleated half eggs begins one division earlier and in quarter eggs two divisions earlier than in whole eggs, indicating that desynchronisation is correlated with cell size (Kobayakawa & Kubota, 1981). In *Rana temporaria* desynchronisation starts relatively later at a lower temperature (Chulitskaia, 1970). The hypothesis of Newport & Kirschner (1982a) was further supported by injecting into the egg a plasmid containing a cloned gene coding for yeast leucine tRNA; this is inactive before MBT but becomes transcriptionally active at MBT. The pre-MBT suppression can be reversed by adding so much competing DNA that its total amount comes to equal that present after 12 cleavage cycles (Newport & Kirschner, 1982b; see also Laskey, 1983). It is of interest to mention that Landström, Løvtrup-Rein & Løvtrup (1975) suggested that the desoxyriboside triphosphates of the *Xenopus* egg support synchronous cleavage by preventing the cells from undergoing differentiation. Injection of desoxynucleotides prolongs synchrony and inhibits RNA synthesis.

Cleavage waves

The fact that the initiation of cleavage differs among the individual blastomeres leads to a spatial pattern of cleavage denoted by the term 'cleavage wave'.

Løvtrup (1965 *a**) stated that in the early amphibian embryo dorso-ventral polarity is expressed in a higher rate of cell division in the dorsal than in the ventral blastomeres, the former thus becoming smaller than the latter. However, although cleavage does not occur simultaneously in the various blastomeres (see below), there is no such difference in average cleavage rate between dorsal and ventral blastomeres (Hara, 1977). In reality the difference in average cell size between dorsal and ventral blastomeres, which is already visible at the 4- and 8-cell stages, is due to the slightly eccentric, more dorsal, position of the cleavage plane during second cleavage.

In side-view timelapse cinematography Hara (1977) observed that in *Ambystoma mexicanum* seven animal–vegetal 'cleavage waves' pass over the egg surface from the 5th cleavage – the earliest time that a cleavage wave is discernible – to the 11th cleavage, with a gradual shift in phase, so that the 10th cleavage of the slowest vegetal cells more or less coincides with the 11th division of the fastest animal cells. Satoh (1977) found that in *Xenopus laevis* cleavage waves usually start in the animal hemisphere near the equator on the dorsal side, and travel towards the animal as well as the vegetal pole, ending at the latter. In a thorough timelapse-cinematographic analysis of cleaving *Xenopus laevis* eggs Boterenbrood, Narraway & Hara (1983) observed that the direction of the cleavage waves initially varies in individual embryos but begins to change from the 10th cycle onwards, resulting in all embryos in a ventral/left to dorsal/right direction prior to gastrulation. This implies that the lengthening of the cleavage cycles is ultimately more pronounced dorsally than ventrally, which may correspond to the predominance of mRNA synthesis in the dorsal half of the embryo as observed by Bachvarova & Davidson (1966). The ultimate direction of the cleavage waves also reflects an asymmetry in the lengthening of the cleavage cycles in the animal hemisphere, lengthening being more pronounced on the right than on the left (Boterenbrood *et al.*, 1983).

Cleavage cycles and onset of gastrulation

In fish and amphibians Dettlaff (1964*) found a correlation between the duration of the [synchronous] cleavage cycles, the number of divisions and the onset of gastrulation. This holds for diploid as well as haploid and polyploid embryos within a normal temperature range.

Nuclear inactivation by X-rays in the loach during the period from fertilisation till the mid-blastula always leads to an arrest of development before the onset of gastrulation (Neyfakh, 1959). This is the developmental period that precedes the onset of nuclear function. It demonstrates the importance of nucleocytoplasmic interactions for the preparation of gastrulation during the subsequent period of asynchronous divisions (see Dettlaff, 1964*). Chulitskaia (1970) found in sturgeons and frogs that, although the beginning of desynchronisation starts later in the succession of cleavage cycles at lower temperature, the ratio between the length of the period from fertilisation to the onset of gastrulation and the duration of the synchronous cycles does not change.

Satoh (1977) observed that in *Xenopus laevis* gastrulation always starts during the same cleavage cycle; i.e. around the onset of the 15th cycle in the animal blastomeres. Boterenbrood *et al.* (1983) noted that the 13th cleavage cycle seems to be a special one; it deviates from both the preceding and the succeeding ones in that its duration shows an inverse correlation with the time elapsing from its beginning to the onset of gastrulation. The 13th cleavage cycle may therefore be of particular significance in the preparation of the gastrulation process, e.g. with regard to mRNA synthesis.

Blastocoel formation

In *Xenopus laevis* the earliest manifestation of blastocoel formation occurs during first cleavage in the form of an expansion of a localised area at the furrow tip (Kalt, 1971 *a*), but a true cleavage cavity begins to form during third cleavage in the centre of the egg, where vacuoles unite at the furrow tips (Motomura, 1960 in *Rhacophorus* embryos). The cavity enlarges during succeeding cleavages. This intercellular cavity is 'sealed off' from the environment by junctional complexes formed at the boundary of pre-existing and nascent membrane (Kalt, 1971 *b*). According to Tuft (1961 *a*) water is the only constituent that is taken up from the environment in appreciable quantities during early development. It is transported to the blastocoelic cavity. Tuft thinks that a net influx of water occurs through the roof of the blastocoel and a smaller net efflux through the cells of the vegetal region, the balance of the two fluxes accounting for the accumulation of the blastocoelic fluid. Tuft (1962) assumed that blastocoel formation is caused by an energy-dependent water-pumping mechanism, but it is more likely that it is due to a passive, osmotic water flow following active Na^+ pumping (see below; see also Zotin, 1965*).

C. Slack, Warner & Warren (1973) measured the K^+ and Na^+ ion concentrations in the early embryos of *Ambystoma mexicanum* and *Xenopus laevis* and found that the total internal K^+ concentration remained constant at nearly 60 mM l^{-1}, while the intracellular K^+ concentration was close to 100 mM l^{-1} cell water. The total internal Na^+ concentration remained constant at nearly 50 mM l^{-1}, while the intracellular Na^+

concentration fell steadily from 80 to 30 mM l^{-1} cell water between fertilisation and the onset of gastrulation. The intercellular fluid contained 100 mM l^{-1} Na^+, but only 1 mM l^{-1} K^+. These ions must have been transferred from the cells to the intercellular spaces, ending up in the blastocoelic cavity. The blastocoelic fluid has a pH of 8.5–8.8 (Holtfreter, 1943b) and contains only a low Ca^{2+} concentration (0.5 mM l^{-1}). At the late blastula stage, when the blastocoelic cavity has reached its maximal size, it occupies about 20% of the egg volume in *Xenopus*, containing 350 nl of fluid. Løvtrup (1965a*) stated that the blastocoelic fluid is initially isotonic with the blastomeres but later decreases in tonicity due to water uptake. The water uptake is likely to be an osmotic effect, since blastocoel formation can be suppressed by 100 mM sodium chloride or by 200 mM sucrose in the external medium (Holtfreter, 1943a; see also Tuft, 1962; Morrill, Kostellow & Murphy, 1974). C. Slack & Warner (1973) showed that the formation of the blastocoelic cavity can be blocked by injection into the cavity of ouabaine, an inhibitor of Na^+/K^+ ATPase. Gerhart (1980*) states that blastocoel formation is caused by the special ion-transporting properties of the new intercellular membranes, the low ion permeability of the outer pre-existing membrane and the close association of the blastomeres through tight junctions, keeping the embryo an osmotically closed system. Blastocoel formation is the automatic consequence of these physiological properties.

Morrill, Kostellow & Watson (1966) noted that the blastocoelic fluid of *Rana pipiens* embryos is 35–40 mV positive with respect to an external 0.1-strength Ringer solution. Replacement of external Na^+ by choline ions leads to a decrease in potential, which suggests that the potential is the result of Na^+ transfer from the medium to the intercellular space. Dicaprio, French & Sanders (1976) found no detectable electrical differences between the blastocoel and the perivitelline fluid in the early *Xenopus* embryo. This could not be confirmed by de Laat, Barts & Bakker (1976). They also measured the coupling ratios between pairs of blastomeres at the 4-, 8- and 16-cell stages in *Xenopus laevis* and observed that adjacent cells were electrically coupled but that there was no direct pathway between non-adjacent cells, so that the blastocoel hardly plays a role in the transport of ions from cell to cell. The blastocoel therefore (at least electrically) isolates the 'ectodermal' blastomeres of its roof from the endodermal vegetal yolk mass cells which form its floor, except in the peripheral region, where the two moieties are in direct contact with each other (cf. Nieuwkoop, 1973*).

Electrical coupling and cell junctions between blastomeres

The membrane potential (MP) of *Rana ridibunda* eggs increases during the cleavage period. Up to the 16-cell stage, K^+ ions seem to play an important role in the generation of the MP, but after the 16-cell stage it hardly depends on the external K^+ concentration any longer (Kvavilashvili *et al.*, 1977). The same holds for the eggs of the loach, *Misgurnus fossilis* (Kvavilashvili *et al.*, 1972).

In electrical measurements on *Triturus pyrrhogaster* eggs S. Ito & Hori (1966) noted that the resistance of the junctional membranes of the blastomeres is low compared to that of the external plasma membrane. Vilain (1974) made similar observations on *Pleurodeles* eggs. All three authors came to the conclusion that the blastomeres of the early embryo are electrically coupled. Palmer & Slack (1970) found a tighter electrical coupling in *Triturus* than in *Xenopus*. C. Slack & Warner (1969) noted that the specific membrane resistance of isolated blastomeres in *Xenopus laevis* is about one-tenth that in *Triturus pyrrhogaster* as determined by S. Ito & Loewenstein (1969). However, de Laat & Barts (1976) point out that in *Xenopus* the very high resistance of the outer membrane of the egg may lead to an exaggerated impression of interblastomeric conductance.

Sanders & Zalik (1972) and Sanders & Dicaprio (1976*b*) observed that in *Xenopus laevis* embryos junctions resembling gap junctions begin to develop between apposed membranes at the 8-cell stage and are fully formed at the early blastula stage. They concluded that during early development coupling is mediated by specialised junctions. S. Ito & Loewenstein (1969) found that in isolated blastomeres of *Triturus pyrrhogaster*, which show low ion permeability, junctional membrane regions are no longer discernible. When such cells come into contact with each other, cell communication is rapidly re-established by reformation of junctional membrane regions with high permeability. S. Ito, Sato & Loewenstein (1974*a*) noted that cells are capable of developing and maintaining cell-to-cell coupling during the entire cell cycle.

Colchicine and cytochalasin B, which both block cytokinesis, do not prevent the development and maintenance of cell coupling. Treatment with dinitrophenol, CN^-, $HgCl_2$, arsenite and *N*-ethylmalcimide breaks down established junctions and thus interferes with coupling. Metabolic inhibitors and SH-blocking agents do not affect coupling (S. Ito, Sato & Loewenstein, 1974*b*; S. Ito, Yamashita & Ahsako, 1977*a*). According to Ito and coworkers these facts are most easily explained by assuming that the permeability of junctional membrane channels depends on cytoplasmic free Ca^{2+} in such a way that elevation of the free Ca^{2+} concentration leads to closure and lowering to opening of channels, as already suggested by Loewenstein (1967*a, b*). Gap junctions in embryos resemble those in adults but are smaller. Due to the unusually high resistance of the outer membrane, coupling in early embryos partially seems to occur via non-junctional pathways, but the presence of low-resistance junctions is required for the observed permeability of interblastomeric membranes, which allows the exchange of ions, metabolites and nucleotides (Sheridan, 1976*).

S. Ito, Yamashita & Ahsako (1977*b*) observed coupling across a

Nuclepore filter as soon as cell processes traversing the filter are capable of forming permeable membrane junctions. Turin & Warner (1977) found that carbon dioxide reversibly abolishes ionic communication between cells of early amphibian embryos. They suggest that it is the change in intracellular pH occurring during uncoupling – from pH 7.7 in 0% to pH 6.4 in 100% carbon dioxide – that controls the junctional permeability, rather than the intracellular Ca^{2+} concentration. Bennett, Spray & Harris (1981) noted that the conductance of gap junctions that is responsible for the coupling of blastomeres can be controlled by a modest increase of H^+ ions and by a relatively large increase of free Ca^{2+} ions. Electrical coupling through gap junctions is such a common occurrence in embryos that their role in intercellular transport of information is rather plausible, although gap junctions are not always present at sites of embryonic interaction (see Gilula, 1980*). Palmer & Slack (1969) observed that the anaesthetic 'halothane' reversibly slows down or blocks cleavage, and drastically lowers the coupling ratio.

Blastomere coupling has also been studied by the transfer of dyes, but this method has its limitations. C. Slack & Palmer (1969) noted that the high ion permeability of junctional membranes does not hold for the fluorescein anion, which has a molecular weight of only about 300. In *Xenopus* embryos fluorescein does not spread beyond the injected blastomere. This must be due to particular properties of the dye, which may be strongly adsorbed to cytoplasmic components. In isolated blastomere pairs of *Ambystoma*, *Rana* and *Xenopus* embryos which are electrically coupled, the junctional permeability of the dye Lucifer yellow is markedly and reversibly lowered by moderate transjunctional polarisation in either direction, making junctional conductance a plausible mechanism of intercellular communication (Spray, Harris & Bennett, 1979; see also Sheridan, 1976*). (See further postscript on p. 94).

Cell adhesion

Letourneau, Ray & Bernfield (1980*); Steinberg & Poole (1981, 1982*) and Poole & Steinberg (1981, 1982) state that cell adhesion is the major controlling factor in cell migration, cell shaping, cell division and cell differentiation. Cell adhesion is based on non-covalent chemical bonds between cells and between cells and extracellular matrix. Desmosomes, zonulae adherentes and synapses play a role in cell adhesion, but gap or tight junctions do not. The basal lamina of cells consists of glycoproteins and sometimes proteoglycans and hyaluronic acid, as well as fibronectin, while the extracellular matrix may contain proteins, glycoproteins, proteoglycans, glycosaminoglycans and gangliosides.

Apart from the general notion of the role of cell-specific chemical patterns on cell surfaces, Curtis (1961) has called attention to the temporal specificity of cell adhesion as a possible additional mechanism of cell sorting in reaggregates.

Reaggregation of *Xenopus laevis* blastula cells obtained by dissociation in Ca^{2+}-free medium evidently depends on the addition of extracellular Ca^{2+}. According to Jumah & Stanisstreet (1982) and Stanisstreet & Jumah (1982) Ca^{2+} binding agencies and Ca^{2+} inhibitors affect cell reaggregation by modifying cell surface structure. Atsumi & Takeuchi (1980) and W. A. Thomas *et al.* (1981) distinguish between tight, Ca^{2+}-dependent and loose, Ca^{2+}-independent adhesion. Grunz (1969) found that reaggregation does not require new RNA synthesis, since actinomycin D does not interfere with it during the first 30 h. Reaggregation does need new protein synthesis, since cycloheximide and puromycin interfere with reaggregation. This was confirmed by Shiokawa *et al.* (1982), who concluded that translation of maternal mRNA plays an important role in cell adhesion of reaggregating cells of early embryos.

Roberson, Armstrong & Armstrong (1980), studying the adhesive properties of cells at the 32–64-cell stage in *Rana pipiens*, found that contrary to the apical or outer cell surface, the lateral and basal or inner cell surfaces are strongly adhesive. Whereas the outer surface shows convolutions but is devoid of microvilli, the inner ones lack convolutions but have scattered microvilli.

Lectins bind to active groups on the cell surface, thus influencing adhesion between cells. Nosek (1978) observed that in blastulae of *Xenopus laevis*, lectin binding occurs on the inner side of surface layer cells and on deeper cells as well. H. L. Harris & Zalik (1982) demonstrated the presence of endogenous carbohydrate-binding proteins in soluble extracts of cleavage, gastrula and neurula stages of *Xenopus laevis* embryos. β-D-galactoside-binding lectins may actually play a role in the adhesiveness of blastula cells (see also Ranzi, 1962*; Hubbert & Miller, 1974*; Overton, 1974*; Barendes, 1980*; as well as Chapter 12, p. 165).

DNA and RNA synthesis during cleavage

The fertilised amphibian egg contains a large store of nucleotides needed for DNA replication during cleavage. Baltus & Brachet (1962) stated that the embryo contains a constant DNA nucleotide pool until the late blastula stage, when rapid new DNA nucleotide synthesis begins. Deuchar (1966*) suggests, however, that in *Xenopus* measurable new DNA nucleotide synthesis already starts during cleavage. The rate of DNA nucleotide synthesis increases during the first few cell divisions due to a stockpile of maternal DNA polymerases, reaching a maximum at the mid-blastula stage. The rate declines rapidly at the end of cleavage and continues to do so in later stages (Gurdon, 1968*b**). These changes are reflected in the rate of uptake of ^{3}H-thymidine and its subsequent incorporation into DNA (Loeffler & Johnston, 1964).

The results of actinomycin D treatment demonstrate that cleavage is independent of new mRNA synthesis up to a late blastula stage. Protein synthesis during early development involves, among other things, the synthesis of large amounts of histones, although a stored pool is provided by histone synthesis during oogenesis. All histone fractions found in the chromatin of differentiated somatic cells of *Xenopus laevis* are already present or are newly synthesised beyond the 16-cell stage. Important changes occur at the mid-blastula stage, leading to a partial decondensation of the chromatin as prerequisite for new mRNA synthesis (Destrée, 1975). Histone protein present in the yolk platelets cannot play a role in early development, since yolk breakdown does not really start before neurulation (Brachet, 1965*). However, Nakamura, Hayakawa & Yamamoto (1966) observed yolk decomposition in the dorsal marginal zone of blastulae of *Triturus pyrrhogaster*. Wright (1978*) could demonstrate that cleavage proteins are exclusively of the maternal type. This tallies with the fact that actinomycin D treatment does not interfere with their synthesis (Kubo & Wright, 1977).

Although in *Xenopus* the period of synchronous cleavage is chiefly given over to DNA replication and synthesis of nuclear and cell membrane proteins, some RNA synthesis occurs in the form of 4S RNA needed for terminal tRNA turnover, and of some new mRNA with a moderately long half-life (Brown & Littna, 1964; E. H. Davidson *et al.*, 1965). However, according to Bachvarova *et al.* (1966) tRNA synthesis does not start before stage 9. mRNA synthesis takes place particularly at later cleavage stages (Gross, 1967*a*, *b*), following the increasing DNA content of the embryo (Brown, 1966*). This new mRNA is probably preserved for later development, notably for the process of gastrulation (Gurdon, 1968*b**). During synchronous cleavage new synthesis of rRNA and 5S RNA could not be demonstrated (see also Gurdon & Woodland, 1968*, 1969; Woodland & Gurdon, 1968).

The period of synchronous cleavage is characterised by the absence of nucleoli; these first form after the appearance of the interphases. Nakamura & Yamada (1971) found that in *Xenopus laevis* a prenucleolar body is formed at stage 7 but is replaced by the primary nucleolus at stage 9. The latter develops into the true nucleolus at stage 12 (Nieuwkoop & Faber, 1975).

Ignatieva & Rott (1970) observed that lengthening of the interphases of the cell cycle in the cleaving teleost egg coincides with the onset of morphogenetic nuclear function, detectable by X-ray nuclear inactivation. This nuclear function is not strictly temporally correlated with the onset of RNA synthesis, however. During asynchronous cleavage an enormous increase (10–20-fold) in DNA-like RNA synthesis takes place, representing mainly mRNA and some tRNA.

The end of rapid synchronous cleavage at about the mid-blastula stage is extremely important for early cell diversification in the embryo (Gurdon, 1968 b*). According to Bachvarova *et al.* (1966) in *Xenopus* a critical period immediately precedes the onset of gastrulation. Old mRNAs are now rapidly broken down and new ones are synthesised. It is at this time that the arrest of development of embryos with lethal genes and of lethal hybrids occurs, and the lethal effect of X-irradiation as well as the arrest caused by actinomycin D treatment become manifest (see also the reviews by Brachet, 1965*, 1967 a*, b*, 1980*; Denis, 1974*; E. H. Davidson, 1976*; Gurdon, 1974*, 1977*; Laskey, Gurdon & Trendelenburg, 1979*).

Regional differences in metabolism and in RNA and protein synthesis

Landström (1977 a) observed a higher metabolism and energy consumption in dorsal than in ventral blastomeres of amphibian embryos. In contrast, Thoman & Gerhart (1979) failed to find a dorso-ventral metabolic gradient in *Xenopus laevis*. However, Shiokawa & Yamana (1979) found that rRNA synthesis starts 4 h earlier in dorsal than in ventral cultured blastomeres isolated from 4-cell *Xenopus* embryos.

Amoebae of the slime mould *Dictyostelium discoideum* show a stronger chemotactic response to dorsal than to ventral cells of the late blastula of *Ambystoma mexicanum*, suggesting a higher cAMP production by dorsal cells (Nanjundiah, 1974).

Capco & Jeffery (1981) observed an uneven distribution of maternal poly(A)$^+$ RNA in the oocyte cytoplasm and in developing embryos of *Xenopus laevis*; injected vegetal pole poly(A)$^+$ RNA ultimately exhibiting a vegetal–animal gradient. C. R. Phillips (1982) examined animal/vegetal, dorsal/ventral and right/left portions of *Xenopus* embryos and found approximately four times more total RNA in animal than in vegetal regions, and a 1.5-fold higher RNA content in the most dorsal region than in the other five regions. Poly(A)$^+$ RNA showed temporally and regionally specific changes in concentration during development, with a higher concentration in the most animal region. Mikawa & Hiroshe (1982) studied the protein composition of the four pairs of blastomeres of the 8-cell *Xenopus* embryo by means of two-dimensional gel electrophoresis combined with a highly sensitive silver staining method: although the overall protein pattern is very similar, a few proteins segregate into one particular pair of blastomeres along the animal–vegetal or the dorso-ventral axis. These observations demonstrate that some maternal proteins become spatially segregated in the very early embryo.

The vegetal yolk cleaves more slowly than the animal, yolk-poor cytoplasm, leading to the first synthetic differences within the embryo

(Gurdon, 1968*b**; Woodland & Gurdon, 1968). 4S RNA and mRNA synthesis begins in the presumptive endoderm and in the dorsal inner equatorial region of the *Xenopus* gastrula, and only later in the ventral inner equatorial region and the animal presumptive neurectoderm (Bachvarova *et al.*, 1966).

Opinions differ about the onset of rRNA and 5S RNA synthesis during embryonic development. According to Bachvarova *et al.* (1966) and Denis (1968*) new rRNA synthesis in *Xenopus* embryos is first detectable at the onset of gastrulation, but Gurdon (1968*b**) states that rRNA synthesis is not detectable before the late gastrula/early neurula stage. Anyway, rRNA synthesis starts several hours earlier in the vegetal endodermal cells than in the other cells of the embryo, indicating that cell diversification begins in the endodermal moiety of the embryo (Flickinger *et al.*, 1966, 1967*b*; Flickinger, Moser & Rollins, 1967*c*; Flickinger, 1970*a*; Flickinger & Daniel, 1972).

Cleavage in the meroblastic anamnia and in the amniotes

(See Chapter 7, p. 64.)

Early regional diversification in chordate eggs and embryos

A spatial segregation of presumptive organ anlagen occurs in the fertilised egg of uro- and cephalochordates prior to cleavage in the form of a spatial localisation of different ooplasms. During cleavage the various ooplasms become localised in different blastomeres (see Conklin, 1905*a*, *b* in *Cynthia* and Conklin, 1932 in *Branchiostoma lanceolatum*; Fig. 13, p. 122). T. C. Tung, Wu & Tung (1962*a*) determined the localisation of the various ooplasms at the 8- and 32-cell stage of *Branchiostoma belcheri*. The presumptive yellow crescent region of the ascidian egg contains a contractile plasma membrane lamina and a cytoskeletal structure. Contraction phenomena may be responsible for the ultimate localisation of the yellow crescent in the ventral vegetal hemisphere of the egg (Jeffery & Meier, 1983). From (among other things) injection of muscle cell cytoplasm into epidermal blastomeres, Whittaker (1979, 1982, 1983) concludes that a quantitative control of muscle acetylcholinesterase synthesis determines the genetic expression of the presumptive muscle cells. Cytochalasin B which arrests cleavage, does not affect the normal transcriptional and translational control mechanisms of muscle acetylcholinesterase synthesis.

The egg of amphioxus has a much greater capacity for regulation than does the ascidian egg (T. C. Tung, 1934). T. C. Tung, Wu & Tung (1958) investigated the developmental potencies of the various blastomeres of *Branchiostoma belcheri* at the 2–16-cell stages. Any blastomere(s) containing all five formative ooplasms may regulate into a complete embryo. Thus the amphioxus egg is not a mosaic egg, as Conklin (1933) thought it to be. Rotation through 90° or 180° of the animal blastomeres at the 8-cell stage does not lead to a reversal of polarity, so that at that stage the dorso-ventral polarity of the egg must mainly reside in the vegetal hemisphere (T. C. Tung, Wu & Tung, 1960*b*).

The totipotent blastomeres of the teleost blastodisc become segregated from the liquid yolk by an interjacent yolk-syncytial layer. This primary segregation probably plays an important role in the formation of the embryonic anlage (see Chapter 10, p. 99).

Blastocoel formation in the amphibian embryo leads to a spatial segregation of the animal, cup-shaped ectodermal moiety from the vegetal endodermal yolk mass, except in the equatorial peripheral region of the embryo, where the two moieties remain in contact.

The segregation of the primary, endodermal hypoblast from the totipotent epiblast forms the first spatial diversification in the reptilian and avian blastoderm (Eyal-Giladi & Kochav, 1976). A similar segregation into a double-layered embryo takes place in the mammalian embryo after the formation of the amniotic cavity (see Chapter 7, p. 68 and Fig. 7e on p. 67; see further Chapter 10, p. 116).

Evaluation

Cleavage in the amphibian egg starts with a period of synchronous cell divisions during which DNA replication takes precedence almost entirely over RNA synthesis. This is followed by an asynchronous cleavage phase during which the cell cycles of individual blastomeres lengthen differentially as a result of the appearance of G_1 and G_2 phases, in which RNA transcription starts.

During the fragmentation of the egg into cells, cleavage waves can be distinguished starting in the animal hemisphere and spreading towards the vegetal pole region. These waves are the expression of a more pronounced lengthening of the cell cycles in the vegetal than in the animal hemisphere. The eccentricity of the initiation points of the later cleavage waves indicates the very early existence of dorso-ventral polarity as well as left/right asymmetry. In the amphibians, the 13th cleavage cycle seems to be of particular significance for the initiation of the gastrulation process. The transition from synchronous to asynchronous cleavage, called the mid-blastula transition, is thought to be linked up with the exhaustion of a cytoplasmic factor by the nuclear replication process, and leads to the appearance of a new developmental programme with renewed mRNA synthesis, preparing the embryo for the important morphogenetic event of gastrulation. The spatial segregation of the embryo into a totipotent animal moiety and an already determined endodermal yolk mass, partially separated from each other by the blastocoelic cavity, is the prerequisite for the spatial interaction of the two components during mesoderm induction.

The special characteristic of the cleavage period is that it is a preparatory period for subsequent cellular interactions. Cleavage is only a special form of cell division conditioned by the enormous dimensions of the egg. Blastocoel formation is the natural consequence of the properties of the newly formed intercellular membranes and those of the outer egg membrane, only the former being comparable to those of cell membranes of differentiated cells.

POSTSCRIPT TO PAGE 94

When antibodies to the major protein of rat liver gap junctions are injected into one of the dorsal blastomeres of the 8-cell stage *Xenopus* embryo both dye transfer and electrical coupling between the progeny cells are disturbed, which leads to abnormal neural development. This demonstrates that intercellular communication through gap junctions is essential for normal development (see also Warner, Guthrie & Gilula, 1984; Slack, 1984).

10

The induction of the meso-endoderm

The ecto-, meso- and endodermal germ layers of the chordate embryo have long been viewed as its three primary moieties, since the three anlagen were thought to be topographically and functionally distinguishable very early in development. This rather preformistic notion was based on, among other things, the visible cytoplasmic segregation of ecto-, meso- and endoplasms in uro- and cephalochordate eggs (Conklin, 1905 a, b and 1932 respectively), as well as on the formation of the grey or light crescent as a putative precursor of the dorsal axial mesoderm in lower vertebrate eggs (phenomena which take place shortly before first cleavage). The first hint that meso- and endodermal structures could be epigenetic in origin was obtained when it proved to be inducible in competent amphibian gastrula ectoderm by bone marrow (Toivonen, 1953); the idea was definitively proved by Nieuwkoop and coworkers in the late 1960s. The induced meso- and endodermal structures will be called 'meso-endoderm'; we shall therefore speak of 'meso-endoderm induction'. In the amphibian embryo the induced endoderm should however be clearly distinguished from the vegetal, endodermal yolk mass (see p. 96) and in the avian embryo from the primary and secondary, endodermal hypoblast (see p. 116).

Among the vertebrates much work has been done on meso-endoderm formation in the amphibians, and during the last decades also in birds, so that we shall principally concentrate on these two groups as representatives of the anamnia and amniotes, respectively.

The induced meso-endoderm, which represents a separate moiety in the early vertebrate embryo, ultimately differentiates into a large number of different cell types and contributes to the formation of a large number of different organ systems. In this chapter we shall restrict ourselves to the initial induction of the meso-endoderm and the subsequent spatial segregation of the mesoderm, leaving their further development and differentiation for separate discussion in the chapters on organ formation (see Chapters 17, p. 228, and 18, p. 249).

The anamnian embryo

The epigenetic origin of the meso-endoderm

After having observed the formation of spino-caudal structures under the influence of kidney tissue implanted into early gastrulae of *Triturus* (Toivonen, 1938, 1940; Chuang, 1938–40), Toivonen (1953) noted that purely mesodermal structures can be induced in early gastrula ectoderm by guinea pig bone marrow.

Nakamura & Matsuzawa (1967), Nakamara & Takasaki (1970) and Nakamara, Takasaki & Mizohata (1970) isolated groups of blastomeres from the presumptive dorsal marginal zone at successively older stages of *Xenopus laevis* and found that mesodermal differentiation tendencies are expressed in cultured isolates from the 32–64-cell stage onwards. The mesodermal differentiations increased in size with advancing donor stage and changed from ventral-mesodermal into dorsal-mesodermal and endo-dermal structures, so that the mesoderm must be of epigenetic origin. These observations were confirmed by Koebke (1976, 1977). He also observed that in *Ambystoma mexicanum* a dorso-ventral gradient in mesodermal differentiation manifests itself in isolated marginal zone material from an early blastula stage onwards.

Nieuwkoop (1969*a*) noted that animal and vegetal blastomeres of middle to late blastulae of *Ambystoma mexicanum*, when cultured sepa-rately, only formed 'atypical ectoderm' and 'endoderm', respectively. When recombined they formed a whole range of mesodermal structures next to ecto- and endodermal structures, and (in the most successful cases) developed into nearly complete, normal embryos. Using xenoplastic recombinates of animal and vegetal blastomeres of blastulae of *Ambystoma mexicanum* and *Triturus alpestris* and recombinates of ^{3}H-thymidine-labelled and unlabelled axolotl blastulae, Nieuwkoop & Ubbels (1972) could convincingly demonstrate that all the mesodermal as well as some endodermal structures arise from the animal, 'ectodermal' moiety of the embryo. This conclusion was corroborated by a quantitative analysis of the various structures formed in animal/vegetal recombinates of blastulae of *Xenopus laevis* (Sudarwati & Nieuwkoop, 1971). Nieuwkoop (1973*)* therefore concluded that the entire presumptive mesoderm as well as the presumptive pharyngeal endoderm, localised above the extending blasto-poral groove of the gastrula, are derived from the totipotent animal moiety of the embryo through an inductive influence emanating from the yolk mass, the vegetal moiety of the embryo.

The induction of meso-endoderm in competent gastrula ectoderm apparently implies the suppression of ectodermal differentiation tendencies. Grunz (1973, 1976*) and Grunz *et al.* (1975) observed in TEM and scanning electron microscope (SEM) studies that the animal moiety of the early gastrula manifests tendencies towards epidermal differentiation in the form of autonomous formation of cilia, but these tendencies are suppressed during mesoderm induction with purified 'vegetalising factor' (see also below under heterogenous inductors, p. 110).

Independently of Nieuwkoop and coworkers, Ogi (1967, 1969) made similar recombinates in *Triturus pyrrhogaster* and obtained comparable results, but interpreted them as a 'vegetalisation' of the embryo, in analogy with Hörstadius' (1939*) double-gradient hypothesis for early sea urchin development. He assumed that the mesoderm was formed from both components of the recombinate. This notion was further developed by Nakamura and coworkers (see, among others, Nakamura, Takasaki & Ishihara, 1971). Nakamura & Takasaki (1971*b*) claimed that the amount of differentiated endoderm was larger in *Xenopus laevis* embryos from which both the most animal and the most vegetal octets of blastomeres were removed at the 32-cell stage, than after removal of the latter only. After removal of the most animal octet they claimed to have observed a weak animalisation. They ascribed these phenomena to animal–vegetal regulation in the operated embryos. However, they did not make a quantitative analysis of their recombinates and did not take a possible interference with gastrulation into account, which may have influenced the differentiation of endodermal structures. As already mentioned, the xenoplastic recombinates made by Nieuwkoop & Ubbels (1972) showed not only induced mesodermal but also induced dorsal endodermal structures.

Unfortunately, Nieuwkoop (1973*) also used the term 'vegetalisation' for this phenomenon, which Nakamura (1978*) then interpreted in the sense of Hörstadius' double-gradient hypothesis. Nakamura considers mesoderm *induction* as described by Nieuwkoop to be an *abnormal* phenomenon, while *normal* mesoderm formation occurs under the influence of oppositely directed animal and vegetal gradients. In order to prevent further misunderstanding, we shall only use the terms *meso-* and *endodermisation*. In our opinion there is a very valid argument against Nakamura's application of Hörstadius' double-gradient hypothesis to early amphibian development: Agents such as monoiodine acetate (CH_2ICO_2Na), sodium thiocyanite (NaSCN) and zinc chloride ($ZnCl_2$), which have a strong animalising action in sea urchin development (Runnström, 1967), have no animalising effect on amphibian embryos when used at moderate concentration, and at higher concentrations only become toxic; these agents do not convert presumptive endoderm into meso- or ectoderm (P. D. Nieuwkoop, unpublished observations; see Nieuwkoop, 1973*).

We must nevertheless briefly discuss some arguments raised against the epigenetic origin of the meso-endoderm in amphibian embryos. The first argument is based on the differentiation of the animal quartet of blastomeres isolated after the completion of third cleavage. In cases where a horizontal third cleavage plane divides the *Rana* egg into a smaller animal quartet of about one-third of the egg volume and a larger vegetal quartet, the animal quartet forms only atypical ectoderm (Ancel & Vintemberger, 1948*). In cases where the third cleavage plane is slightly lower or slightly oblique, some additional mesenchyme is formed (Bustuoabad & Pisanó, 1971). Grunz (1977) in a number of cases even found more extensive mesodermal structures forming from isolated animal quartets of *Triturus alpestris* 8-cell stages. He concluded that the state of regional determination is more or less rigidly fixed at the 8-cell stage. It should be realised, however, that the animal quartet of blastomeres may easily contain some coarse yolk material, e.g. from the dorsal vitelline wall, which originally formed part of the vegetal yolk mass. Inclusion of such yolk material may be responsible for the induction of small mesodermal structures in embryos with a slightly lower or oblique third cleavage plane, which is a common phenomenon in cleaving axolotl and newt eggs.

The second argument is based on the initial suggestion of Motomura (1967), taken up by Ave, Kawakami & Sameshima (1968) and Nakamura & Aochi (1970), that the ectoderm of the early gastrula should consist of a mixed cell population. They observed that dissociated early gastrula ectoderm placed in an electric field segregates into three separate bands, two of which would correspond to similar bands of dissociated mesodermal cells and the third one to that of dissociated neural cells. Mesodermal and neural inductive actions would respectively stimulate the development of one cell type and eliminate the other. This rather revolutionary notion is first of all in flagrant contradiction of the fact that massive cell necrosis has never been observed during mesodermal and neural induction. It was finally disproved by S. Dasgupta & Kung-Ho (1971), who showed that during development the initially heterogeneous cell population becomes homogeneous without loss of cells by migration or selective death (see further Nieuwkoop, 1973*).

Summarising, we may say that *meso-endoderm formation is indeed an epigenetic phenomenon and is caused by an inductive action emanating from the already firmly determined vegetal, endodermal moiety and acting upon the still pluripotent animal moiety of the amphibian blastula.* This conclusion has been confirmed by Asashima (1975), Grunz (1975*) and Maufroid & Capuron (1977); see also T. Yamada (1981*).

Mesoderm induction represents the first large-scale interaction between the two primary moieties of the embryo, i.e. the totipotent animal and the endodermal vegetal moiety. Moreover, it is an interaction between cells, since the egg by that time has been subdivided into a rapidly increasing number of blastomeres separated by newly formed permeable intercellular membranes. Although in normal development cleavage and blastocoel formation are prerequisites for mesoderm formation, it may possibly also occur in activated, non-cleaving anuran eggs (see under pseudogastrulation, Chapter 11, p. 138).

The evocation of a meso-endoderm-inducing centre in the dorsal yolk mass

In the chapter on symmetrisation we came to the conclusion that symmetrisation of the amphibian egg may be the consequence of an interaction between the cytoplasm-rich animal moiety and the yolk-rich vegetal moiety of the egg, an interaction which is enhanced in the region of the dorsal vitelline wall. The initial dorso-ventral polarisation manifests itself primarily in the animal hemisphere of the fertilised, uncleaved egg. In the blastula, however, dorso-ventral polarity is exclusively found in the vegetal hemisphere in the form of a strong mesoderm-inducing capacity in the dorsal portion and only weak inducing capacities in the lateral and ventral portions of the yolk mass (Nieuwkoop, 1969 a; Boterenbrood & Nieuwkoop, 1973). Consequently, at some time the dorso-ventral polarity must be transferred from the animal to the vegetal moiety.

What do we know about the timing and mechanism of this transfer? Ancel & Vintemberger (1948*) found that 180° rotation of the animal quartet of blastomeres performed at the 8-cell stage of *Rana fusca* does not affect the dorso-ventral polarity of the embryo, so that in *Rana* the dorso-ventral polarity must already be localised in the vegetal blastomeres at that stage. However, Milan recently observed that the same operation in *Xenopus laevis* embryos led to a rather high percentage of reversals of the dorso-ventral axis. When performed at the beginning of the 16-cell stage, the percentage of reversals was lower (Cardellini, Milan & Sala, 1982). The discrepancy between Ancel & Vintember's results in *Rana* and Milan's in *Xenopus* may of course be due to species differences, but may also be the result of a delay in the transfer of polarity in Milan's experiments owing to the use of a Ca^{2+}- and Mg^{2+}-free 0.67 M phosphate buffer for disaggregation. These experiments anyhow suggest that the transfer of the dorso-ventral polarity from the animal to the vegetal moiety of the embryo occurs around the 8-cell stage. This is in agreement with the recent observation by Takasaki that regional differences in mesoderm-inducing capacity between the dorsal and ventral portions of the vegetal yolk mass are first discernible at the 16–32-cell stage (H. Takasaki, personal communication).

In the meroblastic teleost egg, in which no actual invagination occurs during 'gastrulation' (Ballard, 1966 a, b, c, 1973 a, b, c*), the syncytial periblast seems to play an important role in mesoderm formation (Luther, 1937, 1938; Oppenheimer, 1947*; Devillers, 1961*). In our opinion the syncytial periblast may act as a mesodermal inductor, like the yolk mass in amphibians.

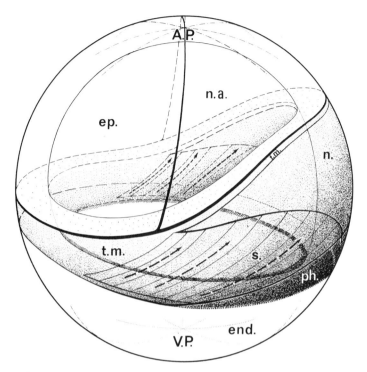

Fig. 9. Diagrammatic representation of meso-endoderm induction in an amphibian blastula/early gastrula shown in spherical perspective (central yolk mass omitted). A.P., animal pole; presumptive anlagen of end., endoderm; ep., epidermis; n., notochord; n.a., neural anlage; ph., pharyngeal endoderm; s., somites; t.m., tail mesoderm; V.P., vegetal pole; end., endodermal yolk mass.

The role of the blastocoelic cavity in meso-endoderm induction

In amphibians the blastocoelic cavity forms at the boundary between the animal and vegetal moieties of the cleaving embryo. It restricts the contact between these moieties to the equatorial, peripheral region of the blastula, where they remain attached to each other (Nieuwkoop, 1969 a, 1973*). This leads to the formation of an equatorial, ring-shaped zone of mesoderm, the so-called marginal zone. Collapse of the blastocoelic roof caused by centrifugation (Pasteels, 1953 a, b, 1954 a), can result in additional mesoderm formation. The blastocoelic cavity therefore plays an important *negative morphogenetic role* in mesoderm formation. It should be realised that the presumptive mesoderm is not a regular ring-shaped structure of equal width but occupies a broad zone on the dorsal side and only a narrow region on the ventral side, with intermediate width in the lateral regions (Fig. 9).

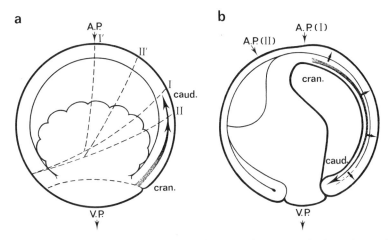

Fig. 10. Diagram of amphibian gastrulation, showing reversal of cranio-caudal axis. a, Early gastrula: with I and II, extension of marginal zone in urodeles and anurans, respectively; I' and II', extension of cranial border of presumptive neural area in urodeles and anurans, respectively. b, Advanced gastrula, with position of animal pole (A.P.) in urodele and anuran embryos, respectively. V.P., vegetal pole; small arrows, inductive interaction.

The development of the regional pattern of the meso-endoderm

For a proper understanding of the formation of the mesodermal mantle it is desirable to distinguish between its regional differentiation along the dorso-ventral and along the cranio-caudal axis. In the essentially tubular mesodermal anlage the dorso-ventral axis finds its expression in the regional differentiation of the mesoderm from the dorsal midline towards the ventral side, while the cranio-caudal axis is established by the regional differentiation of the marginal zone from its boundary with the vegetal moiety in the direction of the animal pole. It should moreover be realised that the latter axis reverses during the gastrulation process (see Fig. 10; see Nieuwkoop, 1977*).

It should be emphasised that the formation of the mesodermal mantle is a process extending over a long period of time. Meso-endoderm induction probably begins already during the fourth or fifth cleavage, while the final pattern of the meso-endoderm does not seem to be established before an advanced neurula stage. In this long-lasting process a number of steps may be distinguished: first the gradual extension of the inductive influence into the animal moiety of the blastula, which may be called early mesoderm induction (see p. 102), then the subsequent interactions within the mesoderm during gastrulation (see p. 104), and finally the late interactions within the mesodermal mantle and between the mesoderm and the overlying neural anlage and underlying endoderm (see p. 105; see also Toivonen, 1978*).

The dorso-ventral differentiation of the mesoderm during early meso-endoderm induction

By recombining animal caps with vegetal yolk masses of middle to late blastulae of *Ambystoma mexicanum*, rotating the former through 0°, 90° or 180°, Nieuwkoop (1969b) could demonstrate that in the blastula the dorso-ventral polarity resides in the vegetal yolk mass and shows no relation with the location of the grey crescent in the animal cap. Boterenbrood & Nieuwkoop (1973), who tested the inductive capacity of dorsal, lateral and ventral portions of the yolk mass of the axolotl blastula, found that ventral and lateral portions induced only weakly and evoked exclusively ventral mesodermal structures such as lateral plate, blood islands and primordial germ cells (PGCs), while dorsal portions induced much larger formations of predominantly dorsal, axial meso-endoderm in the form of notochord, somites and pharynx endoderm (demonstrating the existence of a dorsal meso-endoderm-inducing centre). They moreover concluded from these experiments that purely quantitative differences in inductive capacity are responsible for the regional differentiation of the mesoderm.

T. Yamada (1950a) showed that ventral mesoderm can be converted into dorsal axial mesoderm under the influence of ammonia. As discussed in Chapter 6, ammonia treatment probably causes an elevation of the intracellular pH. A gradual shift from ventral to dorsal mesodermal and further to endodermal structures can also be brought about by a gradual increase in the concentration of a purified vegetalising factor isolated from 13-day-old chicken embryos, acting upon early gastrula ectoderm (Grunz, 1975*).

Nieuwkoop (1973*) made a three-dimensional reconstruction of the regional induction of the meso- and endoderm in the amphibian blastula starting from a single inductive factor (see Fig. 9, p. 100). On the basis of these data Weijer, Nieuwkoop & Lindenmayer (1977) developed a computer programme for a spherical diffusion model with a single initial source.

The cranio-caudal differentiation of the marginal zone during early meso-endoderm induction

Early meso-endoderm induction involves partial endodermisation of the dorsal marginal zone leading to the formation of the pharyngeal endoderm (Nieuwkoop & Ubbels, 1972; see Fig. 11). In normal development endodermisation also seems to occur, though to a much lesser extent, in the lateral and ventral marginal zone in direct contact with the yolk mass endoderm (Koebke, 1976). Gebhardt & Nieuwkoop (1964) observed a strong endo- and mesodermisation of axolotl gastrula ectoderm under the

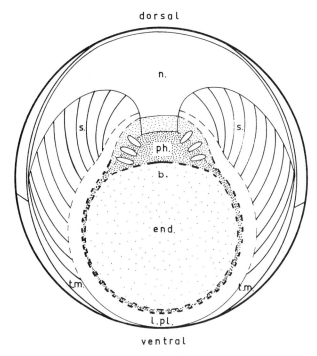

Fig. 11. Fate map of endo- and mesodermal organ anlagen in very early urodele gastrula, seen from vegetal side. b., Initial blastoporal groove; end., endodermal yolk mass; l.pl., lateral plate mesoderm; n., notochord; ph., pharyngeal endoderm; s., somitic mesoderm; t.m., tail mesoderm: Stippled area represents induced endoderm (after P. D. Nieuwkoop & G. A. Ubbels, 1972 and Koebke, 1977).

influence of lithium chloride (see further below, p. 112). Endodermisation was particularly pronounced when blastula ectoderm was used. This was confirmed by Engländer & Johnen (1967) in *Ambystoma* and *Triturus*.

Tseng (1963 a) suggested from her induction experiments with guinea pig bone marrow that a single inductive factor is responsible for the regional dorso-ventral differentiation of the mesoderm as well as for the induction of endodermal structures, both as a function of the intensity (duration × concentration) of the inductive action (Tseng, 1963 b). Kocher-Becker, Tiedemann & Tiedemann (1965) and Asahi *et al.* (1979) found that purified vegetalising factor injected into the blastocoelic cavity of early *Triturus* gastrulae caused a massive endodermisation of the animal moiety, which led to a complete evagination of the gastrula. Asahi *et al.* (1979), using a high concentration of the vegetalising factor, induced predominantly endodermal structures in early gastrula ectoderm. Vegetalising factor diluted with non-inducing protein induced dorsal mesodermal structures as well. From these observations both Kocher-Becker and Asahi concluded that the primary action of the vegetalising factor is an endodermisation of the ectoderm, while mesoderm formation represents a secondary phenomenon evoked by the induced endoderm. M. Minuth & Grunz (1980) confirmed this conclusion. It should however be emphasised that in these experiments rather high concentrations of vegetalising factor were used, which may easily have led to a predominance of endodermal structures. Moreover, a competition between endo- and mesodermal competences has never been observed, as is the case e.g. between neural and mesodermal ones. Recently Grunz (1983) convincingly demonstrated that an increasing concentration of vegetalising factor first induces ventral mesodermal, then dorsal mesodermal and finally also endodermal structures. It may therefore be concluded that *meso- and endodermisation on the basis of purely quantitative differences in a single inductive action is still the most plausible interpretation of the dorso-ventral and cranio-caudal differentiation of the marginal zone.*

In the double-layered anuran embryo the thin outer epithelial layer of the animal moiety seems to be exclusively endodermalised, while the thicker inner layer is predominantly or exclusively mesodermalised during meso-endoderm induction (Sudarwati & Nieuwkoop, 1971). However, Asashima & Grunz (1983) observed that vegetalising factor can induce dorsal mesodermal structures in the isolated outer as well as inner layer of *Xenopus* gastrula ectoderm.

Intermediate inductive interactions in mesoderm formation
At the early gastrula stage the primary mesoderm-inducing action has not yet extended throughout the entire marginal zone (Ban-Holtfreter, 1965). Kaneda & Hama (1979) found that at an early gastrula stage the presumptive trunk organiser possesses hardly any mesodermal differentiation tendencies and no inductive capacity, but that it acquires notochordal and somite differentiation tendencies and spinal-cord-inducing capacity when approaching the dorsal blastoporal lip. The Japanese school of T. S. Okada, T. Hama, H. Takaya and K. J. Kato already in the 1940s and 1950s observed that the dorsal lip of an early gastrula differentiates into notochord and somites and induces rhombencephalon and spinal cord in competent ectoderm. The same material, when tested directly after its invagination around the dorsal blastoporal lip, differentiates into prechordal mesoderm and pharyngeal endoderm of the head organiser and induces archencephalic neural structures (see Hoessels, 1957; Kato, 1957, 1958, 1959, 1963a, b; Nieuwkoop, 1973* and Takaya, 1978*).

There is a striking temporal and spatial relationship between the changes in differentiation tendencies and inductive capacity in the newly invaginated head organiser and the changes in the presumptive trunk organiser. This suggests a two-way interaction between the two regions of the presumptive archenteron roof for the duration of their transient apposition during the invagination process (Nieuwkoop, 1973*; Kaneda, 1980). However, Kaneda (1981) proposes that the influence of the head inductor on the presumptive trunk inductor is mainly achieved by tangential induction. Kaneda & Hama (1979) concluded from isolation and recombination experiments that the mesodermisation of the trunk organiser occurs in two steps: the first step involves a weak mesodermisation with the appearance of ventral mesodermal differentiation tendencies, as a consequence of early mesoderm induction by the dorsal yolk mass. The second step involves an intensification of mesoderm induction when the trunk organiser approaches the dorsal blastoporal lip and finds itself in apposition to the head organiser, leading to the appearance of dorsal mesodermal differentiation tendencies and a corresponding increase in neural inductive capacity. Kaneda & Suzuki (1983) observed mesodermisation within 12 h of early gastrula ectoderm implanted in the region of the trunk organiser of *Cynops pyrrhogaster*, with concomitant development of neural inducing capacity. Miyagawa & Suzuki (1969) found that, analogous to the head organiser, trunk organiser aged *in vitro* shows a decrease in caudal and an increase in cephalic differentiation tendencies, with a corresponding change in neural inductive capacity.

Ignatieva (1960 *a*, *b*) reported similar changes in differentiation tendencies and inductive capacity in the head organiser of *Acipenser* embryos. She also found an increase in notochordal differentiation tendencies and inductive power in the trunk organiser during gastrulation (Ignatieva, 1961, 1962).

A protein fraction from the germinal vesicle content of *Rana pipiens* oocytes, when injected into the blastocoel, causes a considerable enlargement of the brain, which Malacinski (1972) denoted as 'supercephalisation'. This may result from an enlargement of the prechordal plate due to a further transformation of trunk into head inductor (see also Nieuwkoop & Sutasurya, 1983 and Chapter 12, p. 161).

Late inductive interactions in the regional differentiation of the mesodermal mantle

T. Yamada (1937, 1939 *a*, *b*, *c*, 1940), in transplantation and recombination experiments made with neurula mesoderm, showed that the dorso-ventral regional differentiation of the mesodermal mantle depends on an inductive influence emanating from the notochordal anlage and spreading dorso-ventrally with decrement.

An enhancement of notochordal differentiation as a result of concomitant neural induction was first reported by Yamada (1939 *b*), Toivonen & Saxén (1966) and Toivonen (1967) as well as by Kurrat (1974, 1977 and 1978). Nieuwkoop & Weijer (1978), in a quantitative analysis, could demonstrate that during the neural induction process the neural anlage exerts an inductive action upon the underlying mesoderm, enhancing notochordal differentiation tendencies particularly in the caudal portion of the archenteron roof. Reciprocally, in the absence of neural induction and after dis- and reaggregation the dorsal mesoderm forms more ventral structures (Forman & Slack, 1980). This was confirmed by Suzuki, Mifune & Kanéda (1984).

The tail somites develop from the most caudal portion of the neural plate (Bijtel, 1931, 1936) under an inductive influence emanating from the underlying archenteron roof (Spofford, 1945, 1948, 1953; see further Chapter 12, p. 158).

Chuang & Tseng (1956a, b), Tseng (1958) and Muchmore (1957a, b, 1958) called attention to the fact that certain differentiations of the mesodermal mantle are subject to local inductive influences from particular regions of the underlying endoderm, e.g. the blood islands, the heart and the splanchnic mesoderm (see further Chapter 17, p. 241). Capuron & Maufroid (1981) showed that the differentiation of blood cells and primordial germ cells (PGCs) in the ventral mesoderm requires a complementary interaction of the latter with the caudal endoderm. This had already been suggested by Nieuwkoop (1946) for the PGCs.

Homoiogenetic induction of meso- and endoderm

Spemann & H. Mangold (1924) were the first to observe so-called assimilatory induction of dorsal mesoderm after grafting the dorsal blastoporal lip into the ventral side of a host embryo. Nishijima *et al.* (1978) and Kurihara & Sasaki (1981) demonstrated that the mesoderm-inducing action of carp swim bladder can spread from mesodermalised ectoderm into adjacent (vitally stained) competent ectoderm. Asashima (1980) studied the extension of both mesodermal and neural induction in competent gastrula ectoderm of *Ambystoma mexicanum*, using grey crescent material from the uncleaved egg up to the blastula as inductor. It is evident that *in primary mesoderm induction the inductive action spreads tangentially from cell to cell.*

Sasaki, Iyeiri & Kurihara (1976a) found that in the case of homoiogenetic induction the differentiation of the inductor material can be affected by an excess of competent ectoderm, which changes its regional differentiation from a dorsal to a more ventral type. It is not clear whether this is a dilution effect or is due to an unknown influence of the non-induced ectoderm. In recombinates of dorsal and ventral marginal zone the latter could be partially dorsalised but the former apparently could not be ventralised (J. M. W. Slack & Forman, 1980). This may be the result of the more strongly determined state of the dorsal mesoderm. In these experiments the observed effect required contact between the two parts for at least 48 h, showing that it is a late effect.

Summarising, it may be concluded that *a single mechanism may be responsible for the early induction of the meso-endoderm and its subsequent differentiation into head and trunk organiser.* However, different mechanisms may be functioning in the later interactions between the neural plate and the underlying archenteron roof, and in the local interactions between specific regions of the endo- and mesoderm. We think that the most fundamental aspect of mesoderm induction is its spreading from cell to cell over a considerable distance and during a relatively long period of time; a spreading accompanied by a spatial decrease in intensity (see Chapter 12, p. 159, and Chapter 19, p. 287).

The temporal aspects of meso-endoderm induction

The development of the inductive capacity of the endoderm

Asashima (1975) noted that vegetal yolk material from uncleaved eggs and 2–4-cell stages induced only ventral mesodermal structures in competent ectoderm, such as blood cells and endothelium. Yolk material taken from later stages induced notochord, somites and pronephros. Endodermal structures were induced by yolk material taken from the 8-cell stage onwards. The yolk mass of the blastula showed maximal meso-endoderm-inductive capacity. Takasaki recently found that differences in inductive capacity between dorsal and ventral yolk material became discernible between the 16- and the 32-cell stage (H. Takasaki, personal communication). In our opinion the gradual acquisition of inductive capacity by the vegetal yolk mass points to an inductive action from the animal moiety of the egg upon the vegetal yolk mass (see further Chapter 11, p. 130). Boterenbrood & Nieuwkoop (1973) observed that the inductive capacity of the dorsal endoderm declines first, starting at stage 9 and dropping to almost zero at stage 10^-, while the decline in the lateral endoderm begins at stage $9\frac{1}{2}$ and that in the ventral endoderm not before stage 10^-.

The meso- and endodermal competence of the reacting ectoderm

Using lithium as a meso- and endodermalising agent on presumptive ectoderm of *Ambystoma* and *Triturus* embryos, Grunz (1968) noted that mesodermal competence was maximal at the mid-to-late blastula stage. Endodermal differentiations decreased in late blastula/early gastrula ectoderm. While lithium evoked caudal trunk and tail structures in *Ambystoma* ectoderm, it evoked middle and posterior trunk structures in *Triturus* ectoderm (Engländer, 1962a; Engländer & Johnen, 1967), so that the reacting ectoderm also plays a role in regional meso-endoderm induction, probably by species-specific quantitative differences in competence.

Using carp swim bladder as a strong inductor of meso-endoderm, Sasaki *et al.* (1976b) showed that in *Triturus pyrrhogaster* mesodermal competence appears in the presumptive ectoderm 10–16 h before the beginning of gastrulation and increases rapidly over the next few hours. Schmidt (1979) found maximal mesodermal competence in axolotl ectoderm at a late blastula/very early gastrula stage. The decline of mesodermal and neural competences of the reacting ectoderm was first investigated by Leikola (1963, 1965), using guinea pig bone marrow as inductor. He noted that in isolated early gastrula ectoderm of *Triturus alpestris* the mesodermal competence rapidly declines after about 15 h of cultivation *in vitro* and has completely vanished after about 18 h, which corresponds to stage $12\frac{1}{2}$ (Harrison). Neural competence disappears after about 24 h of cultivation (see further Chapter 12, p. 156 and Fig. 19, p. 284). The two competences, which markedly overlap each other in time, are apparently in some way antagonistic, since neural competence was suppressed as long as the mesodermal competence was sufficiently strong and could only express itself when the mesodermal competence declined. Sasaki & Iyeiri (1972*b*) showed that the mesoderm-inducing capacity of guinea pig bone marrow can be counteracted by a subsequent treatment with liver extract containing a potent neuralising agent, when applied less than 7 h after the first treatment. Leikola (1963, 1965) observed that ageing of the ectoderm *in vitro* and *in vivo* is essentially similar, although it is slightly slower *in vitro*. Sasaki, Iyeiri & Tadokoro (1975*a*) found that *Triturus pyrrhogaster* ectoderm of stage 12 (Okada & Ichikawa, 1947) begins to lose its competence for dorsal mesodermal structures after 7 h of cultivation *in vitro*, and that for ventral mesodermal structures after about 24 h.

Differentiation of meso- and endoderm from early gastrula ectoderm of *Triturus* and *Ambystoma* under the influence of guinea pig bone marrow becomes discernible after 3 h of contact with the inductor, already shows the full pattern of mesodermal differentiation after 7 h of contact and reaches its maximum after 9 h of contact (Engländer, 1962*b*). A shortening of the duration of contact leads to the appearance first of spino-caudal, then of deuterencephalic and finally of archencephalic neural structures, with gradual concomitant disappearance of meso- and endodermal structures. Engländer concluded that both neuralising and mesodermalising factors must be present in the bone marrow, the former being suppressed by the latter during longer-lasting contact. A 'subliminal' contact with bone marrow of 0.5–1 h, which is in itself insufficient to evoke mesodermal differentiation in competent ectoderm, nevertheless leads to induction of spino-caudal structures when the ectoderm is subsequently brought under the inductive influence of liver tissue, which acts as a neural inductor, so that a sensitisation for mesoderm formation must have taken place during the short contact with the first inductor (Noda, Sasaki & Iyeiri, 1972*a*, *b*, *c*). Similar observations have been made by Sasaki & Iyeiri (1972*a*). It is of interest to mention that Katoh (1962) could already obtain mesoderm induction in competent gastrula ectoderm after 15 min of contact with purified guinea pig bone marrow extract prepared according to T. Yamada

& Takata (1961). (See also Sasaki *et al.*, 1976 and Kawakami & Sasaki, 1978*).

Sutasurya & Nieuwkoop (1974) failed to find any dorso-ventral differences in mesodermal competence in axolotl blastula ectoderm but observed a decrease in mesodermal competence from the equator towards the animal pole of the blastula.

Toivonen, Vainio & Saxén (1964) found that actinomycin D inhibits mesoderm formation induced by guinea pig bone marrow. Tiedemann, Born & Tiedemann (1967) could show that actinomycin D treatment does not affect the inducing capacity but interferes with the competence of the reacting cells. Cycloheximide also interferes with mesoderm formation (Tiedemann, 1969*). These findings emphasise once more the significance of the reaction system in the inductive interaction.

Using the reversible inhibitor of protein synthesis, cycloheximide, Grunz (1970) could lengthen the period of mesodermal competence of gastrula ectoderm by the duration of treatment. From this observation he concluded that a loss of competence of the ectoderm requires protein synthesis. In our opinion the dependence on protein synthesis holds for the entire ageing of the ectoderm, including the decline of mesodermal competence (see also Kawakami & Sasaki, 1978*).

Double-embryo formation

A few words must be said about double-embryo formation, which is closely connected with mesoderm induction since it represents the formation of two separate mesodermal axis systems. It was first observed in amphibians by Schultze (1894) after inversion of *Rana fusca* eggs slightly compressed between glass plates. Penners & Schleip (1928 *a*, *b*) found that both blastopores always appeared at the edge of the displaced yolk mass. Pasteels (1938, 1939, 1940 *a*, *b*, 1941 *a*, *b*) repeated Penners & Schleip's experiments and observed that in double embryos one blastopore appeared near the original grey crescent and the other at some other site, usually starting slightly later. Dalcq & Pasteels (1937, 1938) attributed blastopore formation to an interaction between an animal–vegetal yolk gradient and a dorsal cortical morphogenetic field, but Penners & Schleip already suggested that additional axis formation may be primarily connected with the displacement of the yolk mass. Gerhart (1980*) and the present authors believe that double-embryo formation in inverted eggs is the result either of splitting of the yolk mass or of the formation of a second dorsalising centre, e.g. a so-called Born's crescent, by the displaced yolk mass at a time when the first dorsal centre is already firmly established. Double-embryo formation no longer occurs after rotation or inversion of 8-cell stages (Penners & Schleip, 1928 *a*, *b*; Curtis, 1962 *a*), when displacement of the yolk is interfered with by the third cleavage plane separating the animal from the vegetal cytoplasm (see further Nieuwkoop, 1973*; Gerhart, 1980* and Chapter 7, p. 60).

In constriction experiments Dollander (1950) regularly observed normal dorsal axis formation in isolated dorsal halves of *Triturus* eggs, but occasionally also in ventral halves. Gerhart (1980*) suggests that in these ventral halves the animal and vegetal cytoplasm are brought into intimate contact by the constriction procedure, causing the formation of a new dorsalising centre.

Double-embryo formation can also be achieved by the implantation of a dorsal blastoporal lip into the ventral side of an early gastrula. This experiment was first performed by Spemann & H. Mangold in 1924 and extended by Spemann (1931). They observed assimilatory (homoiogenetic) mesoderm induction by the graft. Nieuwkoop (1947) analysed the inter-action between the two developing axis systems in such double embryos and observed a mutual 'attraction' of the two anlagen, which often led to their partial fusion in the caudal region. Cooke (1972a, b, c) found in similar experiments that the two organiser regions set up competing fields for cellular orientation and migration. While removal of the dorsal blastoporal lip from a normal early gastrula often led to complete regulation, removal of the host organiser markedly enhanced the activity of the implanted second organiser (Cooke, 1973a, c). Cooke explained the behaviour of the two organiser regions in the sense of Wolpert's positional information hypothesis, but the present authors are inclined to ascribe it to the high adhesiveness of mesodermalised cells, leading to a mutual attraction of the two centres during gastrulation.

Meso-endoderm induction by so-called heterogenous inductors

A very extensive literature exists about heterogenous inductors. We shall restrict ourselves to the more significant facts and interpretations. For more detailed information the reader is referred to the articles and reviews by Saxén & Toivonen (1962*), Yamada (1962), Tiedemann (1968a, b), Saxén & Kohonen (1968*), Tiedemann (1975*) and Grunz (1975*).

Toivonen (1938, 1940) and Chuang (1938–40) described the induction of spino-caudal structures, consisting of notochord and somites with accompanying spinal cord, in *Triturus* early gastrula ectoderm by guinea pig kidney. In 1953 Toivonen discovered an almost purely mesodermal inductor in guinea pig bone marrow. C. Takata & Yamada (1960) noted that guinea pig bone marrow not only induces mesodermal but also endodermal structures in early gastrula ectoderm of *Triturus pyrrhogaster*. This was confirmed by Kocher-Becker & Tiedemann (1971). Saxén, Toivonen & Vainio (1961) showed that HeLa cells have a strong meso-endoderm-inducing capacity, which is destroyed by heat treatment. Eng-länder (1962a) observed that the mesoderm-inducing effect of guinea-pig bone marrow gradually declines with decreasing duration of contact and is replaced by weak neural induction. He concluded that meso-endoderm induction requires a much longer period of contact than neural induction.

It was soon recognised that deuterencephalic and spino-caudal inductions were the result of a combined action of neural and mesodermal inductors. This was elegantly demonstrated by Toivonen, Saxén & Vainio (1963), who implanted the two different inductors side by side in the same host.

Iyeiri & Kawakami (1962) tested the regional inductive capacity of a large number of rat tissues upon early gastrula ectoderm of *Triturus pyrrhogaster* and observed a range of induced arch- and deuterencephalic, spino-caudal and meso- and endodermal structures, without any relationship between the inductive specificity and the embryonic origin or physiological function of the adult organs and tissues. This was confirmed by Saxén & Toivonen (1962*). Kawakami *et al.* (1977) found that carp swim bladder contains a very potent meso- and endodermalising factor. Whereas Saxén & Toivonen (1957), Kawakami (1958) and Chiang (1964) found mammalian liver to be a strong neural inductor, Chuang (1963) and Y.-H. Wang, Mo & Shen (1963) found liver to be a potent inductor of meso- and endodermal structures. This shows that the inductive specificity of a particular organ may not be a constant phenomenon.

Purification and characterisation procedures for neural and mesodermal factors have been developed by the Finnish school of S. Toivonen & L. Saxén, the Japanese schools of T. Yamada and of I. Kawakami, and the German school of H. Tiedemann. While Toivonen & Saxén and Yamada started from mammalian liver and bone marrow, Tiedemann isolated both factors from 9–12 day old chick embryos. The latter material was also used as the starting point by N. Sasaki and coworkers, while Kawakami and coworkers used carp swim bladder for the isolation of the mesodermal factor. For isolation procedures and characterisation of the active factors the reader is referred to the following articles and reviews: Saxén & Toivonen (1962*), Saxén & Kohonen (1968*), T. Yamada & Takata (1961), Kawakami *et al.* (1966*), Kawakami *et al.* (1977), and Tiedemann (1966*a, b*, 1975*, 1978*).

The general outcome of these studies is that the active factor in neural-inducing tissues is likely to be a ribonucleoprotein, the nucleic acid component of which is not essential however, while the mesodermalising or vegetalising factor seems to be a protein with a molecular weight (MW) of 30000–32000. Substances of high MW that inhibit the action of these inducing factors have also been isolated. Faulhaber (1972) and Faulhaber & Geithe (1972) isolated active factors from early amphibian embryos. Tiedemann (1975*, 1978*) mentioned that a mesoderm- and endoderm-inducing factor prepared from amphibian oocytes and early developmental stages has properties that are similar but not identical to those of the vegetalising factor isolated from chick embryos. However, the principal problem is not the *similarity* or *identity* of active factors isolated from adult tissues and early embryos, as discussed by Saxén & Toivonen (1962*), Croisille (1963*), Tiedemann (1975*) and others, but the role of these substances in the *normal* process of induction. Bagnara (1961*) already stated that there is no evidence for the identity of heterogenous inductors and inductive factors acting in normal development. This fundamental question becomes even more significant in the light of the endo- and mesodermalising action of the Li^+ ion (an ion that is completely foreign to the embryo).

The meso- and endodermalising action of Li^+ and other cations

Masui (1960*a*, *b*, *c*, 1961) and Ogi (1961) were the first to discover the mesoderm-inducing action of lithium chloride on *Triturus pyrrhogaster* early gastrula ectoderm. Gebhardt & Nieuwkoop (1964) observed meso-dermalisation as well as endodermalisation of *Ambystoma mexicanum* blastula ectoderm under the influence of Li^+ ions. Masui (1966) noted that the neuralising effect of Zn^{2+}, NH_4^+ and urea inhibits the mesodermalising action of Li^+ ions. Under the influence of Li^+ ions, competent gastrula ectoderm of *Ambystoma mexicanum* forms spino-caudal, neural, meso- and endodermal structures, while *Triturus vulgaris* ectoderm produces almost exclusively meso- and endodermal structures, with predominance of the latter (Engländer & Johnen, 1967). Previous treatment of the ectoderm with an archencephalic inductor (alcohol-treated guinea pig liver) markedly enhanced endodermal differentiation evoked by lithium chloride (Johnen & Engländer, 1967). Using lithium chloride as meso-endo-dermal inductor (see below), Johnen & Albers (1978) showed in sandwich experiments that the regional differentiation of induced mesodermal structures can be influenced by the amount of available competent ectoderm. While lithium chloride treatment of *Ambystoma* ectoderm evokes mainly meso- and endodermal differentiation, sodium thiocyanite in combination with lithium chloride acts as a neuralising agent. Pretreat-ment with lithium chloride enhances the archencephalic induction evoked by sodium thiocyanite. In the reverse experiment the endodermal differ-entiations evoked by lithium chloride are enhanced, which demonstrates the reciprocal sensitising effect of the two ions when acting successively (Johnen, 1970). Masui (1961) and Mager (1972) found that a lowering of the pH intensifies the Li^+ effect. M. Suzuki (1981) noted that foetal calf serum induces mesodermal structures in competent *Cynops pyrrhogaster* gastrula ectoderm pretreated with Ca^{2+}- and Mg^{2+}-free salt solution, which itself has a neuralising effect.

The great similarity between the sensitising or inhibitory effects of successive actions of *different ions*, of neural and mesodermal *heterogenous inductors*, and of neural and mesodermal *natural inductive actions*, in our opinion strongly favours the conclusion that the various actions cannot be based on identity of inducing factors, the more so since the embryo does not contain any Li^+ or Zn^{2+} ions. L. G. Barth (1965) and L. G. Barth & Barth (1972, 1974) suggested that this resemblance may be based on an identical elementary action of ions and proteins. They assumed that proteins act as inductors by increasing the intracellular concentrations of cations, which in turn would bind to phosphate groups in the DNA, thus lifting the inhibitory action of histones. In our opinion the great similarity in the actions of ions, heterogenous and natural inductors rather points to the great significance of the reaction system in induction processes: heterogenous inductor studies may tell us something about the nature of the reaction system rather than of the inductors.

Vainio *et al.* (1962) used fluorescent antisera against heterogenous inductors and observed a substantial transfer of high MW antigenic material from the heterogenous inductor to the reacting ectoderm. Tiedemann, Born & Tiedemann (1972) suggested that the vegetalising factor isolated from chick embryos may act directly upon the genome of the ectodermal cells, since its action is inhibited by combining it with chick or *Xenopus* DNA.

While Ranzi (1975) ascribes the vegetalising action of Li^+ to its stabilising effects on proteins, and Cigada, Maci & de Bernadi (1968) to a non-specific inhibitory effect on transcription and translation in the reacting ectoderm, Duncan (1979) proposes that Li^+ treatment, like the action of UV and of ouabaine, raises the intracellular Ca^{2+} concentration, with concomitant morphogenetic effects at critical stages of development.

The inductive capacity of extracellular matrix material

By means of toluidine blue and lanthanum staining K. E. Johnson (1977*a*) detected the presence of a small amount of extracellular material already in the amphibian blastula; the staining increased dramatically during gastrulation, so that this material could be involved in mesoderm induction.

Landström & Løvtrup (1979) observed mesoderm formation after treatment of early gastrula ectoderm of *Ambystoma mexicanum* with heparan sulphate. This substance could indeed be demonstrated to be present in endodermal Ruffini or flask cells (Løvtrup, Landström & Løvtrup-Rein, 1978*). The latter differentiate in the peripheral yolk mass which acts as mesodermal inductor. Flickinger (1980) found that heparin can cause dorsalisation of mesoderm by stimulating DNA and RNA synthesis, which points to a late action.

Kawakami *et al.* (1978) reported that reptilian liver with its perisinusoidal basement membrane is a potent mesoderm-inducing agent, whereas chick and guinea pig liver, which lacks this type of basement membrane, induces only neural structures. Tanaka *et al.* (1976) observed that glomerulus basement membrane and dentine matrix of 23 day old rabbit fetuses are potent mesodermal inductors. Hoperskaya *et al.* (1984) found that basement membrane of eye tissues, particularly Bruch's membrane and *lamina vitrea*, not only affects periocular mesenchyme but also acts as a strong mesodermal inductor on early gastrula ectoderm, inducing a broad range of mesodermal cell types. They conclude that the action of [vegetalising] heterogenous inductors is based on the presence of extracellular basement membrane components, and that the diversity of mesodermal cell types induced depends on the competence of the reacting cells.

W. W. Minuth (1978) placed a Nuclepore filter between bone marrow and competent ectoderm and observed induction by diffusible substances passing through the filter pores, without cellular contact between inductor and reacting tissue. Similar results were obtained by Kawakami (1976) and Kawakami *et al.* (1978), using swim bladder as the mesodermal inductor in transfilter experiments. However, the filters may have permitted the transfer of extracellular matrix material (see further Chapter 12, p. 169).

These findings not only emphasise the significance of the reaction system, the competence of which changes with the progression of development, but may actually point to a different interpretation of the spatial propagation of the inductive action. The inductor may act through the formation of extracellular matrix, which activates its own formation in adjacent cells, thus spreading from cell to cell. There still remains the problem of how the inductive action can spread *with decrement*. As we shall see under neural induction, p. 162, this may be based on a loss of competence due to the ageing of the reaction system. However, at an early gastrula stage, when mesodermal competence is still high, mesodermal differentiation tendencies decline from the equator of the embryo towards the animal pole and are still chiefly confined to the presumptive head organiser. This decline cannot simply be accounted for by a loss of competence (see further Chapter 19, p. 287).

Some biochemical data relevant to meso-endoderm induction

We have seen in Chapter 9, pp. 90, 92, that mRNA synthesis already starts during cleavage and blastula formation and that the endoderm is the first region of the embryo where RNA synthesis is initiated (Gurdon, 1968 b*; Bachvarova *et al.*, 1966). This mRNA has a relatively long half-life and seems to be preserved for later development, probably for mesoderm formation and gastrulation. Brachet (1965*) pointed out that the latter processes undoubtedly require new transcriptional activity in the form of mRNA synthesis. There is a dramatic increase in DNA-like RNA synthesis during the mid-to-late blastula stage, comprising mRNA and possibly tRNA (Gurdon, 1968 b*). Gurdon & Woodland (1969) and Flickinger,

Daniel & Greene (1970) suggest a direct antagonism between DNA and RNA synthesis, allowing gene expression as soon as DNA replication slows down. DNA/RNA competition–hybridisation experiments by E. H. Davidson, Crippa & Mirsky (1968) and Denis (1968) demonstrated that the RNA synthesised at cleavage stages is non-competitive with gastrula RNA and thus represents an entirely different molecular population. Using hybrids of two *Rana* species characterised by different electrophoretic protein variants, Wright (1978*) could show that the proteins synthesised during cleavage and blastula stages are still of maternal origin (see also Gerhart, 1980*).

Noda & Kawakami (1976) reported that during the first 4 h of mesoderm induction in *Triturus pyrrhogaster* ectoderm by rat bone marrow mitosis is repressed, the cells being arrested in the S phase of their cell cycle. This is followed by accelerated cell division up to 12 h, after which cell proliferation returns to normal. Nakamura & Takasaki (1971*a*) found that actinomycin D and actinomycin S_3 treatment of presumptive dorsal marginal zone material blocks further development when applied at stages 7 and 8 (O. & I.), but not when applied at stage 9 and older. This indicates that the RNA synthesized at stages 7 and 8 is essential for *mesodermal differentiation*. It does not mean, however, that new RNA synthesis is required for meso-endoderm induction that precedes it.

Grunz & Tiedemann (1977) were unable to evoke meso- and endodermal differentiation in amphibian early gastrula ectoderm with cyclic nucleotides (see further Chapter 12, p. 165). Tiedemann & Born (1978) and J. Born *et al.* (1980) found that, contrary to the neural inductor, the vegetalising factor loses its inductive capacity when bound to activated CH-sepharose. They concluded that the vegetalising factor must be internalised before being able to exert its action. As far as we know these results have not been confirmed so far (see also Chapter 12, p. 166).

Neufang *et al.* (1978) noted that a proteoglycan isolated from chick embryos counteracts the inducing capacity of Tiedemann's vegetalising factor. This proteoglycan is in turn inactivated by hyaluronidase.

J. Born *et al.* (1972) noted that the first visible effect of a vegetalising inductive action is a change in cell affinity of the reacting cells, which develop a high affinity for endodermal cells. This was confirmed by Grunz (1972) and Grunz & Staubach (1979*b*), who observed a lowering of the cell surface charge of *Triturus* gastrula ectoderm cells treated with vegetalising factor for 24 h which is correlated with a change in cell affinity. Grunz (1976*) found that the plant lectin Con A supports cell aggregation, probably by binding sugar residues of glycoproteins at apposing cell surfaces. Cell affinity among ectodermal and among mesodermal cells is likewise enhanced by Con A, so that the change in cell affinity observed during mesoderm induction must be based on binding sites other than those occupied by Con A. K. Y. Yamamoto *et al.* (1982) observed that ferritin-Con-A and ferritin-DBA (horse gram agglutinin) bind weakly to the extracellular matrix of the outer cell surface of the ectodermal cells but more intimately to that of internal cell surfaces; DBA inhibits mesoderm induction.

Tseng (1982) observed a dispersion of chromatin in the nuclei of early gastrula ectoderm of *Cynops orientalis* that has been mesodermalised with guinea pig bone marrow extract.

Meso- and endoderm formation in the amniotes, studied in particular in birds

The origin of the embryonic and extraembryonic meso- and endoderm

In the flat avian blastoderm, mesoderm formation finds its expression in the development of a primitive streak in the epiblast layer. Pasteels (1937*a*, *b*) and Spratt (1955) still thought that the streak area only contains mesoderm, which moves towards the streak, invaginates, and spreads anteriorly and laterally. Vakaet (1962), using carbon marking, was the first to find indications of endoderm ingression through the streak. Isotopic cell marking showed that endodermal cells ingress through Hensen's node and the anterior streak region and give rise to the entire embryonic endoderm, a process which is completed at the head-process stage (Modak, 1965, 1966; Nicolet, 1965, 1970*a*, 1971*; Rosenquist, 1966). Since primitive streak formation thus involves the formation of both mesoderm and embryonic endoderm, Hara (1978*) proposed the term 'streak induction' instead of mesoderm induction for the organised formation of endo- and mesoderm, with the concomitant establishment of an antero-posterior axis.

We have already discussed in Chapter 7, p. 65, that the primary hypoblast segregates from the initially single-layered blastoderm by poly-invagination progressing from back to front (Eyal-Giladi, Kochav & Yerushalmi, 1975*a*; Eyal-Giladi & Kochav, 1976; Kochav, Ginsburg & Eyal-Giladi, 1980). Removal of the peripheral zone of the area pellucida prevents regeneration of the primary hypoblast (Azar & Eyal-Giladi, 1979). The secondary hypoblast is formed from the posterior margin of the blastoderm, the so-called Koller's sickle (Vakaet, 1962). It pushes the primary hypoblast towards the anterior and lateral margins of the blastoderm, while finally the ingressing tertiary hypoblast (embryonic endoderm) displaces the primary and secondary hypoblast towards extra-embryonic blastoderm areas (Modak, 1965, 1966; Nicolet, 1965, 1970*a*, 1971*; Rosenquist, 1966; Eyal-Giladi *et al.*, 1975*a*; see also Fig. 12).

Epiblast from primitive-streak-stage rat embryos formed structures of all three germ layers in homografts to the kidney capsule, whereas epiblast from head-fold stages formed only ecto- and mesodermal structures. This suggests that the embryonic endoderm is formed from the epiblast, as in

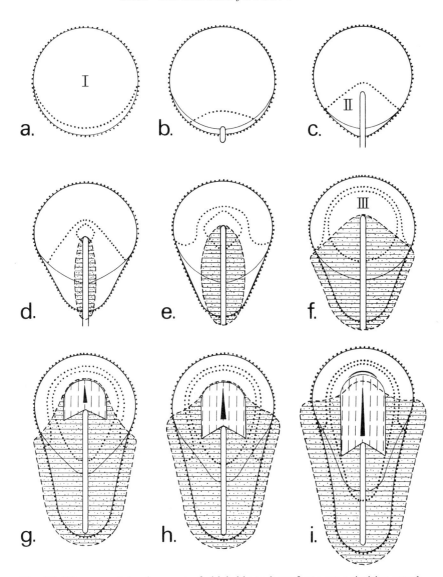

Fig. 12. Diagram of development of chick blastoderm from pre-primitive streak till head-fold stage: formation of primary (I), secondary (II) and tertiary (III) hypoblast (dotted lines); development of mesoblast (broken lines, horizontal hatching and stippling) and notochord (black); formation of neural anlage in epiblast layer (continuous lines and vertical hatching). (Redrawn after L. Vakaet, 1970.)

birds (Levak-Svajger & Svajger, 1974). Otherwise little is known about the nature of primitive streak formation in the mammalian embryo, although the process strongly resembles streak formation in birds.

The induction of the primitive streak

Waddington (1930, 1932, 1933) and Waddington & Schmidt (1933) were the first to show that the hypoblast plays an important role in primitive streak formation. Reversal of the [primary and secondary] hypoblast with respect to the epiblast can lead to a partial or complete reversal of the antero-posterior axis of the primitive streak. Lutz (1962) concluded from his blastoderm bisection experiments that the orientation of the embryonic axis is due to [secondary] endoblast formation from the posterior margin of the blastoderm. Rotation of the hypoblast at stage XIII† led to primitive streak formation in accordance with the orientation of the hypoblast, while the same operation at successively older stages showed a gradual change in streak orientation to that of the epiblast (Azar & Eyal-Giladi, 1981). The retention of hypoblastic cells underneath the primitive streak anlage in normal development allows for a prolonged inductive action (Azar & Eyal-Giladi, 1983). Mitrani & Eyal-Giladi (1981) found that hypoblast reconstituted after dis- and reaggregation can induce a primitive streak in competent ectoderm, but its orientation always follows that of the epiblast. Contrary to the situation in the amphibians, where in the blastula dorso-ventral polarisation is restricted to the endoderm, in the avian blastoderm both hypoblast and epiblast are polarised, although the polarity of the hypoblast normally prevails.

Lutz *et al.* (1963) observed that the blastoderm of the unincubated duck egg is still equipotential, any fragment being able to form a primitive streak. This was confirmed by Eyal-Giladi & Spratt (1964, 1965) for winter eggs of the chick, whereas in slightly more advanced summer eggs, in which the process of hypoblast formation had started, only marginal fragments, particularly from the posterior margin, were able to form a primitive streak. At a later stage, when the blastoderm consisted of epi- and hypoblast, central portions were again capable of streak formation. These findings were corroborated by experiments involving folding of early blastoderms either parallel or perpendicular to the antero-posterior axis (Eyal-Giladi, 1969, 1970a, b). These experiments in addition demonstrate that the antero-posterior polarity of the double-layered blastoderm is not yet firmly determined and can be altered by experimental procedures.

Gallera & Nicolet (1969) noted that the frequency of induction of a

† Eyal-Giladi & Kochav (1976) extended to Hamburger & Hamilton's (1951) normal table of the chick to earlier stages, distinguishing 14 stages prior to their stage 2, designated stages I–XIV.

secondary streak decreases at the mid-primitive-streak stage, while meso-dermal competence is lost at the definitive primitive-streak stage. In contrast to the area pellucida, streak induction is hardly possible in the area opaca (Gallera, 1971).

A Hensen's node grafted into the posterior portion of a young primitive streak becomes entirely endodermalised and induces another primitive streak in the overlying epiblast (Gallera, 1972). Vakaet (1965) found that Hensen's node induces only neural structures in competent epiblast while the middle portion of the primitive streak can induce a secondary primitive streak. Gallera & Nicolet (1969) showed that a primitive streak can be induced by middle and anterior thirds of the primitive streak taken from a mid-streak stage. Vakaet (1973) even obtained streak inductions in competent ectoderm by the posterior end of the streak, which by itself forms only extraembryonic mesoderm. These observations indicate the homoiogenetic character of mesoderm induction, which spreads from cell to cell. Homoiogenetic mesoderm induction was also observed by Eyal-Giladi (1969, 1970a) in transversely folded blastoderms up to mid-primitive-streak stage. Mesoderm induction is replaced by neural induction in folded definitive-streak embryos (see for further details Leikola, 1976a*; Hara, 1978*).

By inserting a millipore filter between the epiblast and hypoblast Eyal-Giladi & Wolk (1970) showed that a general [weak] inductive influence can pass through the filter, but that direct cellular contact is required for the formation of a mature primitive streak complete with Hensen's node. It is not clear to us what this means. Contrary to Gallera & Nicolet (1969), Eyal-Giladi & Wolk (1970) could demonstrate that the competence of the epiblast for streak formation outside the normal streak area appears at an early streak stage and disappears at the head-process stage, while neural competence appears at a late streak stage. Raveh, Friedländer & Eyal-Giladi (1971) found that the streak-inducing capacity of the secondary hypoblast manifests itself only after complete segregation of epi- and hypoblast. Azar & Eyal-Giladi (1981) concluded from experiments involving translocation of the hypoblast with respect to the epiblast that the streak-inducing capacity is mainly localised in the posterior region of the hypoblast. Mitrani, Shimoni & Eyal-Giladi (1983) found that the posterior [secondary] hypoblast is inductive, whereas the anterior [primary] hypoblast is not (see Fig. 12, p. 117). At stage 2 (Hamburger & Hamilton, 1951) the competence of the epiblast is already mainly restricted to the posterior half of the blastoderm, showing a rapid further antero-posterior restriction at later stages. The anterior end of the primitive streak topographically corresponds to the most anterior part of the secondary hypoblast that is still inductive and to the most anterior region of the epiblast that is still competent. The antero-posterior extension of the primitive streak therefore seems to be controlled by both the inductive capacity of the hypoblast and the loss of competence of the epiblast.

Data relevant to the possible nature of the induction process

The hypoblast acquires specific antigenic properties during its cytoplasmic segregation from the epiblast. The latter does not show any specific antigens before stage XIII, characterised by full [primary and secondary] hypoblast development (Wolk & Eyal-Giladi, 1977). The [secondary] hypoblast is characterised by the synthesis of particular proteins in the form of two main peaks on the electropherogram after ^{3}H-phenylalanine incorporation. Later two similar peaks arise in the epiblast (Eyal-Giladi *et al.*, 1975*b*). These authors suggest that one of these proteins may be responsible for induction. In our opinion, these peaks may just as well express an enhanced synthetic activity first occurring in the hypoblast and then in the epiblast. The burst of synthetic activity in the hypoblast coincides with the onset of its inductive activity (Eyal-Giladi *et al.*, 1975*b*). Blocking of sulfhydryl groups prevents streak formation. Cysteine stimulates post-nodal fragments to differentiate into axial mesodermal structures (K. V. Rao, 1969), which is comparable to dorsalisation of mesoderm in the amphibians.

Although Sherbet & Lakshmi (1967) suggest that FSH of the anterior pituitary may be the natural neural inductor, neural inductions observed with FSH-treated posterior primitive-streak mesoderm as inductor (Sherbet & Mulherkar, 1965) can in our opinion be adequately explained by a dorsalisation by FSH of the inductor used.

Local application of cAMP to the blastoderm leads to a bending of the embryonic axis, while high concentrations cause a disruption of the blastoderm (Gingle & Robertson, 1979). Bromodeoxyuridine (BUdR) which neither interferes with hypoblast formation nor with the development of its inductivity, affects the competence of the epiblast for streak induction (Zagris & Eyal-Giladi, 1982). This inhibitory effect of BUdR can be eliminated by methionine (Lee & Redmond, 1975).

Streak induction by heterogenous inductors and the possible role of extracellular matrix material

Streak formation can be evoked by different heterogenous inductors (McCallion & Leikola, 1967). Morphogenetically active substances have been claimed to include sulfhydryl groups (Waheed & Mulherkar, 1967; K. V. Rao, 1973), mRNAs from different sources (Butros, 1963*a*, *b*, 1965; Lee & Niu, 1973*); etc.). Lee & Niu (1973*) reported the induction of a secondary axis in definitive streak chick embryos by calf testis mRNA, but not by heart mRNA. Sherbet & Mulherkar (1963) found that treatment with FSH leads to a dorsalisation of embryonic and extraembryonic mesoderm. Waheed & McCallion (1969), who pretreated posterior primitive streak with lithium chloride, observed differentiation of prospective extraembryonic mesoderm into axial mesodermal structures (dorsalisation) and the induction of corresponding neural structures. Noto (1967) noted metaplasia in addition to malformations in early chick embryos treated with lithium chloride.

Mitrani & Eyal-Giladi (1982) found that both hypo- and epiblast of stage XIII (see note on p. 118) form a basement-membrane-like structure when cultured *in vitro*, which suggests a possible role of extracellular matrix material in mesoderm induction.

Rostedt (1968, 1971) states that the nature of the heterogenous inductors used is of minor importance, the response depending primarily upon the intrinsic potencies of the reacting cells.

Mesoderm formation in uro- and cephalochordates

As shown in Chapter 7, p. 63, the various regions of the ascidian and cephalochordate egg are equipotential before fertilisation. After fertilisation a process of cytoplasmic segregation starts, leading to a spatially segregated egg with separate regions of animal ectoplasm, dorsal neuro- and chordoplasm, ventral mesoplasm and vegetal endoplasm (Conklin, 1905 *a*, *b* in *Cynthia* and Conklin, 1932 in *Branchiostoma lanceolatum*; see Fig. 6, p. 63). During cleavage these different plasms become localised in separate blastomeres (see Fig. 13).

T. C. Tung (further to be noted as Tung) noted as early as 1934 that the egg of *Branchiostoma belcheri* has a much higher capacity for regulation than the ascidian egg, and consequently is not a 'mosaic' egg. Tung, Wu & Tung (1958) observed that isolated longitudinal halves of 4-cell stages regulate into complete embryos when the first cleavage plane more or less coincides with the plane of bilateral symmetry of the egg. Testing different longitudinal, transverse and oblique halves of 8–16-cell stages, Tung concluded that any group of blastomeres which contains all five organ-forming ooplasms may regulate into a complete embryo.

Isolated tiers of blastomeres of 32-cell stages developed more or less according to their prospective significance (Tung, Wu & Tung, 1959, 1960*a*, 1962*a*). However, recombinates of different animal–vegetal tiers of blastomeres formed much more than the mere sum of the prospective fates of the components; for instance, the recombinates $an_1 + veg_2$ and $an_2 + veg_2$ gave rise to more or less complete larvae (see Fig. 13). Tung, Wu & Tung (1960*a*) concluded that the veg_2 blastomeres must be

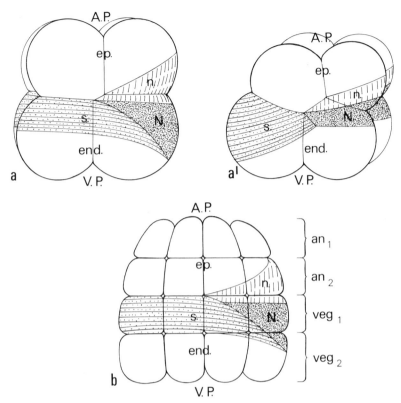

Fig. 13. Localisation of organ anlagen at the 8-cell stage: a, in *Branchiostoma belcheri* (after T. C. Tung *et al.*, 1962*a*); a′, in *B. lanceolatum* (after E. G. Conklin, 1932); b, localisation of organ anlagen at the 32-cell stage in *B. belcheri* (after T. C. Tung *et al.*, 1962*a*). A.P. animal pole; end., endoderm; ep., epidermis; n., neural anlage; N., notochord; s., somites; V.P., vegetal pole; an_1, an_2, veg_1 and veg_2, successive animal–vegetal tiers of blastomeres.

able to differentiate into notochord and somites under the influence of an_1. This was corroborated by Wu & Cai (1964), who also noticed self-differentiation of muscle tissue.

Grafting of prospective ectodermal cells of an_1 or an_2 into vegetal regions led to an endodermisation of the graft, whereas vegetal blastomeres grafted into animal regions invaginated and usually maintained their endodermal nature (Tung, Wu & Tung, 1961). Thus endodermal cells are far less readily converted to other cell types than ectodermal cells. Y. F. Y. Tung, Luh & Tung (1965*b*) removed increasing amounts of vegetal material from blastulae or early gastrulae and concluded from the results that the embryo has high regulative capacity for endoderm formation. Testing recombinates of various numbers of animal and vegetal blastomeres taken from 8–16-cell stages, T. C. Tung *et al.* (1965*a*) found that one 8-cell vegetal blastomere can still form a harmonious embryo when combined with a quartet or octet of animal blastomeres, provided it contains presumptive notochordal material.

Comparing these recombination experiments with similar experiments involving various animal–vegetal regions of blastulae of *Ambystoma mexicanum* made by Nieuwkoop (1969a, 1970) and Nieuwkoop & Ubbels (1972), Nieuwkoop & Sutasurya (1979*) suggested that also in *Branchiostoma* the mesoderm may be induced in the animal tiers of blastomeres by the vegetal, endodermal blastomeres. The development of the cephalochordate embryo may not only be far less mosaic-like than previously assumed by Conklin (1933), but may even be highly epigenetic, notwithstanding the early segregation of the different ooplasms.

Recently, Nakauchi & Takeshita (1983) could obtain nearly normal tadpoles from *Styela* half embryos which hatched and metamorphosed, thus demonstrating a considerable regulative capacity. Unfortunately nothing is known about possible interactions between animal and vegetal blastomeres during mesoderm formation in ascidians, so that the epigenetic nature of mesoderm formation cannot at present be extended to the entire phylum of the chordates (see Nieuwkoop & Sutasurya, 1979*; see also Chapter 2, p. 9).

Evaluation

In the telolecithal amphibian egg with holoblastic cleavage, meso-endoderm induction constitutes the first large-scale cellular interaction in the developing embryo, taking place between the two primary moieties of the egg, the animal, totipotent 'ectodermal' moiety and the vegetal, already firmly determined 'endodermal' one. The induction process spreads slowly from cell to cell into the animal cap of the blastula, reaching about half-way its ultimate extension at the early gastrula stage. The process is completed stepwise during gastrulation and neurulation through additional interactions between transiently apposed parts of the archenteron roof, and between the latter and the overlying neural plate as well as the underlying endoderm. Meso-endoderm induction starts very early in development, probably already around the fourth to fifth cleavage. No new transcription seems to be required for its initial steps, since the newly synthesised proteins are all of maternal character.

The induction of the meso-endoderm very probably consists of a large number of successive steps, each of which may be susceptible to different agents. Meso-endoderm induction involves the suppression of the initially present ectodermal differentiation tendencies of the animal moiety.

Meso-endoderm induction can also be evoked by high molecular weight substances isolated from adult or embryonic tissues, as well as by Li^+, a cation that is completely foreign to the embryo. This strongly emphasises both the non-specificity of the inductive factor(s) involved and the great significance of the competence of the reacting cells. The recent discovery that components of the extracellular matrix may play a role in the induction process suggests that the interaction may be the direct result of the divergent cellular differentiation pathways of the two interacting cell populations.

It seems plausible that in amphibians purely quantitative differences in the intensity of the inductive action are responsible for the cranio-caudal and dorso-ventral segregation of the marginal zone and its succeeding endo- and mesodermal differentiation. After invagination around the blastoporal lip, the marginal zone forms the pharyngeal endoderm and the various structures of the mesodermal mantle. The notochordal anlage, as a medio-dorsal differentiation, plays an additional role in the dorso-ventral segregation of the mesodermal mantle (see further Chapter 17, p. 230).

Although meso-endoderm formation follows a different spatial course in the megalolecithal avian embryo, which shows meroblastic cleavage, the processes involved may be very similar if not identical to those taking place in the amphibian embryo. This seems also to hold for the cephalochordate egg. However, it is not clear whether the notion of meso-endoderm induction as presented above can also be extended to the urochordates.

11

The gastrulation process

Ballard (1976*) states that no adequate general definition has ever been given for the process of gastrulation, the evolutionary divergence being too great for any adequate generalisation. Uro- and cephalochordates show a relatively simple indentation of the vegetal hemisphere of the blastula, followed by a gradual inward translocation of particular cell types. A tubular, sheet-like invagination of endo- and mesoderm takes place in the amphibian embryo. In reptiles, where the endoderm first segregates from the epiblast, possibly as in birds by poly-invagination, the mesodermal archenteron invaginates in the form of a tubular mesodermal cell sheet. In birds and mammals individual cells and cell groups migrate and ingress through the primitive streak. In the teleost embryo, however, no invagination of cell sheets or individual cells occurs but a multilayered embryo is formed by a complex process of cell segregation. Gastrulation is therefore characterised by a great variety of cellular processes, which ultimately lead to the formation of a triple-layered embryo, composed of an upper or outer neuro-ectodermal, an intermediate mesodermal and a lower or inner endodermal layer.

Gastrulation is a very complex process which can be divided into a number of sub-processes, but it must be realised that the latter are so accurately coordinated in space and time that the whole process is in effect a single event. Although archenteron formation in the anamnia and reptiles has several features in common with primitive streak formation in birds and mammals, the temporal and spatial relationships are markedly different. We shall therefore separately discuss gastrulation in the amphibian embryo and primitive streak formation and regression in the avian embryo, these being the best-studied groups.

The amphibian embryo

Invagination of the meso- and endoderm is preceded by an expansion of the animal cap material, a process called 'epiboly', and a contraction of the vegetal area, called 'polar ingression'. These processes taken together are known as pregastrulation movements (T. M. Harris, 1964). Superimposed upon these movements, which continue during gastrulation, bottle or flask cells are formed along the periphery of the endodermal yolk mass (Holtfreter, 1943 *b*, 1944 *a*), while a sheet-like migration of the invaginating meso- and endoderm takes place along the inner surface of the blastocoelic wall.

In the urodeles mesoderm invagination starts at the dorsal surface of the embryo and continues around the dorso-ventrally extending blastoporal lip. Mesoderm and endoderm invagination go hand in hand, the floor of the archenteron consisting of endoderm and its roof of mesoderm. The gastrulation process is completed by a delamination of the endo- and mesodermal portions of the archenteron roof and the ventro-dorsal closure of the endoderm into the enlarging enteric cavity.

In the anuran embryo mesoderm migration occurs exclusively or predominantly around an internal blastoporal lip (so-called cryptic gastrulation; Keller, 1976, 1981; Dettlaff, 1983; J. C. Smith & Malacinski, 1983) while the strictly endodermal archenteron wall invaginates through the enlarging blastopore. Mesoderm migration precedes endodermal archenteron formation, but the extending archenteron later catches up and finally overtakes the migrating mesoderm (Nieuwkoop & Florschütz, 1950; Keller, 1975, 1976; see p. 128). Subsequently the endodermal archenteron widens at the expense of the blastocoelic cavity.

In his classical 1929 study on amphibian gastrulation, Vogt established the first fate maps of the urodele early blastula and early gastrula and anuran early blastula. These maps have been supplemented by Pasteels (1942) and Nakamura (1942). Nakamura & Kishiyama (1971) made an additional fate map for the 32-cell stage of *Xenopus laevis*. In 1975 Keller performed a very accurate vital dye mapping of the superficial layer, and in 1976 of the deeper layer of the double-layered *Xenopus* gastrula, confirming Nieuwkoop & Florschütz's (1950) histological observations. These observations were recently also confirmed by Dettlaff (1983). Nakatsuji (1974 *a*) provided a detailed staging of blastopore formation in *Xenopus* (see further the general reviews on cell movements by Trinkaus, 1976*, 1982*).

Pregastrulation movements

Epiboly
From a mid-blastula stage onwards the cells of the animal moiety begin to expand, whereas the cells of the vegetal moiety contract. At the time of appearance of the blastoporal groove the animal moiety has enlarged by approximately two-thirds of its original area and the vegetal moiety has contracted by about the same amount. Expansion occurs more or less uniformly without marked differences between dorsal and ventral regions. It continues after blastopore formation. Epiboly was first described by

Holtfreter (1943a), who attributed it to special properties of a putative extracellular surface coat, a notion which has now been replaced by that of a filament meshwork situated directly underneath the plasmalemma and having contractile properties (Merriam & Sauterer, 1983).

At the mid-blastula stage, when epiboly starts in the animal hemisphere (Keller, 1978), tight junctions between the superficial blastomeres are replaced by fully developed desmosomes (Sanders & Zalik, 1972). Monroy, Baccetti & Denis-Domini (1976) observed locomotory activity in the cells lining the roof of the blastocoel prior to the onset of epiboly. From the mid-blastula stage onwards, a stage considered to be critical for development (Newport & Kirschner, 1982a, b; see Chapter 9, p. 83), isolated blastomeres begin to show locomotory behaviour in the form of blebs, ruffles and pseudopods (Hara, 1977). The various types of movement of isolated blastomeres were first described by Holtfreter (1943a, b, 1944a, 1946, 1947a) as 'circus' movements of large lobopodia, slow creeping movements characterised by directional rhythmic peristaltic contractions, and formation of contractile filo- and lamellipodia. Motility is attributed to the newly formed interblastomeric cell walls.

Satoh, Kageyama & Sirakami (1976) and K. E. Johnson (1976) found that locomotory activity increases rapidly from mid-blastula to late blastula and early gastrula stages and is highest in the presumptive mesoderm and head endoderm, particularly shortly before gastrulation. Locomotory activity is intermediate in the presumptive ectoderm and lowest in the endoderm of the blastocoelic floor. Johnson ascribes cell motility to an interaction of membrane proteins with contractile and other cytoskeletal elements. At the mid-blastula stage, when cell division begins to slow down and RNA synthesis starts, the cells apparently acquire new surface properties, which are expressed in an increased adhesiveness of the cells and in the acquisition of cellular motility. K. E. Johnson (1976) noted that the surface activity is well correlated with the decrease in mitotic index of the cells. Gerhart (1980*) considers the motility of the newly formed cell wall, as observed in isolated cells, to be a prerequisite for true locomotion during gastrulation.

EDTA-dissociated gastrula cells show circus movements depending on the stage of isolation; whereas very few pre-gastrula cells show this activity, in the early gastrula cells of the dorsal blastoporal region are the most active ones (K. E. Johnson & Adelman, 1981). During gastrulation the proportion of the cell population performing circus movements increases progressively. These authors consider the circus movements as an artefact of dissociation, but at the same time as an expression of the capacity of these cells to participate in morphogenetic movements. H. Y. Kubota (1981) observed that isolated endodermal cells perform creeping movements by means of directional flow of cell surface material and antero-posterior propagation of constrictions.

The expansion of the animal moiety leads to a thinning of the blastocoelic roof, while the contraction of the vegetal area results in a thickening and elevation of the floor of the blastocoel. The thinning of the blastocoelic roof is brought about by a reduction of the number of cell layers by means of cell interdigitation. This holds both for urodeles and anurans; in the urodeles the interdigitation concerns the entire wall of the blastocoel while in the anurans, where the wall is double-layered, interdigitation is restricted to the deeper layer (Keller & Schoenwolf, 1977). Whereas the cells of the deeper layer are attached to each other by cellular protrusions, the cells of the superficial layer are joined together by circumferential, apical tight junctions and desmosomes. In the anuran *Xenopus laevis* no migration of cells occurs from the superficial to the deeper layer or vice versa (Keller, 1978). Epiboly therefore implies flattening of the superficial cells and interdigitation of the cells of the much thicker, deeper layer (Keller & Schoenwolf, 1977). In *Xenopus* epiboly is isotropic in the animal region but anisotropic in the marginal zone, which begins to constrict circumferentially below the equator. A fast expansion takes place initially in the animal region and later during invagination in the dorsal marginal zone. At the onset of invagination the deep cells of the dorsal marginal zone become elongated and interdigitate, forming fewer cell layers (Keller, 1980). Keller assumes that epiboly is brought about by the active force-producing process of interdigitation and the subsequent shortening of the deep cells, the superficial layer being under tension and the deep layer under compression.

As a consequence of the different configuration of the blastular wall in urodeles and anurans, the presumptive mesoderm occupies part of the outer surface in the urodele embryo but is situated internally in the anuran embryo, where it forms the so-called 'internal marginal zone' and is covered by presumptive ecto- and endoderm. In *Xenopus* the presumptive mesoderm is located exclusively in the internal marginal zone (Nieuwkoop & Florschütz, 1950; Keller, 1976), but in other anuran species the delamination of the blastocoelic wall into two separate layers seems to be less complete, so that some mesoderm may be present on the surface of the embryo. The conclusion of Landström & Løvtrup (1979) that all amphibian embryos are double-layered, with only presumptive ecto- and endoderm at the surface and the presumptive mesoderm exclusively located in the interior, is certainly an unjustified generalisation.

Brachet (1962) showed that epiboly can be blocked by mercaptoethanol and lipoic acid, agents which break disulfide bonds. Tseng *et al.* (1982) emphasise the role of tight junctions in maintaining the cohesiveness of the ectoderm during epibolic extension.

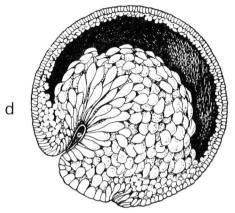

d

Fig. 14. Slightly schematised sagittal section through an early yolk plug stage of a urodele gastrula, showing flask-cell formation around the dorsal archenteron and in the ventral blastoporal groove, the flask cells protruding deeply into the yolk mass. d., dorsal side (after J. Holtfreter, 1943 *b*).

Vegetal ingression

Vegetal ingression was first observed by Schechtman (1934) in *Triturus torosus* but is a universal phenomenon in amphibian development. It has been described as multipolar ingression, but seems mainly to represent an inward flow of subcortical cytoplasm along the cleavage furrows (M. B. Clayton & Dixon, 1975), since cells apparently do not detach from the surface. The subsequent occurrence of tangential cleavages may be responsible for the upward movement of vegetal pole cells into the vicinity of the blastocoelic floor (Nieuwkoop, 1946; Züst & Dixon, 1975).

Blastopore formation

Blastopore formation is due to a local surface contraction and the formation of flask-shaped cells, also called 'Ruffini cells' (Holtfreter, 1943 *b*, 1944 *a*). Balinsky (1961) observed in anuran embryos that the external surface of the cells lining the blastoporal groove is thrown into folds and deep crypts. Baker (1964, 1965) showed in *Hyla regilla* that the distal surface of the flask cells bears many microvilli and is underlain by an electron-opaque layer of fine granular material and microfilaments. Perry & Waddington (1966) noticed that many microtubules are oriented parallel to the main axis of the flask cells. These observations were confirmed by Nakatsuji (1975 *a*), among others. The cells lining the groove are tightly joined together at their distal ends, so that the dense layer can act as a supracellular contractile system. Proximally, however, the cell bodies containing the nuclei are separated by wide intercellular spaces and protrude more or less freely into the interior of the embryo (Holtfreter,

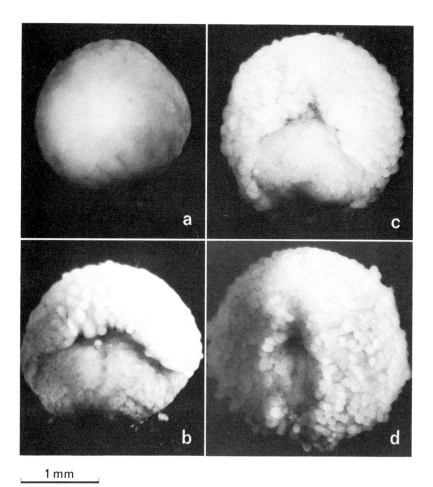

1 mm

Fig. 15. Progressively spreading flask-cell formation around the vegetal yolk mass isolated at successively older stages of development in *Ambystoma mexicanum*. a, Stage $7\frac{2}{3}$ (H.), no flask cell formation; b, $8\frac{1}{2}$, dorsal flask cells only; c, stage 9, dorsal and lateral flask cells; and d, stage 10^-, flask-cell formation all around the circumference (after M. H. M. Doucet-de Bruïne, 1973).

1943b; see Fig. 14; see also Bluemink, 1978*; Burgess & Schroeder, 1979*; Monroy, 1979).

Nieuwkoop (1969b) isolated the yolk mass of *Ambystoma mexicanum* embryos at successively younger stages. Isolated late blastula endoderm formed flask cells all around the yolk mass, starting on the dorsal side and spreading to the ventral side, as in normal development. Flask-cell formation was much more extensive on the dorsal than on the ventral side, and intermediate laterally. At that stage the endoderm is apparently already fully programmed for flask-cell formation. Endoderm isolated at a mid-to-late blastula stage developed flask cells dorsally and laterally only, and to a much smaller extent. Isolated mid-blastula endoderm only formed a small number of flask-like cells on the dorsal side, while still younger endoderm did not show any flask-like cell formation (see Fig. 15). The earlier the endoderm was isolated, the longer the time that elapsed between isolation and flask-cell formation (P. D. Nieuwkoop, unpublished observations).

The retrogressive behaviour of the endoderm isolated at successively younger stages cannot be ascribed to a deleterious effect of the culture medium, since the isolated yolk mass remained healthy for several days. Nieuwkoop therefore concluded that the appearance of these special 'differentiation tendencies' in the peripheral cells of the yolk mass must be based on an interaction of the yolk mass with the animal moiety of the embryo before isolation. Doucet-de Bruïne (1973) actually found that marginal zone material exerts a weak stimulating influence upon flask-cell formation by the endoderm, mainly on the dorsal side, but could not demonstrate a correlation between the amount of induced mesoderm and the incidence of flask-cell formation in recombinates. She deduced that flask cell formation is a direct consequence of the dorso-ventral polarisation of the endoderm, without intervention by the induced mesoderm.

Doucet-de Bruïne (1973) also found that the dorso-ventral spreading of flask-cell formation is due to an inductive influence propagating from cell to cell, since isolated ventral endoderm of a late blastula did not form flask cells, whereas flask-cell formation occurred around the entire circumference in the isolated entire yolk mass. H. Y. Kubota (1983), studying gastrulation in anuran amphibians, came to the conclusion that flask-cell formation does not play an essential role in gastrulation, the main function being performed by the deep marginal cells. We still think, however, that flask-cell formation is important for the initiation of the gastrulation process.

Holtfreter (1944a) assumed that during gastrulation the flask cells are ultimately discarded into the gut lumen, but Nieuwkoop noticed a retransformation of flask cells into cuboidal cells and subsequently into flattened epithelial cells (P. D. Nieuwkoop, unpublished observations), showing that flask-cell formation is a transient change in cell shape and does not represent a particular type of permanent cellular differentiation, as suggested by Løvtrup, Landström & Løvtrup-Rein (1978*).

Landström, Løvtrup-Rein & Løvtrup (1976) isolated small groups of cells from the vegetal surface of *Ambystoma mexicanum* blastulae not far from the pigment boundary and observed the formation of Ruffini cells, which they erroneously called 'fibroblasts'. They could suppress this spontaneous cell 'differentiation' by lactate and various inhibitors of RNA synthesis, thus demonstrating that new transcription is required for the transient cell change. Malpoix, Quertier & Brachet (1963) found that β-mercaptoethanol inhibits epiboly and mesoderm ingression but hardly affects flask-cell formation.

What determines the onset of gastrulation? One-half and one-quarter fragments of *Cynops pyrrhogaster* eggs containing the nucleus give rise to embryos which start gastrulation at about the same time as whole embryos, though having a smaller number of blastomeres, in proportion to the diminished egg volume (Kobayakawa & Kubota, 1981). Comparing embryos of *Rana temporaria, Xenopus laevis, Ambystoma mexicanum* and *Triturus vulgaris* reared at different temperatures, Valough, Melichna & Sladeček (1970) found that the number of cells at the beginning of gastrulation is temperature dependent but also shows species-specific differences.

Beetschen (1970) studied the mutant *ac* of *Pleurodeles waltlii,* which exhibits a syndrome of disturbances of gastrulation, leading in extreme cases to complete evagination of the endodermal yolk mass. He described the formation of additional pseudoblastopores in the animal cap. The syndrome could be prevented by injection of the contents of the germinal vesicle or of cytoplasm of fertilised uncleaved eggs, but surprisingly also by simple puncturing of the animal hemisphere (Fernandez, 1979). After observing mutant embryos, Nieuwkoop has the impression that a disturbance of blastocoel fluid production or its leakage from the embryo may be the primary cause of the strongly reduced blastocoel and the severely disturbed gastrulation process.

Mesoderm migration

In urodeles and anurans the presumptive mesoderm moves around the outer or inner blastoporal lip, respectively, and then moves upwards towards the animal pole along the inner surface of the blastocoelic wall. In *Xenopus laevis* the dorsal mesoderm, after involution around the inner blastoporal lip, shows dorsal convergence and cranio-caudal extension, accompanied by rearrangement of the multilayered marginal zone into the single cell layer of the mesodermal mantle by means of interdigitation (Keller, 1981, 1984; Lundmark *et al.,* 1984). Cooke (1979*a, b*) found that cells generally maintained their relative positions during the formation of the mesodermal mantle, which is formed by posterior recruitment of cells. Mesodermal mantle formation is characterised by a succession of wave-like movements which pass obliquely across the mantle in antero-posterior direction (compare mesoderm ingression through the primitive streak, p. 145). K. E. Johnson (1970) stressed the role of cell surface adhesiveness in the gastrulation process: for instance, the presumptive mesodermal cells of the early gastrula show strong adhesiveness to each other as well as to a substrate, such as glass. In the early gastrula of *Bufo japonicus* the cells of the pharyngeal endoderm and prechordal mesoderm form pseudopodia, whereas cells of the endodermal mass and the ectoderm rarely do so (Nakatsuji, 1974*b*). In *Cynops pyrrhogaster* invaginating cells of the dorsal, lateral and ventral marginal zone form pseudopodia and establish contact with the blastocoelic wall by means of lobo-, filo- and lamellipodia. This behaviour suggests that the invagination of the pharyngeal endoderm and prechordal mesoderm is at least partially brought about by active migration along the inner surface of the ectoderm (Nakatsuji, 1975*b*).

In TEM and SEM studies on *Bufo*, *Xenopus* and *Rana*, Nakatsuji (1976) observed that filopodial processes of the advancing endo- and mesodermal cells, which contain microfilaments, make focal close contacts with the inner surface of the blastocoelic wall, probably as an expression of altered adhesive properties. K. E. Johnson (1976) noticed similar cell movements of *Rana pipiens* gastrula cells *in vitro*. With time-lapse cinematography H. Y. Kubota & Durston (1978) actually observed mesodermal cell migration along the inner surface of the ectoderm in opened *Ambystoma mexicanum* gastrulae.

The lectin Con A binds to the cell surface of blastula and gastrula cells and affects cell adhesiveness. This can be counteracted by methyl-D-mannopyranoside (K. E. Johnson & Smith, 1976, 1977). Cytochalasin B, which influences the microfilament system, strongly inhibits migration, whereas colchicine and other microtubule-disrupting drugs have little effect (Nakatsuji, 1979).

Nieuwkoop & Florschütz (1950) observed that in *Xenopus laevis* gastrulae the rolling-in and upward migration of the dorsal mesoderm markedly precedes the formation of the endodermal archenteron. At the large circular-blastopore stage the extension of the archenteron begins to catch up with the invaginating mesoderm and soon overtakes it, leading to the formation of the pharyngeal endoderm in front of the prechordal mesoderm. Nakatsuji (1975*b*) found that the dorsal blastoporal groove initially deepens at a rate of only 1 μm min^{-1}. At the circular-blastopore stage the prechordal plate and chordomesoderm move at a rate of about 6 μm min^{-1}, while the endodermal archenteron elongates at a rate of 9 μm min^{-1}. Ventrally, the invagination of the archenteron and the movement of the mesoderm are slower.

Boucaut (1973) injected ^{3}H-thymidine-labelled ecto-, meso- and endodermal cells into the blastocoelic cavity of *Pleurodeles* embryos at an advanced blastula stage and observed selective adhesion to corresponding host tissues in 80%, 65% and 95% respectively.

In the early gastrula of *Rana pipiens* Leblanc & Brick (1981) noted differences in spreading and adhesive properties *in vitro* between ectodermal, mesodermal and endodermal cell populations; within each population all cells had similar properties. At the mid-to-late gastrula stage each of these populations splits into two subpopulations: neural and epidermal, notochord and other mesoderm, and head and trunk endoderm, respectively. Leblanc & Brick ascribe this phenomenon to tissue interactions and to the subsequent appearance of different tissue-specific kinetic properties, which are apparently related to their morphogenetic activity *in vivo*. Morphogenetic processes such as gastrulation and neurulation are based on the appearance of cell-specific properties.

The role of the extracellular matrix in mesoderm migration

Sirakami (1963) found hyaluronidase-sensitive, PAS-positive intercellular matrix material in blastula ectoderm. Motomura (1967) observed the secretion of mucosubstances by ectoderm of early gastrulae. Cells of the blastoporal lip showed very active secretion both on the outer surface and into intercellular spaces. Motomura postulates functions for this extracellular material in cell adhesion, as a lubricant in gastrulation movements, and in the filling-up of embryonic cavities. From the retention of polarity in reversed dorsal marginal regions of the gastrula Cooke (1972 c) concluded that gastrulation movements are primarily controlled by the intrinsic properties of the invaginating mesodermal cells and not by cell guidance along the ectoderm (however, see the work of Nakatsuji and Johnson described below).

In an SEM study on *Ambystoma mexicanum* Moran & Mouradian (1975) observed the appearance of external cell surface material in the blastoporal region. In a TEM study of *Xenopus laevis* blastulae and gastrulae, K. E. Johnson (1977d) noticed the presence of lanthanum-staining material on the internal cell surfaces and in intercellular spaces. Stainability starts at the late blastula stage on about 3% of the cell periphery, increases rapidly to about 29% of the cell periphery at the beginning of gastrulation, and reaches a value of about 82% at a late gastrula stage, when an appreciable accumulation also occurs in intercellular spaces. K. E. Johnson (1977a) found a dramatic increase in extracellular matrix material at the beginning of gastrulation in the marginal zone, particularly in the dorsal blastoporal lip region. The material seems to be formed in Golgi complexes (K. E. Johnson, 1977b). It has a high MW, is not degraded by hyaluronidase, is not collagen, but is degraded by pronase (K. E. Johnson, 1977c, 1978). In an SEM study, Karfunkel (1977) observed fine filamentous extracellular material on the inner surface of the ectoderm at a late gastrula stage. Nakatsuji & Johnson (1982a) found that presumptive mesodermal cells of *Xenopus* gastrulae manifest substantial locomotion when cultured on a glass surface coated with collagen and foetal bovine serum. The migration *in vitro* strongly resembles normal migration during gastrulation. There is contact inhibition between mesodermal cells but no contact inhibition between mesodermal and ectodermal cells. Presumptive mesodermal cells of the blastula do not yet show migratory activity, but mesodermal cells of the gastrula migrate actively. Gastrula ectodermal cells have no migratory activity. In *Ambystoma mexicanum* gastrulae migrating mesodermal cells are strongly oriented towards the animal pole. Nakatsuji & Johnson (1982b) and Nakatsuji,

Gould & Johnson (1982) observed an extracellular network of fibrils covering the inner surface of the ectodermal layer. Filopodia of the mesodermal cells attach to this network and the mesodermal cells follow the fibrils as a guiding substratum. The fibrils exhibit a statistically significant alignment in the direction of the animal pole.

Nakatsuji & Johnson (1982c) were able to transfer the extracellular fibrils from fragments of *Ambystoma maculatum* gastrula ectoderm to plastic coverslips. Mesodermal cells attach to such a substratum and migrate actively along the fibrils, which in the original gastrula ectoderm were predominantly aligned in the direction of the animal pole. The mesodermal cells showed a slightly stronger tendency to migrate towards the original animal pole side than towards the blastoporal side, a phenomenon which is not yet understood. The fibril network on the inside of the ectoderm can be artificially aligned by exerting mechanical tension, leading to directed movements of mesodermal cells (Nakatsuji & Johnson, 1984). Nakatsuji (1983) suggests that morphogenetic movements occur by contact guidance in combination with contact inhibition. Extracellular matrix material was recently observed in early blastulae and more abundantly in early to late gastrulae of *Pleurodeles waltlii* and *Cynops pyrrhogaster* by Boucaut & Darribere (1983a) and Komazaki (1982, 1983) respectively. Boucaut & Darribere (1983b), using immunofluorescence, detected fibronectin in the cell surface as well as in the extracellular matrix of the blastocoelic roof and in the blastopore region of *Ambystoma mexicanum* and *Pleurodeles waltlii* early gastrulae, suggesting its involvement in early morphogenesis.

Archenteron formation

Little attention has been paid to the transformation of the blastoporal invagination cavity into the tube-like archenteron. P. D. Nieuwkoop concluded from unpublished observations that, as soon as the extension of the invaginating archenteron begins, the bottle cells are retransformed into cuboidal and subsequently into flattened cells. This was confirmed by Keller (1980). In *Xenopus laevis* bottle-cell formation initiates blastoporal groove formation, but according to Keller (1981) and Lundmark *et al.* (1984) the archenteron is carried inward by the involuting mesoderm. The constriction of the ring-shaped blastopore may also facilitate the inward movement of the meso- and endoderm.

Mercaptoethanol treatment leads to persistence of the blastocoelic cavity by blocking the transfer of fluid from the blastocoel to the archenteron. This transfer is therefore not simply due to an osmotic gradient but must be based on active transport of fluid by polarised transport mechanisms located in the cells separating the two cavities (Tuft, 1961a, b, c, 1962). Herkovits, Bustuoabad & Pisanó (1977) called attention to the pseudopodial activity of the yolk cells of the blastocoelic floor, which may facilitate the mechanical collapse of the blastocoelic cavity and thereby the widening of the archenteron.

Hama (1978*) stresses the fact that the length of the archenteron is a much more reliable criterion for the progression of gastrulation than the shape of the blastopore, which may vary considerably. Hama makes a clear distinction between the behaviour of the so-called head organiser, which is represented by the prechordal endo- and mesoderm and expands in medio-lateral direction after invagination, and the trunk organiser, which is represented by the chordomesoderm and elongates in cranio-caudal direction both before and after invagination.

Curtis (1962*b*) and Cooke (1975*a*) observed normal gastrulation some time after removal of the dorsal blastoporal lip from *Xenopus laevis* and *Bombina orientalis* early gastrulae. This regulation was prevented by treatment with the chelating agent EDTA, from which they concluded that it requires intercellular communication.

Malacinski & Spieth (1978*) found that embryos of the *ovum deficient* mutant, *o*, of *Ambystoma mexicanum* (discovered by Humphrey in 1966) are arrested at gastrulation regardless of the genotype of the sperm, showing a slowing-down of DNA synthesis and cell division after 24 h of development.

Exogastrulation

Exogastrulation was described by Holtfreter (1933*b*) as an autonomous segregation of the ectoderm from the endo- and mesoderm. It occurs in decapsulated *Ambystoma mexicanum* and *Rana fusca* eggs kept in hypertonic solution. Exogastrulation can be produced in *Xenopus laevis* embryos by lithium treatment (J. L. Smith, Osborn & Stanisstreet, 1976; Stanisstreet & Smith, 1978) and in *Bufo arenarum* by treatment of the opened blastula/early gastrula with tunicamycin, a glycoprotein glycosylation inhibitor (Sanchez & Barbieri, 1983). Apart from the evagination of endo- and mesoderm *per se*, the morphogenetic movements in exogastrulae, such as epiboly, blastoporal groove formation, dorsal convergence and ventral divergence are very similar to those in the normal gastrula.

Biochemical and biophysical data relevant to gastrulation

Lohmann (1972) found stage- and regional-specific DNA amplification in the neurectoderm, the chordomesoderm and the endoderm of *Triturus* embryos which was correlated with cyto- and histodifferentiation (Lohmann & Vahs, 1969). Lohmann & Schubert (1977) observed differential DNA replication during gastrulation in nuclei being in the G_1 phase. Desnitsky (1974), working with early gastrulae of *Ambystoma* and *Triturus*, noted a temporary block of the cell cycle in the G_1 phase in the presumptive chordomesoderm, and subsequent synchronisation of cell division after the onset of invagination. The dorsal blastoporal lip region is characterised by a very low mitotic index in comparison with other regions (Desnitsky, 1978).

Flickinger (1969, 1970*a*, *b*, 1972) found that during gastrulation rRNA synthesis begins sooner in the dorsal ecto- and mesodermal than in the ventral endodermal portion of *Rana pipiens* embryos. More DNA-like RNA is formed in the former than in the latter. Moreover, there is a higher rate of DNA synthesis in [induced] dorsal than in [non-induced] ventral ectoderm. Whereas RNA synthesis per cell is equivalent, RNA synthesis calculated on the basis of the total amount of protein is higher in the former than in the latter.

Wright (1978*), using hybrids of two *Rana* species characterised by different electrophoretic protein variants, could show that some gastrula-specific proteins are of maternal and others of paternal character, which demonstrates that their formation partially requires new gene activation.

The morphogenetic events of gastrulation and neurulation are not marked by overall changes in enzyme patterns. This does not exclude regional changes, however. Enzyme activity in the late gastrula to early neurula decreases in the following order: presumptive neurectoderm > presumptive epidermis > dorsal marginal zone > ventral marginal zone and endoderm (Barth & Sze, 1951, 1953; Sze, 1953). The higher glycogen metabolism of the dorsal blastoporal lip seems to be correlated with its intense morphogenetic movements (Heatley & Lindahl, 1937; see also Duspiva, 1962*a**, *b**).

Brick *et al.* (1974) observed no differences in electrophoretic mobility among cells of different presumptive regions in the late blastula of *Rana pipiens*, but differences arise at the transition from blastula to early gastrula. In the gastrula the cells of the archenteron floor possess the highest surface charge and those of the inner neural ectoderm the lowest, while dorsal lip cells show an intermediate value. The greatest rise in surface charge occurs in the cells which arrive at the dorsal blastoporal lip. During gastrulation the alterations in surface charge result in a hierarchy in which endoderm is more negative than chordomesoderm, which again is more negative than ectoderm. This coincides with Steinberg's (1964*, 1970) hierarchy in cell adhesiveness. Brick *et al.* (1974) suggest that an increase in negativity is correlated with a diminution of intercellular contact within a cell layer by increasing the energy of repulsion.

B. E. Schaeffer, Schaeffer & Brick (1973*a*) found a positive correlation between surface charge and morphogenetic movements, i.e. a rapid increase in surface charge coinciding with the initiation of morphogenetic activity. The correlation between surface charge and adhesiveness is negative, higher surface charge coinciding with lower adhesiveness and vice versa. Cytochalasin B treatment produced a significant reduction in electrophoretic mobility of disaggregated endodermal and chordomeso-dermal cells of the late gastrula but did not significantly alter that of ectodermal cells (H. E. Schaeffer, Schaeffer & Brick, 1973*b*). The electro-phoretic mobility of *Rana pipiens* gastrula cells is dependent on pH over a wide pH range, but different cell populations behave differently, which also suggests the presence of quantitative differences in cell surface components (H. E. Schaeffer *et al.*, 1973*c*; see also Monroy, 1976*).

S. Ito & Ikematsu (1980), studying inter- and intra-tissue communication in *Cynops pyrrhogaster*, observed electrical coupling among the cells of the same tissue from gastrulation up to the closure of the neural tube. Presumptive chordomesodermal cells and presumptive neurectodermal cells are coupled during the initial stages of gastrulation but coupling is reduced between stages 15 and 16 and disappears at stages 22 to 23 (see also Jaffe & Nuccitelli, 1977*).

UV irradiation and gastrulation

Grant & Youngdahl (1974) noticed that UV irradiation of the vegetal hemisphere of fertilised *Rana pipiens* eggs causes abnormal gastrulation.

In *Xenopus laevis* Beal & Dixon (1975) and Züst and Dixon (1975) observed inhibition of cytokinesis but normal nuclear division after UV irradiation, so that blastomeres with multiple nuclei were temporarily present in the vegetal part of the embryo. UV irradiation leads to a delay in the initiation of gastrulation and subsequent archenteron formation. Grant & Wacaster (1972) and Malacinski, Allis & Chung (1974), Malacinski, Benford & Chung (1975) and Malacinski, Ryan & Chung (1978*b*) found that UV irradiation of the vegetal pole of uncleaved *Rana pipiens* eggs produced abnormalities in gastrula morphogenesis and neural plate formation. UV damage could be remedied by incubating the embryos at low temperature (10 °C) (Malacinski *et al.*, 1974), by injection of oocyte homogenate or a protein component of the germinal vesicle content (Malacinski, 1972, 1974), by replacing the dorsal blastoporal lip of an irradiated embryo by a normal one (H.-M. Chung & Malacinski, 1975), or by rotation of the egg after irradiation (Kirschner *et al.*, 1981; see also Chapter 7, p. 60 and Brachet, 1977*).

Pseudogastrulation

Holtfreter (1943*b*) was the first to describe the phenomenon of pseudo-gastrulation: unfertilised eggs of *Rana pipiens* may show characteristic features of 'gastrulation' movements in the form of animal–vegetal surface streaming (comparable to epiboly) and vegetal contraction (comparable to vegetal ingression) followed by blastoporal groove formation at the boundary between the pigmented and unpigmented egg regions, and finally an inward movement of pigmented cytoplasm (comparable to invagination). This was confirmed by L. D. Smith & Ecker (1970 *a**) and Baltus *et al.* (1973) in *Rana pipiens* and by Malacinski, Ryan & Chung (1978 *b*) in *Rana nigromaculata*. It has also been occasionally observed in other amphibian genera such as *Discoglossus* (E. C. Boterenbrood, personal communication). Holtfreter (1943 *b*) mentioned that dorso-ventral polarity was less evident in pseudogastrulae than in normal gastrulae, but L. D. Smith & Ecker (1970 *c*) obtained pseudogastrulae with characteristic pseudo-yolk-plug formation, while the eggs could even reach a pseudo-slit-blastopore stage with clear dorso-ventral polarity. Malacinski *et al.* (1978 *b*) observed a similar completion of pseudogastrulation in *Rana pipiens*.

Malacinski and coworkers noted that pseudogastrulation regularly occurs when *Rana pipiens* oocytes are induced to undergo maturation *in vitro* by progesterone treatment, a phenomenon taking place also after removal of the nucleus. This is in agreement with L. D. Smith & Ecker's (1970 *a**) conclusion that pseudogastrulation must be programmed in the egg cytoplasm. These authors noticed that pseudogastrulation is reversibly inhibited by anaerobiosis and thus requires energy. Malacinski *et al.* (1978 *b*) found that continuous exposure to a sodium-chloride-containing salt solution is a prerequisite for pseudogastrulation. Whereas UV irradiation and actinomycin D treatment had no effect, cytochalasin B, which affects the microfilament system, and colchicine and vinblastine, which influence the microtubular system, blocked pseudogastrulation; this demonstrated the involvement of the cytoskeleton. Puromycin also prevented pseudogastrulation, so that protein synthesis is also involved, but no new RNA synthesis is required, as shown by actinomycin-D-insensitivity and independence of nuclear presence.

Malacinski *et al.* (1978 *b*) could not demonstrate any mesodermal- or neural-inductive capacity of the lip-like structure of the pseudogastrula when they implanted it into the blastocoel of a host late blastula. Although this may be a significant observation, it is possible that the acellular graft disintegrated in the blastocoel (see also Malacinski, 1984*).

In our opinion pseudogastrulation, which may continue up to a slit-blastopore stage, cannot be explained without postulating an interaction between the two primary moieties of the egg, leading to some initial steps of 'mesoderm formation' in the uncleaved egg. Such mesoderm formation might be comparable to the segregation of mesoplasm in the uncleaved uro- and cephalochordate egg (see below). It would therefore be very interesting to test this hypothesis with antisera made against gastrula mesoderm. Huff & Preston (1965) found that artificially activated eggs of *Rana pipiens*, which do not undergo cleavage, are capable of producing a cleavage-initiating factor that is normally only detectable from the beginning of gastrulation. This suggests that some intracellular interactions proceed normally in uncleaved eggs (see Brachet, 1977*; Gerhart, 1980*; Malacinski, 1984*).

Hypotheses on amphibian gastrulation

Many hypotheses have been proposed to explain the gastrulation process in amphibians. We shall discuss them only briefly, because in our opinion they are all too one-sided for such a complex process. Holtfreter (1944*a*) emphasised the significance of the high pH and low Ca^{2+} concentration of the blastocoelic fluid, which may lead to a lowering of the surface tension at the internal cell surfaces. He considers flask-cell formation as a cell-specific response to the internal medium. Beloussov, Dorfman & Cherdantzev (1975) related morphogenetic patterns such as gastrulation and neurulation to mechanical stress. They believe that the mechanical stress consists of a passive component that is insensitive, and an active component that is sensitive to cooling, cyanide and cytochalasin B, the active component being involved in cell elongation and cell migration and the passive component in cell-shape preservation. H. M. Phillips & Davis (1978) developed a gastrulation model based on differences in surface tension (γ) of presumptive ectoderm (e), mesoderm (m) and endoderm (n) in the order $\gamma e > \gamma m > \gamma n$. They concluded that control of tissue surface tension constitutes the key morphogenetic mechanism in amphibian gastrulation, combined with active changes in cell shape and changes in intercellular adhesiveness. Løvtrup (1965*b**) considers invagination to be the result of a 'transformation of cell types', viz. from amoebocytes into mechanocytes. We have already pointed out that the formation of bottle or Ruffini cells represents a *transient* change in cell shape and not any form of more permanent cellular differentiation, as Løvtrup, Landström & Løvtrup-Rein (1978) suggest.

According to Løvtrup the dorso-ventral polarity of the invagination process is due to the fact that a time gradient is imposed upon the essentially radially symmetrical process of epiboly and invagination. However, he ignores the pronounced quantitative dorso-ventral differences in the invagination process. As already mentioned, Løvtrup (1975) unjustifiably extends the double-layered nature of the anuran egg to all amphibians. In the urodeles he nevertheless has the mesodermal notochord derive from superficial cells. Løvtrup *et al.* (1978*) assume that the animal–vegetal polarity (a hypothetical gradient of carbohydrate metabolites) is responsible for the level of invagination, while the dorso-ventral polarity (a hypothetical gradient of heparan sulphate) determines its initiation on the dorsal side.

Brannigan & Fabian (1980) developed a gastrulation model based on the diffusion of an animal–vegetal and a dorso-ventral morphogen analogous to Dalcq & Pasteel's (1937, 1938) and Gierer & Meinhardt's

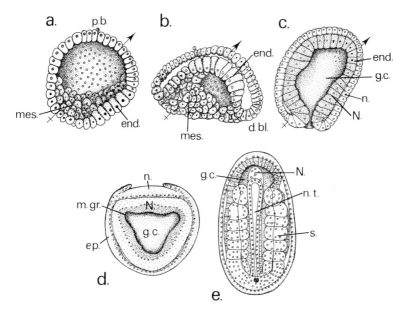

Fig. 16. Normal development of *Branchiostoma lanceolatum*. Median sections through a, early gastrula at $5\frac{1}{2}$ h of development; b, mid-gastrula at 9 h; c, late gastrula at 14 h; d, transverse section through neurula at 16 h; and e, frontal section of early larva at 18 h of development. Arrows indicate future antero-posterior axis. d.bl., dorsal blastoporal lip; end., endoderm; ep., epidermis; g.c., gastrocoel; mes., mesoblast, m.gr., mesodermal groove; n., neural plate; N., notochord; n.t., neural tube; p.b., polar body; s., somites (after E. G. Conklin, 1932).

(1972, 1974) developmental theories. Odell *et al.* (1981) built their gastrulation model on the presence of an apical contractile filament system, which would show elastic properties under moderate stretching and contractile properties when stretched more strongly. In this model coordinated movements arise as a consequence of the local behaviour of individual cells. An expanding contraction wave would be responsible for invagination. Contraction may also be triggered by chemical signalling, e.g. Ca^{2+} release. Zeeman (1975) applied the mathematical catastrophe theory to gastrulation in amphibians and birds.

Gastrulation in uro- and cephalochordates

Conklin (1905 *a*, *b*) extensively described gastrulation in the ascidian *Cynthia* (*Styela*) *partita* and Conklin (1932) the comparable process in the cephalochordate *Branchiostoma lanceolatum* (see Fig. 16). T. C. Tung, Wu & Tung (1959, 1960 *a*) studied the developmental potencies of the animal–vegetal octets of blastomeres of the 32-cell stage of *Branchiostoma belcheri* and concluded that each tier of blastomeres has its own developmental capacity, but that recombinates of animal–vegetal octets show considerable regulative capacity, leading, among other things, to normal gastrulation. Satoh (1979) studied time-specific enzyme synthesis during ascidian embryogenesis. The 'clock mechanism' he deduced is not regulated by cytokinesis or the mitotic cycle, but is related to the cycle of DNA replication (see also Chapter 10, p. 123).

Gastrulation in sturgeons

Ignatieva (1963) compared gastrulation in the sturgeons with that in axolotl embryos and found only minor differences. Ballard & Ginzburg (1980) described the principal morphogenetic movements during gastrulation in *Huso* and three *Acipenser* species and prepared a fate map for the late blastula of the sturgeon. They reported a close resemblance between the fate map and morphogenetic movements in the sturgeons and those in the anuran amphibians. The resemblance with *Xenopus* is greater than with *Rana* (Ballard, 1981).

'Gastrulation' in teleost fish

As already mentioned in the beginning of this chapter, the teleost embryo does not show true invagination but goes through a complex process of spatial segregation leading to the formation of a triple-layered state (Ballard, 1976*). Nevertheless, Kageyama (1980) still erroneously called the expansion of the ring-shaped embryonic anlage over the yolk mass in *Oryzias latipes* 'gastrulation', and its completion near the original vegetal pole of the egg 'closure of the blastopore'. The latter phenomenon occurs by interdigitation and elongation of the component marginal cells (Kageyama, 1982 in *Oryzias latipes;* Keller & Trinkaus, 1984 in *Fundulus heteroclitus*). Trinkaus & Erickson (1981) observed that deep cells of the *Fundulus* blastodisc show low mutual adhesion and rapid movement by means of lamellipodia during 'gastrulation'. Van Haarlem and coworkers studied the process of epiboly in the annual fish genus *Notobranchius*. The blastodisc consists of an outer enveloping layer and a large number of tightly packed deep cells, underlain by the syncytial periblast. During epiboly the enveloping layer spreads over the egg surface, overgrowing the yolk, while the deep cells gradually build up a monolayer of moving and colliding cells. Their dispersion is due both to contact inhibition and to their attachment to the lower surface of the spreading enveloping layer (van Haarlem, 1979, 1983; van Haarlem, van Wijk & Fikkert, 1981). Unfortunately we know very little about the spatial segregation process in the teleost blastodisc.

Gastrulation in reptiles

Gastrulation in reptiles constitutes a very interesting transition between archenteron formation in the spherical amphibian embryo and cell ingression through a primitive streak in the flat avian blastoderm. In the megalolecithal reptilian egg the blastoderm is also flat and forms a mesodermal archenteron by invagination through the posterior blastoporal plug; this then extends in anterior direction between the thick epiblast and the very thin hypoblast (Pasteels, 1957 *a, b*). In *Lacerta vivipara* the dorsal wall of the mesodermal archenteron gives rise to the prechordal mesoderm and the notochord, the lateral walls supply the entire mesoblast consisting of presumptive somite and lateral plate mesoderm, while the ventral blastoporal lip forms the entire posterior and lateral extraembryonic mesoblast (Hubert, 1962). The archenteron soon opens into the subgerminal cavity, after the hypoblast layer has first been broken up. In *Clemmys* the straight or crescent-shaped blastopore separates dorsal and ventral blastoporal lips, so that there is no yolk plug homologous to that of the amphibian embryo. At the end of gastrulation a continuous endodermal layer is reformed underneath the chordomesoblast from the caudal and lateral edges of the original hypoblast (Hubert, 1962). Unpublished observations by P. D. Nieuwkoop on the marine turtle *Chelonia mydas* generally confirm Hubert's observations.

Gastrulation in birds and mammals

Cell ingression through the primitive streak

As we have seen in the chapter on meso-endoderm induction on p. 118, primitive streak formation is due to its induction in the totipotent epiblast under an inductive influence from the underlying hypoblast. From cell density and growth rate studies Spratt & Haas (1962) concluded that in the chick the primitive streak functions as a proliferating bud and cannot be compared to the blastopore of the amphibians. Apart from the fact that cell proliferation is actually highest in the node region and directly adjacent to it (Emanuelsson, 1961), Vakaet (1960, 1962, 1971) and Shieh, Ning & Tsung (1963) disproved Spratt & Haas' notion and showed that the primitive streak is a region of extensive cell ingression and not a growth centre. The elongation of the primitive streak in postero-anterior direction is caused by an active stretching of its middle region (Vakaet, 1962). As soon as the median groove appears in the extending primitive streak, cells begin to move towards the groove, leave the surface of the blastoderm in the groove region and disappear into the interior of the embryonic anlage.

Wakeley & England (1978, 1979) followed the process of hypoblast delamination and the subsequent ingression of endo- and mesoderm

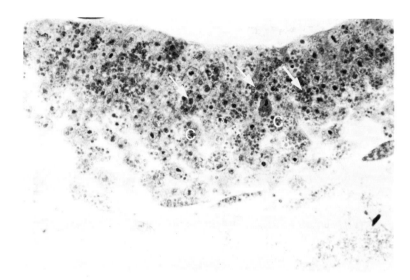

Fig. 17. Transverse section through anterior half of fully developed primitive streak, showing flask-cell formation as indicated by arrows and cells (c.) which have just left the streak (original).

through the streak in a TEM and SEM analysis. The extending groove terminates anteriorly in a pit-like depression, the so-called Hensen's node. Cells approaching the streak elongate and become flask-shaped in the groove area (see Fig. 17). The flask cells have fine filopodia and broad lamellipodia with which they are attached to subjacent cells. Subsequent cell shortening results in an inward movement of cells forming a sort of cell stream (Solursh & Revel, 1978). This holds for both chick and rat embryos. Apical microfilament bundles are present in the cells in the centre of the chick primitive streak as well as in the presumptive neural plate (Vakaet & Vanroelen, 1982). In the primitive groove cells coming from the left and right intermingle, so that in the internal layers they are more or less arbitrarily distributed over the left and right sides.

Cell ingression begins in the anterior third of the streak, soon includes Hensen's node, and subsequently extends posteriorly along the streak. The first cell population to ingress through the anterior third of the streak and Hensen's node becomes inserted into the hypoblast, forming the presumptive embryonic endoderm (Vakaet, 1962; Modak, 1965, 1966; Nicolet, 1965, 1970a; Rosenquist, 1966). Veini & Hara (1975) confirmed the epiblastic origin of the embryonic endoderm by comparing the development of hypoblast-free blastoderms of medium-streak to head-process stages cultured as intracoelomic grafts. Sumiya (1976b) came to the same conclusion when he separately cultured epi- and hypoblast isolated at successively later stages and enveloped in vitelline membrane (Wolff & Haffen, 1952). (See also Leikola, 1978).

After the ingression of the embryonic endoderm the presumptive prechordal mesoderm leaves Hensen's node, moving anteriorly between epi- and hypoblast. This is followed by the ingression of the chordomeso-derm, which is laid down successively in antero-posterior direction as the primitive streak with Hensen's node regresses (see above: Vakaet, Modak, etc.). The presumptive somitic mesoderm is still located in the epiblast layer at the definitive-streak stage (Gallera, 1975). Spratt (1955, 1957a, b) extensively described the regression of the streak and the laying-down of the notochordal and somite anlagen. Spratt & Haas (1960a, b) studied the morphogenetic movements in the hypoblast of reversed early chick blastoderms *in vitro*.

The primitive streak is a very dynamic region possessing high regulative capacity. Shoger (1960) observed pronounced regulation after injuring or stirring the node region. Spratt & Haas (1960c, 1961a, b) analysed the integrative mechanisms acting in chick blastoderm and streak formation. Ballester (1966) and Gallera & Dicenta (1966) found more or less complete regulation after inversion of Hensen's node involving the rejection of the exposed hypoblast. Implantation of a second node directly behind the host node also led to complete regulation when host and graft were of the same age, whereas no regulation occurred when the graft was either younger or older than the host (Gallera, 1974a, b; see also Nicolet, 1971*; Hara, 1978*).

L. F. Jaffe & Stern (1979) detected steady currents of about 100 μA leaving the streak of the chick embryo and returning elsewhere through the epiblast. Hensen's node was their 'epicentre'. Cell movement may be directed by the currents. The authors suggest that ions are pumped into the extraembryonic space by the epiblast.

In a time-lapse cinematographic study Stern & Goodwin (1977) noted that the morphogenetic movements during streak formation in the chick are periodic in nature, with a mean frequency of one pulse every 2.6 min. This periodicity is temperature dependent. Pulses are seen as slow waves starting at the posterior end of the streak and moving towards its anterior end (compare with wave phenomena in the mesodermal mantle during amphibian gastrulation, p. 132).

Neumann (1983) and Neumann, Laasberg & Kärner (1983) demonstrated adenylate cyclase and cAMP phosphodiesterase activity in the apical parts of the lateral surfaces of epiblast cells and in the entire lateral surfaces of primitive streak cells.

Colchicine and vinblastine, which influence microtubule formation, prevent elongation and migration of cells through the primitive streak (Granholm, 1970). Colchicine causes eversion instead of groove formation by transforming columnar cells into polyhedric or squamous cells. Colchicine does not affect intercellular contacts, however (Klika, Myslivečková & Rychter, 1980). In the chick, primitive streak ingression is not affected by transcription inhibitors but is blocked by translation inhibitors (Olszanska & Kludkiewicz, 1983).

The role of the cell membrane and extracellular matrix in cell ingression and migration

T. Yamada (1978*), emphasising the role of morphogenetic movements in development, called attention to alterations in the chemical specificity and fluidity of the cell membrane in association with subsurface structures such as microtubules, microfibrils and actin filaments, alterations which play an important part in gastrulation.

Wakely & England (1979) studied basement membrane structure and composition during and after ingression through the chick primitive streak. The basement membrane of the epiblast consists of a thin amorphous sheet without regional differences, but there are four different regions of fibrils superimposed on it which seem to play an important role in mesoderm migration. The fibrous network is laid down as a result of an interaction between the hypo- and epiblast (see Fig. 18). Zalik, Milos & Ledsham (1983) isolated two endogenous lectins from ingressing chick blastoderms: a soluble one from the entire area pellucida and area opaca and a particle-associated one from the hypoblast.

According to K. M. Yamada, Olden & Hahn (1980*) fibronectin, a cell-surface protein of 210000–250000 MW, represents a common type of adhesive molecule with tissue specificity. Fibronectin is a predominant constituent of mesenchyme cells and of basement membranes *in vivo*. A distinction must be made between 'cell surface fibronectin' and 'cytoplasmic fibronectin'. Gangliosides are effective competitive inhibitors of fibronectin action. Critchley, England & Wakely (1979) found fibronectin in the basement membranes of epi- and hypoblast before mesoderm formation in early chick and mammalian embryos. They suggest that fibronectin is involved in morphogenetic movements both by facilitating cell movement over a substratum and by improving attachment of cells to collagen fibrils. Spreading of chick mesodermal cells *in vitro* is strongly enhanced by the extracellular matrix material (ECM) attached to endoblast or hypoblast used as a substratum (Sanders, 1980). Mesodermal cells have a low level of surface fibronectin as compared to endo- and hypoblast, and therefore a low adhesiveness, which is important for their morphogenetic activity. Vanroelen, Vakaet & Andries (1980) reported that hyaluronate is absent in ingressing cells but rapidly accumulates in the basement membranes of laterally migrating cells. Prior to gastrulation, when only passive cell movements occur, fibronectin is not detectable. It is first found in the

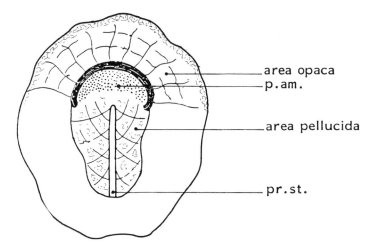

Fig. 18. Regional differences in fibril arrangement and fibronection distribution in the ectodermal basement membrane of a stage-4 (H. & H.) chick embryo, viewed from the inner surface after removal of both endo- and mesoderm. Area opaca with thin, mainly radially oriented fibrils and fibronectin network at outer margin; thick parallel fibrils and high concentration of fibronectin along anterior boundary of area opaca and area pellucida; proamnion (p.am.) with short rod-like fibrils; ectoderm of area pellucida with sparse thin fibrils in grid pattern and sparse distribution of fibronectin. pr.st., Primitive streak (after J. Wakely & M. A. England, 1979 and D. R. Critchley *et al.*, 1979).

basement membrane of the epiblast during streak formation. It disappears where cells sink down through the streak and reappears in the basement membrane of the definitive endoderm after its completion (see Fig. 18). Fibronectin seems to play a role in active cell movement (Duband & Thiery, 1982).

Evaluation

Whereas the amphibian blastula shows only animal–vegetal and dorso-ventral polarity, the advanced gastrula exhibits cranio-caudal, dorso-ventral and medio-lateral axes. During gastrulation the cranio-caudal axis is reversed 180°. These three axes are primarily expressed in the spatial segregation of the mesodermal mantle and secondarily in the pattern of the overlying ectodermal and underlying endodermal layers (see further Chapter 18, p. 249).

Gastrulation in the amphibians is characterised by the transformation of the blastula, which initially consists of an animal 'ectodermal' and a vegetal endodermal moiety, into a triple-layered embryo. The same holds for the flat teleost blastodisc, which initially consists of a multicellular epiblast and a syncytial periblast bordering on the huge semiliquid yolk mass. The transformation into a triple-layered embryo occurs by means of invagination in the amphibians but by means of spatial segregation in the teleosts. Little is known about the latter process.

In the amphibians gastrulation starts with flask-cell formation in the peripheral (blastoporal) region of the yolk mass, followed by mesoderm invagination around an external (urodeles) or internal (anurans) blastoporal lip. Flask-cell formation is apparently due to an interaction between the totipotent animal and the polarised endodermal moiety of the blastula. Mesoderm invagination seems to be the direct result of the changing properties of the presumptive mesodermal cells, properties which have been evoked by the inductive influence emanating from the endodermal yolk mass (see Chapter 10, p. 98).

In the avian embryo the cranio-caudal axis is expressed in the craniad extension of the primitive streak and its subsequent caudad regression, while the medio-lateral axis finds its expression in the lateral migration of the ingressing endo- and mesoderm.

An extracellular fibrillar matrix formed along the inner surface of the ectodermal blastocoelic roof of the amphibian gastrula guides the cells of the actively migrating edge of the mesodermal sheet, and possibly also of the endodermal archenteron, towards the animal pole, pulling the mesoderm and endoderm inwards. Contraction of the ring-shaped blastoporal lip region supports the invagination of endo- and mesoderm. The retransformation of the flask cells through cuboidal to squamous cells may support archenteron formation; the cranio-caudal extension of the latter may mainly be due to inward pulling by its actively migrating anterior tip.

It seems likely that similar cellular properties are responsible for the ingression of endo- and mesodermal cells through the avian and mammalian primitive streak and their subsequent spreading below the epiblast.

It may be concluded that gastrulation is the direct result of inductive interactions between the different moieties of the embryo and of the subsequent altered behaviour of the induced cells. The extracellular matrix, which plays an important guiding role in cellular migration during gastrulation, must be considered as a product of cellular differentiation.

We want to emphasise the great importance of the gastrulation process for embryogenesis. Whereas in the essentially single-layered amphibian blastula interaction is restricted to the peripheral region, where the animal and vegetal moieties are in direct contact with each other, the triple-layered advanced gastrula/early neurula has not only increased its number of moieties from two to three, i.e. the ectodermal, mesodermal and endodermal germ layers, but has also greatly increased the area of its interacting surfaces as a result of the spatial extension of the three layers, which are now in apposition along nearly their entire surfaces. The same holds for the avian and mammalian embryo as it changes from a double- to a triple-layered state.

The mesodermal layer represents the leading germ layer, showing a cranio-caudal and dorso-ventral segregation into various regions with different properties and, as we shall see, also with different inductive capacities. It should moreover be emphasised that the reversal of the antero-posterior axis of the mesoderm during amphibian gastrulation, and the ingression of the endo- and mesoderm through the primitive streak in the avian and mammalian blastoderm, brings into contact parts which were originally far apart, so that the now adjoining regions are very different, which strongly enhances their potential interactions. The first extensive interaction, that between the invaginating archenteron roof and the overlying ectoderm, results in the formation of the neural plate, the anlage of the central nervous system, which thus constitutes the fourth large-size moiety of the vertebrate embryo.

12

The induction of the neural plate

In this chapter we shall restrict ourselves to the formation of the neural anlage and to its primary, cranio-caudal segregation into a number of different regions, such as the arch- or prosencephalon, the rhomb- or deuterencephalon, the spinal cord and the tail somites. The morphogenesis and further development of the central nervous system (CNS) and the accompanying sense organs will be discussed in Chapters 13 (p. 181), 14 (p. 190) and 15 (p. 203).

Due to the different spatial and temporal courses of the gastrulation process in anamnian and amniote embryos, the induction of the neural plate will be dealt with separately for the two groups. For more detailed information the reader is referred to a number of reviews: Holtfreter & Hamburger (1955*), Saxén & Toivonen (1962*), Saxén (1963*), Nieuwkoop (1955*, 1966*, 1967c*, 1973*), S. Cohen (1965*), Sherbet & Lakshmi (1974*), Toivonen, Tarin & Saxén (1976), Grunz (1978*), Hara (1978*), K. Takata & Hama (1978*), Tiedemann (1978*), T. Yamada (1978*), Gerhart (1980*) and Saxén (1980a*).

Before going into the nature of the neural induction process we want first to elucidate the epigenetic origin of the neural anlage and the essentially reciprocal nature of the neural induction process.

The epigenetic origin of the neural anlage

In the predominantly descriptive literature of the last part of the nineteenth and the beginning of the twentieth century the notion prevailed that the nervous system, as all other organs, developed from a part of the egg predetermined for that particular development, so that development would be more or less an unfolding of an already existing but still invisible developmental programme.

Since Spemann & H. Mangold (1924) we have learned, however, that transplantation of an amphibian dorsal blastoporal lip to the ventral side of another gastrula leads to the induction of a secondary neural plate in the overlying ectoderm by the invaginating archenteron formed by the graft. Likewise, Bytinski-Salz (1937) and T. Yamada (1938) observed the

150

induction of a secondary neural plate upon insertion of a dorsal blastoporal lip into the blastocoelic cavity of an early lamprey embryo. The absence of any neural structures in total exogastrulae constitutes additional strong evidence for the epigenetic origin of the neural anlage (see under exogastrulation, Chapter 11, p. 136). Numerous other experiments support the epigenetic origin of the central nervous system in the various groups of the chordates (see further below).

It is therefore remarkable that M. Jacobson (1982*) recently returned to the old preformistic concept by having the central nervous system develop from particular, predetermined blastomeres of the 512-cell embryo, which he considers to be composed of separate 'compartments'. During cleavage the dividing blastomeres may shuffle around slightly but, essentially, they maintain their relative positions (Burnside & A. G. Jacobson, 1968; Cooke, 1979b). Since the first cleavage plane more or less coincides with the plane of bilateral symmetry, marking one of the blastomeres of the 8- or 16-cell stage with horseradish peroxidase may lead to a mirror-image picture of marked and unmarked areas in the left and right halves of the CNS (Hirose & M. Jacobson, 1979; M. Jacobson & Hirose, 1981).

It is not our intention to discuss here all M. Jacobson's arguments in favour of his preformistic notion of development, all of which can in our opinion be adequately explained by inductive interaction. We shall restrict ourselves to a few general remarks. In his 1982 paper he sums up all the possible arguments which seem to support his point of view, but simply ignores the very extensive experimental evidence in favour of an epigenetic origin of the CNS. He criticises Spemann & H. Mangold's classical 1924 experiments by stating that in *his* dorsal blastoporal lip transplantations the secondary nervous system always arose from graft tissue. However, it is common experience to those who more frequently perform dorsal blastoporal lip transplantations that with implantation of slightly older lips, or of young lips which do not heal in properly, the graft may not invaginate normally, so that it does not come to underly host tissue. Under such circumstances the outer cell layer of the graft will develop in an ectodermal direction and will be induced to form nervous tissue by the partially invaginated mesoderm. In any case, one cannot apply this criticism to Spemann & H. Mangold's (1924) and Spemann's (1931) experiments in which pigmented and unpigmented species were used and the nervous system of the secondary embryo unmistakably arose from the ventral host ectoderm. Recently Gimlich & Cooke (1983) and J. C. Smith & Slack (1983) showed convincingly by means of horseradish peroxidase labelling that additional nervous systems induced after dorsal blastoporal lip transplantation in *Xenopus laevis* are undoubtedly derived from host cells, with cell lineages separate from those which form the host CNS.

M. Jacobson (1981 a, b) removed blastomeres at the 16-cell stage and observed normal development. He attributed the results to the still uncommitted state of the blastomeres at this stage. We want to mention Model's (1978) and Model & Wurzelmann's (1982) translocation experiments involving the presumptive mes-rhombencephalic region of the early neural plate, which led to complete regulation and normal outgrowth of Mauthner cells, results which constitute yet another argument against Jacobson's compartment hypothesis.

Very recently M. Jacobson (1984) himself again repeated Spemann & H. Mangold's classical experiments and, obtaining results similar to those of Gimlich & Cooke and Smith & Slack, retracted his previous criticism of the organiser concept.

The reciprocal nature of the neural induction process

The invaginating archenteron roof, which begins to segregate into prechordal plate and chordomesoderm directly after its invagination around the dorsal blastoporal lip, induces the overlying ectoderm to form the anlage of the CNS as a separate differentiation of the outer, ectodermal germ layer. During this interaction, which extends through a rather long period of development (see Chapter 10, p. 101), the developing neural anlage exerts a reciprocal action upon the underlying archenteron roof. This was first noticed by Toivonen & Saxén (1966) and later by Kurrat (1974, 1977, 1978), who observed enhancement of notochordal differentiation tendencies particularly in the more caudal region of the archenteron roof.

Ohara & Hama (1979 *a*, *b*) made recombinates of trunk organiser with ageing exogastrula ectoderm and noticed that the differentiation of the inductor is strongly influenced by the presence or absence of neural competence in the reacting ectoderm. In the absence of neural competence the trunk organiser differentiated into mesenchyme and mesothelia only, whereas in the presence of an induced spinal cord it formed notochord and somites. The same holds for the so-called tail organiser, representing the most caudal region of the archenteron roof.

Nieuwkoop & Weijer (1978) performed a quantitative analysis of recombinates of middle archenteron roof with either competent gastrula ectoderm or non-competent neurula ectoderm of axolotl embryos and observed that notochordal differentiation was markedly enhanced in the former. Ohara (1980) found that head organiser isolated at the slit-blastopore stage formed more mesodermal and fewer endodermal structures in contact with competent than in contact with aged, non-competent exogastrula ectoderm. Leyhausen (1982) could demonstrate that the neural competence of the ectoderm markedly strengthens the self-differentiation and inductive capacity of the chordomesoderm in *Ambystoma mexicanum*. She subjected her data to extensive statistical analysis.

S. Ito & Ikematsu (1980) found that in the embryo of *Cynops pyrrhogaster* neurectoderm cells are electrically coupled to adjacent chordomesoderm cells during gastrulation. The coupling generally decreases at somite stages. It is therefore likely that a reciprocal interaction between archenteron roof and overlying ectoderm occurs over the entire cranio-caudal length of the embryo during the long period of their interaction.

Since inductive interactions involve pluripotential systems, the activation of the new developmental pathway necessarily leads to the suppression of other potential pathways (Holtfreter, 1968*; Nieuwkoop, 1968). This was clearly demonstrated by Grunz (1975*), who observed that formation of cilia, an expression of the epidermal developmental pathway already in progress in competent gastrula ectoderm, is suppressed when mesoderm is induced in the ectoderm by vegetalising factor. The suppression of cilia formation holds for the entire neural plate and the inner side of the neural folds (Kessel, Beams & Shih, 1974). This means that induction implies both derepression and repression of different sets of genes for the activation and inhibition, respectively, of alternate developmental pathways.

Neural induction in the anamnia

Cellular and ultrastructural changes in the reacting neurectoderm

In *Taricho torosa* (Normal Table of *Taricho torosa* by Twitty & Bodenstein, 1962) and *Xenopus laevis* embryos the first visible evidence of neural induction is a thickening of the mid-dorsal ectoderm overlying the median prechordal plate and notochordal anlage, resulting from a reduction of the outer surface area of the individual cells (Burnside & Jacobson, 1968 and Tarin, 1971 b respectively), and a lengthening of the cell bodies perpendicular to the surface. This change is first noticed at stage $11\frac{1}{2}$. Suzuki & Miki (1983) describe similar cell changes in the neurectoderm of *Cynops pyrrhogaster* at stage 13b (O. & I.). In *Xenopus laevis* the brain and spinal cord regions of the neural plate become distinguishable between stages $12\frac{1}{2}$ and 13 (N. & F.) and the outer margin of the neural anlage is clearly discernible at stage 14, while the final regionalisation of the CNS becomes externally visible during stages 18 and 19 (Tarin, 1971 a).

Ultrastructurally neurulation proceeds in nearly identical manner in the pigmented embryos of *Triturus pyrrhogaster* and the unpigmented embryos of *Rhacophorus schlegelii* (Tsuda, 1961), showing that the pigment does not play any role in the induction process. Tseng, Mo & Chang (1965) observed changes in both mitochondria and nucleoli after neural activation by Ca^{2+}-free Holtfreter solution and subsequent cultivation for 8 h, but these changes may have been due to sublethal cytolysis caused by the deleterious medium (Holtfreter, 1944c, 1945, 1947c). Yamazaki-Yamamoto, Yamazaki & Kato (1980) describe changes in the mitotic chromosomes of neuralised ectodermal cells in *Cynops pyrrhogaster*. Otherwise, the cellular and ultrastructural changes which are characteristic for the initial stages of neuralisation remain largely unknown.

A. Suzuki, Kuwabara & Kuwana (1976) state that the presumptive neurectoderm of the early gastrula of *Cynops pyrrhogaster* mainly consists of cells of the 15th cell generation. During gastrulation the number of ectodermal cells nearly doubles, so that at the end of gastrulation most cells belong to the 16th generation; consequently the decision between neural and epidermal development would be made during the 15th or 16th cleavage cycle. In *Cynops pyrrhogaster* the neural competence of the ectoderm decreases sharply after 12–18 h of cultivation *in vitro* from the beginning of gastrulation, during which period the cells divide at least once (A. Suzuki & Ikeda, 1979). According to A. Suzuki & Kuwabara (1974) the first cell cycle after the beginning of interaction is the critical cell cycle for neural induction. However, the fact that neural induction and differentiation can proceed normally, notwithstanding complete blockage of mitoses by colcemid or mitomycin C, when applied not earlier than stage 10^+, seems to contradict this conclusion (Cooke, 1973 *a, b*). Desnitsky (1979) states that induction may accelerate, slow down or not affect cell proliferation; the essential point is that a reprogramming of gene activity takes place in particular phases of the cell cycle in response to inductive stimuli.

Neural induction through permeable membranes

During neural induction the interspace between the inducing chordo-mesoderm and the reacting ectoderm seems to be traversed by cell processes extending from both layers and forming intimate membrane contacts between the two layers (Grunz & Staubach, 1979 *a, b*). These authors failed to find any cell anastomoses, the existence of which had been postulated by Eakin & Lehmann (1957). Grunz & Staubach suggest that in normogenesis the neuralising factor fixes to the cell membrane of the inducing tissue and reacts with receptors in the plasma membrane of the ectodermal target cells in the regions of close membrane contact (see also under heterogenous inductors, p. 166).

Saxén (1961) was the first to obtain neural induction through a millipore filter of 0.8 μm average pore size, using dorsal blastoporal lip or heat-treated HeLa cells as inductor and competent early gastrula ectoderm as reaction system. He concluded that no cellular contact is required for neural induction. Nyholm *et al.* (1962) obtained 50% neural induction through a 'TA' millipore filter in the absence of cytoplasmic contacts. However, it turned out that millipore filters have a spongy structure with the interconnecting spaces varying strongly in diameter, so that cell processes may pass unnoticed through a filter with very small average pore size. By scanning electron microscopy this was actually shown to be the case (Wartiovaara *et al.*, 1972). Toivonen *et al.* (1975) therefore employed Nuclepore filters with constant pore size and straight perforations. They found that the neuralising action traversed filters of any pore size between 0.05 and 1.0 μm during the 24 h the experiment lasted. The best inductions were obtained with pore sizes of 0.1–0.6 μm. The reason for this peak at medium pore size is not clear. They concluded that the morphogenetic signal for neural induction is due to transmissible compounds rather than to cytoplasmic contacts. With a Nuclepore filter of 0.2 μm pore size the first observable effect occurred after 6 h of contact. After 10 h, neuralisation took place in 100% of the cases, but no deuterencephalic structures developed as long as the filter was not penetrated by cytoplasmic processes; this only happened after 15+ h of contact (Toivonen & Wartiovaara, 1976; Toivonen, 1979). Toivonen (1979) also observed neuralisation across a Nuclepore filter of 0.05 μm pore size, as well as through a dialysing membrane which permits the passage of molecules of up to 12000 daltons, which strongly pleads in favour of transmissible compounds (see also Saxén, 1980*b**).

The temporal course of neural induction

The inductive capacity of the archenteron roof
Neural induction in amphibians and holoblastic fishes, as for instance the lamprey and the sturgeon, occurs *during* the gastrulation process. Ignatieva (1968), who defines the period between the beginning of gastrulation and the first appearance of the neural anlage as one developmental unit, states that the movement of the invaginating chordomesoderm towards the animal pole starts at time 0.15. The anterior border of the invaginating prechordal mesoderm comes into contact with the anterior portion of the presumptive neural plate at time 0.6, so that the induction of the definitive forebrain starts at that time. This holds for both axolotl and sturgeon.

In *Triturus pyrrhogaster* the neural inductive capacity of the presumptive organiser region seems to arise in the blastula of stage $6\frac{1}{2}$ (Nakamura *et al.*, 1970, 1971*a*). The inductive capacity gradually increases to reach its full extent by stage 9. Since a brief period of contact between inductor and reacting ectoderm only yielded positive results at stages 9 and 10, these authors suggest that the inducing agent is produced or activated at stage 9 to 10. In our opinion, however, this observation may just be due to different intensities of the inductive action at successive stages. Ohara (1981) found that the head organiser still possesses weak inductive capacity at the slit-blastopore stage. After more than 6 h of subsequent ageing it has lost its inductive capacity. However, the notochord retains its neural inductive capacity up to an advanced tail bud stage (Takaya, 1977). According to Takaya (1977) the segregating mesenchyme rather than the notochord is responsible for neural induction. In our opinion, however, the mesenchyme is important for the further differentiation of the nervous system (see Chapter 14, p. 197).

Minimal induction time

The minimal time required for neural induction varies rather much among different species. Archencephalic induction occurs after 0.5–1 h of contact in *Ambystoma mexicanum*, but needs at least 4 h in *Triturus vulgaris* (Johnen, 1961, 1964*a*). Xenoplastic combinations of ectoderm and archenteron roof have shown that the minimal time required is primarily linked up with the competence of the reacting ectoderm and is only secondarily influenced by the inductive capacity of the archenteron roof (Johnen, 1964*b*). In *Pleurodeles waltlii* 4–5 h of contact are sufficient for archencephalic neural induction (Gualandris & Duprat, 1981), while in *Cynops pyrrhogaster* only 3 h are required (A. Suzuki, 1968*a*, *b*, *c*; A. Suzuki, Kuwabara & Kuwabara, 1975).

Using middle archenteron roof as inductor, brief contact with competent ectoderm results in the formation of archencephalic neural structures, while a longer period of contact leads to the formation of deuterencephalic differentiations. More caudal archenteron roof finally induces spinal cord structures (see under regional neural induction, p. 158). In the axolotl, deuterencephalic differentiation tendencies appear after 4–16 h of contact, and spinal cord tendencies after 12–16 h (Johnen, 1961). *Triturus vulgaris* ectoderm requires at least 8 h of contact for the formation of deuterencephalic, and 12–16 h for spinal cord structures (Johnen, 1964*a*). Similar data were obtained by Goettert (1966) for *Triturus alpestris* and *Ambystoma mexicanum*. A. Suzuki *et al.* (1975) claimed that in *Cynops pyrrhogaster* the entire neural anlage is determined after 14 h of contact between the invaginated archenteron roof and the overlying ectoderm. However, Ohara (1981) found that in this species 15 h of contact are required for the formation of deuterencephalic and 18 h for spinal cord structures.

The neural competence of the reacting ectoderm

Using a short treatment of the ectoderm with Ca^{2+}-free medium, Chuang (1955) found that in *Cynops orientalis* neural competence already appears 24 h before the onset of gastrulation, at the mid-blastula stage.

Kawakami, Watanabe, Ave & Iyeiri (1969) cultured early gastrula ectoderm of *Cynops pyrrhogaster* and observed an increase in neural competence after 6 h. Full competence was still present after 12 h but competence was lost after 16 h.

Goettert (1966) found species-specific differences in the time course and intensity of neural competence between *Triturus alpestris* and *Ambystoma mexicanum* ectoderm.

Nieuwkoop (1958, 1960) studied the decline and disappearance of neural

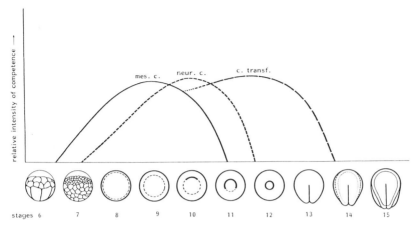

Fig. 19. Diagrammatic representation of relative intensity and duration of meso-endodermal competence (mes.c.), neural competence (neur.c.) and competence for transformation (c. transf.) during axolotl development from morula (stage 6) to open neural plate stage (stage 15, H.).

competence in folds of axolotl ectoderm taken at successive stages of development from mesoderm-free areas of the embryo. These folds were attached to the midline of the neural plate in the presumptive anterior rhombencephalic region of young neurulae. While full competence was present in folds taken from stages 8 to $11\frac{1}{4}$, competence began to decline at stage $11\frac{1}{2}$ and disappeared at stage 12, when only placodal structures were formed (see Fig. 19).

Ohara & Hama (1979a, b), using trunk or tail organiser as inductor in *Cynops pyrrhogaster*, observed that ectoderm of ageing exogastrulae still showed maximal competence after 12 h, while the competence began to decline between 12 and 24 h and had disappeared after 30–6 h at room temperature (c. 20 °C). Sasaki, Iyeiri & Tadokoro (1975b) used guinea pig liver as an archencephalic heterogenous inductor and found that early gastrula ectoderm of *Cynops pyrrhogaster*, when aged *in vitro*, begins to show a reduced rate of neural induction after 12 h and shows complete loss of competence after 24 h.

Toivonen (1967) found that induced presumptive archencephalon could be transformed into rhombencephalon and spinal cord by combining it with trunk mesoderm from stages up to stage 14 (see further p. 161). By transplanting presumptive rhombencephalon on top of exposed prechordal endo- and mesoderm, Nieuwkoop, & van der Grinten (1961) could show that rhombencephalic differentiation tendencies could be reconverted into archencephalic tendencies by a strong archencephalic inductive action at stage 12, and still partially at stage 13. After that the rhombencephalic differentiation tendencies were irreversibly determined.

Regionalisation of the neural anlage

Sala (1955, 1956) demonstrated that the consecutive cranio-caudal regions of the archenteron roof are responsible for the regional differentiation of the neural anlage. These regions pass the dorsal blastoporal lip at successive stages of development; first the head inductor, then the trunk inductor and finally the tail inductor. This implies that the most caudal region of the presumptive neural plate, which will give rise to tail somites (Bijtel, 1931, 1936; Spofford, 1945, 1948, 1953) is the region that comes into contact first with the most anterior region of the invaginating archenteron roof, the prechordal plate, and subsequently with the anterior, middle and posterior chordomesoderm. By interrupting the induction process at successive stages of gastrulation, Eyal-Giladi (1954) could show that the caudal portion of the presumptive neural plate acquires first archencephalic, then rhombencephalic and finally spinal cord differentiation tendencies.

Nieuwkoop *et al.* (1952) implanted folds of competent gastrula ectoderm at different cranio-caudal levels in the midline of the neural plate of host embryos. They observed the formation of regionally segregated nervous systems in the attached folds and concluded that two successive inductive actions occur in the regional segregation of the CNS. The first action was called 'activation'; this not only leads to neuralisation of the ectoderm but also to subsequent archencephalic differentiation. The second action, called 'transformation', is superimposed upon the first and is responsible for the cranio-caudal organisation of the CNS by transforming the initially activated archencephalic differentiation tendencies into those for rhombencephalon and spinal cord, and finally into those for tail somites, with increasing strength of the transforming action. Sala (1955, 1956) showed that the activating action has its maximum in the prechordal plate and anterior notochordal region of the archenteron roof and decreases cranio-caudally. The transforming action is restricted to the notochordal region of the archenteron roof, increases cranio-caudally, and has its maximum in the caudal region of the archenteron roof. Both Nieuwkoop and Sala hold the opinion that the size and shape of the neural anlage are determined by the spatial extension of the activating action in the overlying ectoderm, spreading from the median prechordal plate and notochordal anlage both anteriorly and medio-laterally, while the transforming action is responsible for the cranio-caudal regional pattern of the CNS.

The two – activating and transforming – actions, exerted by the invaginating archenteron roof upon the overlying ectoderm during gastrulation, constitute two successive waves which move caudo-cranially through the neurectoderm (Eyal-Giladi, 1954). The activating wave passes in front and extends farthest anteriorly, followed by the transforming wave, which only reaches up to the posterior boundary of the presumptive

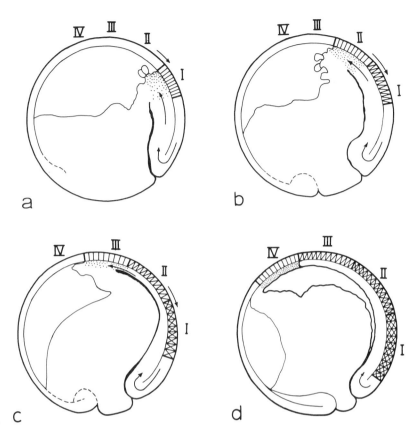

Fig. 20. Diagrammatic representation of four successive stages (a to d) of postero-anterior progression of neural induction during gastrulation. a, Interaction of prechordal endo-mesoderm (stippled) with most posterior region of presumptive neurectoderm; b, c and d, interaction of prechordal endo-mesoderm with successively more anterior regions of presumptive neurectoderm, and simultaneous confrontation of the more posterior regions of the neurectoderm with more and more posterior regions of the archenteron roof, leading to progressively stronger transformation (increasingly dense hatching).

prosencephalon (see also Nieuwkoop, 1962, 1967a; and Fig. 20). The transforming principle is not only responsible for the development of rhombencephalon and spinal cord but also for that of the tail somites which are formed in the most posterior region of the neural plate. Niazi (1969) could actually demonstrate that the latter region still forms neural tissue when isolated at stage $14\frac{1}{2}$ (H.) but differentiates into tail somites when combined with caudal notochord (see also Nieuwkoop, 1973*).

Leussink (1970) tested the inductive capacities of median and more lateral regions of the early neural plate and the underlying archenteron roof at four successive cranio-caudal levels on early gastrula ectoderm, using reciprocal combinations of *Ambystoma mexicanum* and *Triturus alpestris*. He found a very sharp medio-lateral decline in inductive capacity in the archenteron roof at all four cranio-caudal levels, and a slightly more gradual medio-lateral decline in (homoiogenetic) inductive capacity in the neural plate (see also under homoiogenetic induction, p. 162). He concluded that both the activating and the transforming action pass from the archenteron roof to the overlying ectoderm predominantly in the mid-dorsal region and subsequently spread medio-laterally through the ectodermal layer. Both actions show a field-like configuration, but with a different localisation of their maxima. Different threshold values of the transforming action are thought to be responsible for the formation of rhombencephalon, spinal cord and tail somites. Neural crest is formed at the periphery of the neural anlage under the influence of the transforming action (Nieuwkoop, 1962). In a quantitative analysis of the interaction between the archenteron roof and the overlying neural anlage Nieuwkoop & Weijer (1978) found that the size of the neural structures formed is correlated with the total mass of the axial mesoderm and not with that of either notochord or somites.

A rather profound misunderstanding about the regional induction of the CNS arose between the Finnish school of Toivonen & Saxén and coworkers and the Dutch school of Nieuwkoop and coworkers. We have seen in the chapter on mesoderm induction that the various heterogenous inductors can be classified into so-called neural inductors (such as guinea pig liver), which induce archencephalic neural formation and so-called mesodermal inductors (such as guinea pig bone marrow), which induce meso- and endodermal structures in early gastrula ectoderm (see under heterogenous inductors, p. 110). The combined action of the two types of inductor, both present e.g. in guinea pig kidney, elicits rhombencephalic and spino-caudal structures (Toivonen, Saxén & Vainio, 1963). The Finnish authors found that guinea pig liver and guinea pig bone marrow, when implanted side by side, induced additional rhombencephalon and spinal cord. They concluded that the regional induction of the CNS is due to the combined action of a mesodermalising and a neuralising inductor and presented their idea as the 'double gradient' hypothesis (see Saxén & Toivonen, 1962*; Saxén, 1978). Combining heat-treated HeLa cells (an archencephalic inductor) with untreated HeLa cells (a spino-caudal inductor) in proportions varying between 9:1 and 1:9, Saxén & Toivonen (1961) observed a gradual shift from archencephalic through rhombencephalic to spino-caudal induced structures. Similar results were obtained by a successive treatment of competent ectoderm with bone marrow and heat-treated HeLa cells (Toivonen, 1961), with non-treated and alcohol-treated HeLa cells (Toivonen, Saxén & Vainio, 1961*b*), and by preparing mixtures of liver-induced neural cells and bone-marrow-induced mesodermal cells (Toivonen *et al.*, 1963), all supporting Saxén's & Toivonen's double gradient hypothesis.

Unfortunately Saxén and Toivonen had not made a distinction between the *initial induction of the mesoderm in competent blastula ectoderm* and *the subsequent transforming or caudalising action of the differentiating mesoderm* upon the neuralised ectoderm. After Nieuwkoop (1955, 1966*) and Nieuwkoop & van der Grinten (1961) appeared Toivonen & Saxén (1966) investigated the postulated later tissue interactions between neural and mesodermal cells by means of cell reaggregates and actually found that presumptive archencephalic cells of stages 13 and 14 could be transformed into rhombencephalon under the influence of trunk mesoderm, the competence for transformation having disappeared by stage 15 (Toivonen, 1967). Toivonen & Saxén (1968) could subsequently demonstrate that the process leading to the regionalisation of the CNS is actually quantitative in character and can be mimicked stepwise by using different ratios of neuralised cells of stage 13 and cells of the mesodermal archenteron roof, varying between 10:1 and 1:5 (see also Saxén & Kohonen, 1968*). Toivonen (1972) also found that neuralised cells of stage 13 lose their competence for transformation after 10 h of cultivation *in vitro*, while the inducing capacity of the mesoderm is lost at stage 15–16. Finally, Toivonen (1970) showed that the transforming capacity is only a property of archenteric roof mesoderm and not, for instance, of limb-bud mesenchym. Kurrat (1974, 1977) confirmed Toivonen & Saxén's results, using reaggregates of presumptive archencephalon and caudal chordomesoderm, and could demonstrate that the competence for transformation is maximal in the early neurula. Using *Cynops orientalis* grafts and *Ambystoma mexicanum* hosts, Y. C. Wang (1965) found partial regulation after cranio-caudal translocation of various regions of the neural plate at the early neurula stage, which means that the regionality of the CNS is not yet firmly determined at the early neural plate stage.

T. Yamada (1950*a*, 1958*) also proposed a double-gradient hypothesis for the regional differentiation of the CNS. He distinguished a dorso-ventral gradient in morphogenetic potential (P d/v) and a cephalo-caudal gradient in stretching and convergence movements (P c/c). In our opinion, the former is the consequence of the medio-lateral extension of the activating inductive action, while the latter is an expression of the transforming or caudalising inductive action. Therefore, the hypotheses of Nieuwkoop and Yamada are closely related.

Injection of germinal vesicle content into the blastocoelic cavity of *Rana pipiens* early gastrulae can lead to the formation of an enlarged anterior neural plate and cement gland (supercephalisation) as a consequence of an enlargement of the prechordal plate region (Malacinski, 1972; see also Chapter 10, p. 105).

According to Ignatieva (1960*a*) the regional induction of the CNS in the sturgeon is very similar to that in amphibians. The blastoporal lip of large- and small-yolk-plug stages induces spinal cord and a tail-like outgrowth, respectively (Ignatieva, 1960*b*).

The propagation of neural induction in the ectodermal layer, or so-called homoiogenetic neural induction

Leussink (1970), among others, observed induction of large neural structures in the course of testing parts of the neural plate as inductors on early gastrula ectoderm. Deuchar (1970, 1971) showed that ^{3}H-thymidine-labelled neuralised cells are able to neuralise unlabelled competent ectoderm cells. Rollhäuser-ter Horst (1977*a, b*, 1981) observed that anterior neural plate of *Ambystoma mexicanum* can induce neural tissue, particularly retina and lentoids, in adjacent ectoderm. These are clear examples of homoiogenetic neural induction. In our opinion the anterior and medio-lateral extension of the neural plate reflects homoio-genetic induction, i.e. the propagation of the inductive stimulus from the mid-dorsal region directly overlying the prechordal and notochordal anlagen to the ultimate boundary between the neural plate and the surrounding epidermis. B. Albers (unpublished) recently demonstrated that the neural inductive action becomes spatially limited by the loss of neural competence of the ectoderm during its propagation: competent ectoderm implanted outside the lateral boundary of the neural plate, even some distance away from it, also becomes neuralised (see also Holtfreter, 1933*a*). This shows that the inductive action itself spreads beyond the boundary of the neural plate and the adjacent placodal ectoderm, from which the cephalic placodes develop.

Biochemical data relavant to the neural induction process

A wave of DNA synthesis and mitosis passes through the neurectoderm of *Xenopus laevis* (Maleyvar & Lowery, 1973*, 1976); it is correlated with neural induction occurring during invagination of the mesoderm, with a peak in the area overlying the prechordal mesoderm, where as we saw above activation is maximal. The wave is sensitive to 5-bromo- and 5-fluorodeoxyuridine (BUdR and FUdR) and to mitomycin C (agents which are known to disturb cytodifferentiation). It is therefore likely that DNA synthesis is primarily involved in the cellular differentiation of the neurectoderm (Maleyvar & Lowery, 1981). K. Takata, Yamamoto & Ozawa (1981) found that in Con-A-activated ectoderm ^{3}H-thymidine incorporation is lower in the first 30 h and higher in the next 30 h than in non-treated ectoderm.

Using cytophotometry, Lohmann & Vahs (1969) and Lohmann (1972) found regionally specific differences in DNA content in *Triturus vulgaris* neural plate, archenteron roof and endoderm, with short-lasting phase-specific increases in DNA at mid-to-late gastrula and late neurula stages. They take both phases of additional DNA synthesis to be the expression of enhanced gene activity at the onset of cytodifferentiation, possibly in the form of partial gene amplification (as observed also during oocyte maturation). Different melting profiles of the DNA in the mid-to-late gastrula and early neurula support the notion of differential DNA replication during gastrulation (Lohmann & Schubert, 1977). Whereas the methylation pattern of the DNA remains constant during early development, the proportion of DNA cleavage products that reassociate increases significantly from 20% in the gastrula to 30% in the neurula (Schubert & Lohmann, 1982).

T. Yamada (1978*) suggests that DNA transcription, which is required for the cells to enter a new developmental pathway, is regulated by cell-and stage-specific non-histone proteins. Continuous treatment with non-histone protein (NHP) prepared from adult liver inhibits neural induction in recombinates of competent gastrula ectoderm with dorsal blastoporal lip (Duprat, Mathieu & Buisan, 1977). The NHP affects neither the inductive capacity of the blastoporal lip nor the competence of the reacting ectoderm within a period of 4 h, so that NHP may act on the new pathway of differentiation at the transcriptional or post-transcriptional level.

Brachet (1942, 1943) already noticed that a high level of RNA synthesis first occurs in the inducing archenteron roof and later in the reacting ectoderm. This was confirmed by Rounds & Flickinger (1958), Pfautsch (1960) and Vahs (1962). Brachet's suggestion that RNA may be directly involved in the induction process was taken over by Niu (1959*, 1963, 1964), Niu & Sasaki (1971) and Niu & Deshpande (1973), who claimed that RNAs from inducing tissues mediate homoiogenetic inductive actions. Using the same technique as Niu, i.e. culturing small explants in conditioned media, Ajiro (1971) could not confirm Niu's observations, however. Lepanto (1965) found that treatment of the inductor with adenylate cyclase, thymidine or uracil had no effect on the inducing capacity of the archenteron roof, which suggests that the inductor is not a nucleic acid. Moreover, Hayashi (1959*a*, *b*), A. Suzuki (1966) and Tiedemann (1975*, 1978*) showed that the nucleic acid component of the neural inductor isolated from 9–12 day chick embryos, which has been characterised as a ribonucleoprotein, does not play any role in the inductive action. This also holds for the ribonucleoprotein isolated by Faulhaber (1972), Faulhaber & Geithe (1972) and Faulhaber & Lyra (1974) from the microsomal fraction made from yolk platelet coats of early developmental stages of *Xenopus laevis*. The consecutive increase in RNA synthesis first in the archenteron roof and then in the overlying neural plate may simply reflect the sequential initiation of cellular differentiation in the two layers (see also Saxén & Toivonen, 1962* and p. 176).

Lohmann (1979) also noticed that during the early phase of neural induction in *Triturus vulgaris* the RNA content is considerably higher in the inducing system than in the reacting ectoderm. About 10 h later the RNA content of the neurectoderm increases rapidly. Another 10 h later it decreases in both layers, while a second increase occurs during neural tube formation and notochord differentiation. In all stages the RNA content of the endoderm is higher than that of other tissues. Tiedemann, Born & Kocher-Becker (1965) and Tiedemann (1966b*) found the same level of tRNA and rRNA synthesis in non-neuralised and neuralised ectoderm of *Triturus alpestris*, but a somewhat higher level of mRNA synthesis in the latter. Suzuki (1968b) noted an increase in RNA and protein synthesis in the reacting ectoderm of *Cynops pyrrhogaster* after 12 h of contact with dorsal blastoporal lip. In *Xenopus laevis* N. Thomas & Deuchar (1971) observed a more rapid synthesis of high MW RNA in induced neurectoderm than in uninduced ectoderm at stage $12–12\frac{1}{2}$ (N. & F.).

Tiedemann (1978*, 1982*) considers regulation of mRNA transcription and processing to be very important mechanisms in tissue differentiation, but believes that additional mechanisms at the translational level may be involved in inductive interaction. According to Spelsberg (1974*) neural induction may have certain features in common with the binding of steroid hormones to nuclear acidic proteins.

Neural induction can be prevented by treatment with actinomycin D, which blocks RNA synthesis (Toivonen, Vainio & Saxén, 1964). Whereas Denis (1964) concluded that actinomycin D suppresses both the neural competence of the ectoderm and the inductive capacity of the dorsal blastoporal lip by preventing transcription and subsequent protein synthesis, Tiedemann, Born & Tiedemann (1967) could show that actinomycin D in a concentration of 2.5 μg ml^{-1}, which completely inhibits RNA synthesis, does not interfere with the inducing capacity of the dorsal blastoporal lip but does impair the competence of the reacting ectoderm. Løvtrup-Rein, Landström & Løvtrup (1978) found that both lactate and actinomycin D inhibit RNA synthesis during neural induction.

Another drug that affects neural induction is 5-fluorouracil, which inhibits protein synthesis (Toivonen *et al.*, 1961a). W. W. Minuth (1977) compared protein biosynthesis in induced and non-induced gastrula ectoderm and found the first differences in protein pattern after 48 h of cultivation. These were differences in cytoplasmic proteins, while 24 h later differences appeared in the nuclear proteins as well.

Kuusi (1959, 1960, 1961) noted that label transferred from a glycine-1-^{14}C, Na$_2$ ^{35}SO$_4$, ^{32}P- or ^{35}S-methionine-labelled heterogenous inductor to adjacent ectoderm with a lag phase of 1 or 2 days. The transferred label seemed to comprise non-specific compounds. This was confirmed by Vainio *et al.* (1962) using radioactively labelled HeLa cells. A more specific rapid transfer (30 min) of label from the 'inductor' labelled with ^{14}C-uracil, ^{14}C-glycine or ^{14}C-DL leucine to the reacting ectoderm was reported by A. Suzuki (1968*b, c*). After 3 h the label was found in the ectodermal nuclei.

Using antisera against guinea pig bone marrow, Vainio, Saxén & Toivonen (1960) and Vainio *et al.* (1962) observed a transfer of high MW material, in the form of antigenic granules, from inductor to reacting ectoderm in 3–6 h. In *Cynops pyrrhogaster* stage-specific antigen changes and strong inhibition of neural and notochordal differentiation particularly by antisera against gastrula and postgastrula stages were reported by Inoue (1962). Using immunoelectrophoretic methods, Stanisstreet & Deuchar (1972) and Stanisstreet (1975) found that dorsal, neuralised ectoderm contains higher concentrations of the same antigenic compounds than ventral, non-neuralised ectoderm, which they interpreted as evidence for changes in protein synthesis immediately after induction.

Wahn, Lightbody & Tchen (1975) and Wahn *et al.* (1976) observed neural differentiation of explants of early gastrula ectoderm of *Xenopus laevis, Pleurodeles waltlii* and *Ambystoma mexicanum* treated with dibutyryl cAMP, 8-bromo cAMP and cAMP together with theophylline, which suggests that the action of the neural inductor may be mediated via cAMP (see the similar observation by Bjerre (1974) in the chick, p. 177). They emphasise, however, that cAMP is not necessarily the normal agent in neural induction. Grunz & Tiedemann (1977) failed to obtain neural induction in isolated competent ectoderm treated with cyclic nucleotides. Already in 1975 Pays-de Schutter *et al.* suspected that negative results with exogenous nucleotides may be due to their rather high natural concentration in amphibian ectoderm.

Lallier (1960) found that formaldehyde suppresses the inducing capacity of the dorsal blastoporal lip, which is probably due to its crosslinking with proteins.

Hardly anything is known about the mechanism of the transforming action of the mesoderm on neuralised ectoderm,. Tseng (1960) mentioned that ethionine inhibits protein synthesis during the regionalisation of the CNS. Masui (1960*a*) and Ogi (1961) stated that lithium chloride suppresses archencephalic differentiation tendencies, transforming them into rhombencephalic tendencies in the late gastrula and early neurula.

From all these observations it becomes evident that in the entire process of neural induction and subsequent neural differentiation, many at first sight unrelated cellular mechanisms may be involved. The process may start with membrane changes and lead through many steps to ultimate new transcription and protein synthesis for cell-specific differentiation.

Neural heterogenous inductors

Many adult tissues contain neuralising as well as mesodermalising agents, e.g. mouse spleen and HeLa cells (Kohonen, 1963), mouse kidney (Vahs, 1962, 1965), rat spleen and liver (A. Suzuki & Kawakami, 1963 *a*, *b*; A. Suzuki, 1966; Ave, Sasaki & Kawakami, 1968 *a*, *b*) and guinea pig kidney (Kawakami & Iyeiri, 1963). Tiedemann & Tiedemann (1964) and Kawakami & Iyeiri (1964) isolated a neuralising as well as a mesodermalising factor from 9 day chick embryos. Kriegel (1961) detected neural-inducing capacity in *Triturus vulgaris* oocyte nuclei. However, there is no correlation between the neural-inductive action of different heterogenous inductors and their origin from adult tissues (see Saxén & Toivonen, 1962*).

Saxén & Toivonen (1962*) noted that the neuralising factor isolated from various heterogenous inductors is relatively resistant to heat treatment. Its activity is destroyed by proteolytic enzymes but not affected by RNase. The latter observation had already been made by Hayashi (1959 *a*, *b*) and Sasaki (1961), and was later confirmed by Vahs (1965) and Tiedemann (1975*, 1978*).

Basic proteins isolated from alcohol-fixed mouse kidney act as neural inductor (Vahs, 1962). Removal of the basic proteins from the inductor prevents neural induction, whereas treatment with purified RNase has no effect (Vahs, 1965), so that the active component of the ribonucleoprotein must be its basic protein moiety. A. Suzuki & Kawakami (1963*b*) obtained neural induction with ribonucleoprotein particles isolated from rat liver. The neural-inductive capacity of rat liver can be abolished by antisera against the ribonucleoprotein fraction from guinea pig liver or 9 day chick embryos (A. Suzuki, 1966). Ave *et al.* (1968 *a*, *b*) noted that the neural-inductive capacity of a protein fraction prepared from guinea pig liver can be annihilated by modification of its guanidyl or amide groups, or both, and by changing the threonine-hydroxyl and amino groups. Tiedemann *et al.* (1969) found that the neuralising factor isolated from 9 day chick embryos is not affected by thioglycolic acid or mercaptoethanol (agents which block SH groups).

Sasaki, Iyeiri & Tadokoro (1975*a*) subjected competent ectoderm to an initial short treatment with guinea pig liver as neural inductor and noticed that subsequent treatment with guinea pig bone marrow as mesodermal inductor markedly improved neural differentiation of the ectoderm. In Chapter 10, p. 112, we have already discussed the sensitising respective inhibitory effects of successive neural and mesodermal inductive actions, and vice versa.

Tiedemann & Born (1978) and J. Born *et al.* (1980) reported that, contrary to mesodermal induction, neural induction is not affected by the adsorption of the neural-inductive agent to BAC-cellulose or on Br-sepharose beads. They concluded that the neuralising factor acts upon receptor sites at the cell surface. As far as we know these important results have not yet been confirmed (see Chapter 10, p. 115).

The neuralising action of cations

Holtfreter (1947*b*, 1948*, 1955) was the first to observe the neural-inductive action of high and low pH and of Ca^{2+}-free medium and suggested that neural induction is due to a sublethal cytolysis of the treated ectoderm as a result of which neuralising factors are set free. John *et al.* (1984) reported that neural-inducing factors are released from cytoskeletal structures after homogenisation, freezing or ethanol treatment of gastrula ectoderm. According to Janeczek *et al.* (1983), the archencephalon-inducing capacity is particularly found in RNA particles released from the microsomal fraction. Holtfreter (1934*a*, *b*) had already observed that ectoderm killed by heat treatment acquires neural-inducing capacity, which was later confirmed, among others, by Rollhäuser-ter Horst (1977*a*, *b*).

Barth & Barth (1963) obtained differentiation of nerve cells from small ectodermal explants by changing the composition of the medium, particularly the Ca^{2+} and Mg^{2+} concentrations or the pH. Barth (1964) found that neural competence, when tested by Ca^{2+} concentration changes in the medium, arises at the late blastula stage and disappears at the early neurula stage. The late phase of competence may however apply only to certain types of cellular differentiation, such as pigment and neuroglia cells. Barth (1966) believes that sodium chloride is the true neural inductor and suggests that normal induction is brought about by Na^+ ions in the blastocoelic fluid (see also below). This would mean that the invaginating archenteron roof would only enhance neural development in the ectoderm, after it has been sensitised by the ions of the blastocoelic fluid. In our opinion this sounds very unlikely. Gerhart (1980*) suggests that entry of Na^+ ions leads to a transient Ca^{2+} flux and that variations in neural competence are brought about by variations in intracellular Ca^{2+} level. Barth & Barth (1969) noticed that nerve and pigment cell induction in small aggregates of chordomesoderm and ectoderm only occurs at sodium chloride concentrations of 0.88 M or higher.

Tiedemann (1976) believes that neuralisation by means of ions is not incompatible with a membrane-bound neuralising agent, which would be easily released by a change of membrane conformation. According to Grunz (1978*) the neuralising factor may enhance the activity of membrane-bound AMP or GMP cyclase.

Testing stage 11 gastrula ectoderm of various *Rana* species and hybrids for its reactivity to neural induction by lithium chloride Ansevin (1969) observed species-specific differences in incidence of neural induction. Haploid *Rana pipiens* ectoderm did not differ from diploid ectoderm, but in hybrids the response specific for the maternal species was modified by the presence of a foreign set of chromosomes. Ansevin (1966) noted a partial reversal of Li$^+$-induced neuralisation by subsequent cultivation at 4 °C. At low temperature several steps of the neuralisation process may be strongly repressed, which may lead to a return of the ectoderm to a more labile condition. Barth & Barth (1963) found that lithium chloride acting upon small ectodermal explants results in the formation of a variety of cell types, depending upon the duration of treatment. With increasing duration of treatment, first ciliated epidermis, then mucus-producing cells, nerve cells, pigment cells and finally neuroglia appear. Neural induction by lithium chloride or sucrose is accompanied by an increase in the uptake of ^{22}Na$^+$ and the induction by Li$^+$ depends on the Na$^+$ concentration of the medium (Barth & Barth, 1964, 1967; see above, p. 167). A high K$^+$ concentration reverses the Li$^+$ effect by affecting the intracellular Ca^{2+} concentration (L. G. Barth & Barth, 1974; L. J. Barth & Barth, 1974). Barth & Barth (1968, 1972) suggest that ions primarily affect the permeability of the plasma membrane and consequently alter the intracellular ion pool and ion ratios, leading to various types of cellular differentiation (L. G. Barth & Barth, 1974). Currently, H$^+$ ions are also considered as regulators of intracellular metabolic circuits (Epel, 1978*). (See also Chapter 10, p. 112).

The possible role of the cell surface and the extracellular matrix in neural induction

The cell surface and the extracellular matrix seem to play a role in morphogenetic movements as well as in inductive interactions. It is at present difficult to distinguish clearly between the two phenomena, the more so since morphogenetic movements may lead to inductive interactions.

Grobstein (1961) already mentioned that induction requires close association rather than contact between cells, so that contact should be taken in a physiological rather than a physical sense. Insertion of a filter may not affect physiological contact through the extracellular matrices. According to Grobstein (1967 a*, b*) the formation of macromolecular complexes at apposed interfaces may be important in embryonic induction.

Whereas dissociation of gastrulae and neurulae of *Cynops pyrrhogaster* with EDTA or alkali does not affect subsequent cell behaviour during reaggregation, sorting out and cellular differentiation, trypsin treatment disturbs sorting out and affects the differentiation of notochordal cells by causing irreversible changes in cell surface properties (Matsuda, 1980).

According to Landström & Løvtrup (1977) heparan sulphate may be the natural neural inductor, since it is an extracellular constituent of the 'fibroblast-like' Ruffini cells, while epidermal cells are characterised by chondroitin sulphate and mesenchyme cells by hyaluronate as cell surface constituents. In our opinion it is more likely that heparan sulphate plays a role in the migratory activity of the meso- and endodermal archenteron cells.

Løvtrup & Perris (1983) report that small explants of axolotl ectoderm which differentiate into neural crest derivatives under the influence of Li$^+$, also do so after treatment with ouabain, calcium-ionophore A 23 187, the potassium-ionophore valinomycin, cAMP and cGMP, heparan sulphate, prostaglandin and glucagon. These agents in one way or another seem to affect the cell surface of the reacting ectodermal cells.

In transfilter experiments where the passage of cytoplasmic processes was prevented by the small size of the pores, the thickness of the filter or the relatively short duration of the experiment, extracellular material was detectable in the filter pores by TEM or SEM analysis (Kelley, 1969; Tarin, 1972, 1973; Toivonen & Wartiovaara, 1976; see Fig. 21). This may explain the cases of neural induction obtained with Nuclepore filters of very small pore size, but cannot explain those obtained with dialysing membrane (Toivonen, 1979).

Voss (1965) was one of the first to demonstrate the presence of PAS-positive material in the form of intra- and extracellular glycogen in *Ambystoma mexicanum* neurectoderm by the end of gastrulation. Ubbels & Hengst (1978) demonstrated the presence of intercellular glycogen in the blastoporal region of *Ambystoma mexicanum* embryos. Kelley (1969) found that in gastrulae of *Xenopus laevis*, mesodermal and ectodermal cells are separated by large quantities of extracellular material consisting of glycogen, ribonucleoprotein (RNP) particles and fibrillar material. In exclusively ectodermal 'sandwiches' no extracellular material was discernible, but in ectodermal sandwiches containing mesoderm extracellular material appeared after 3 h. During late stages of gastrulation ectodermal cells incorporate fibrillar material from the interspace by means of pinocytosis (Kelley, 1969). Labelled RNA is not transferred from the mesoderm to the ectoderm during gastrulation. The RNA seems to be involved in protein synthesis in the mesodermal cells, possibly for the synthesis of extracellular matrix material.

Intercellular metachromatic material appears during the gastrula stages $10\frac{1}{2}$–13 in *Xenopus laevis* (N. & F.) (Tarin, 1971a, 1973), while the first evidence of neural induction in the form of a thickening of the mid-dorsal ectoderm is discernible at stage $11\frac{1}{2}$. TEM analysis showed that changes occur at the interface between the notochordal anlage and the mid-dorsal ectoderm, where extracellular granules accumulate at stage 12. These are composed of RNA and glycosaminoglycan fibrils (Tarin, 1973). They are replaced by fine fibrils at stage 14 (Tarin, 1972). Tarin does not believe that either the RNA or the glycosaminoglycans are directly involved in the neural induction process; according to him this either occurs through

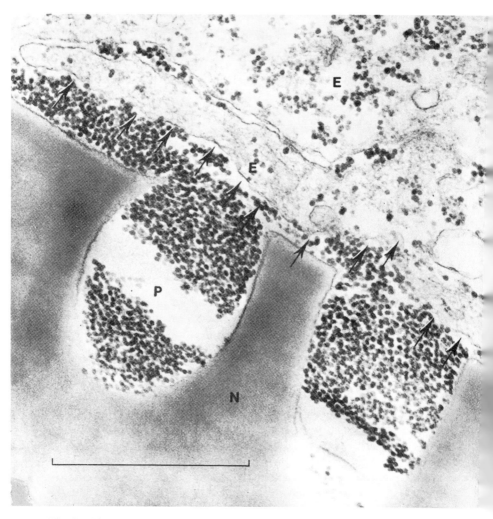

Fig. 21. Electron micrograph of portion of *Triturus* early gastrula ectoderm (E) in transfilter culture with dorsal lip mesoderm (not shown); N = Nuclepore filter. The ectodermal cells traverse the small orifices of the pores (P) and do not enter them. Within the pores extracellular granules can be seen. Cultivation time 24 h. Arrows indicate ectodermal cell boundary. (Courtesy S. Toivonen). Bar = 1 μm.

diffusible substances or through an interaction of the apposed cell membranes. Moran & Mouradian (1975) reported a correlative appearance of cell surface material of mucopolysaccharide nature in tissues undergoing morphogenetic movements and those involved in inductive interactions during gastrulation in amphibians. Moreover, the technique of La^{3+} treatment after glutaraldehyde fixation, which displaces Ca^{2+} from negatively charged moieties in biological membranes, showed that cell surface material is deposited in the developing urodele neural axis (Moran & Rice, 1975). X-ray-probe microanalysis indicates that Ca^{2+} levels are considerably higher in the neural axis, particularly in the neural fold region, than in the adjacent ectoderm (Rice & Moran, 1977).

K. Takata, Yamamoto & Ozawa (1981) reported that the lectins Con A and 'gorse agglutinin' have strong neural-inducing effects, which can be abolished by mannoside or fucose. According to K. Y. Yamamoto *et al.* (1981) the lectin Con A induces changes in the cell surface architecture of early gastrula ectodermal cells of *Cynops pyrrhogaster*, which lead to neural differentiation. Con A also affects cellular adhesiveness, but succinyl Con A (Sl-Con A) and *Dolichos biflorus* agglutinin (DBA) neither change the cell surface nor lead to neuralisation.

However, treatment of competent *Pleurodeles* ectoderm with lectins can also provoke surface modifications which prevent neural induction (Duprat, Gualandris & Rougé, 1982). These authors conclude that the structural integrity of the inner surface of the ectoderm is necessary for neural induction to occur, which suggests an active role of the cell surface in the induction process. K. Takata, Yamamoto & Takahashi (1983) found that Con A binds to asparagin-linked glycoproteins at the cell surface and is subsequently incorporated into the cytoplasm. In this way Con A binding would interfere with neural induction. Gualandris, Rougé & Duprat (1983) observed that the inner surface of isolated gastrula ectoderm becomes more densely labelled by radioactive lectins than the outer surface. (See further postscript on p. 180).

Dissociated presumptive neural cells of late gastrulae of *Bufo arenarum* are less strongly agglutinated by lectins than are epidermal cells, which suggests that the inducing factor generated by the mesoderm primarily affects the surface of the epidermal cells (Barbieri, Sanchez & Del Pino, 1980). Belousov & Petrov (1983) propose that neural induction suppresses extracellular matrix formation of the ectodermal layer, thus favouring cellular interaction. Pretreatment of the ectoderm with colchicine or cytochalasin B (or both) does not alter this behaviour towards lectins, so that the cytoskeleton does not seem to be directly involved in the cell surface alterations (Barbieri *et al.*, 1980). Tunicamycin, which blocks the glycosylation of glycoproteins and causes exogastrulation, does not affect neural induction but leads to an enhanced uptake of glucose, mannose and leucine (Sanchez & Barbieri, 1983).

Although in our opinion pinocytosis of extracellular material formed as a result of the interaction between mesodermal and ectodermal cells (Kelley, 1969) points to the possible role of extracellular matrix components in the neural induction process, a direct demonstration of neural induction by extracellular matrix material, as provided for mesodermal induction by Hoperskaya *et al.* (1984), has not yet been given. However, John *et al.*

(1984) found neural-inductive capacity in the protein but not in the proteoglycan fraction isolated from extracellular matrix material present at the interface of mesoderm and neural plate in *Triturus alpestris*.

It is evident that the extracellular matrix represents an important component of the cell surface, being a cell type-specific product of cellular differentiation. It certainly plays a role in cell locomotion and in the guidance of other cells (see Chapter 11, p. 134 and 146). We feel that the presence of extracellular matrix material in the filter pores in transfilter experiments implies that transfilter experiments are not conclusive for the possible role of cellular contact in induction experiments, since soluble or colloidal precursors may 'diffuse' over considerable distances during extracellular matrix deposition.

UV- and X-irradiation and neural induction

UV-irradiation of the vegetal pole region of uncleaved *Rana pipiens* eggs results in acephalic development which, according to Grant (1969*), is due to lesions in the grey crescent cortex. Transplantation of nuclei from outer cells of the previously irradiated grey crescent region into enucleated recipient eggs gave normal development. Grant concluded that apparently no significant interaction occurs between the purportedly damaged cell cortex and the subjacent nuclei during cleavage and blastulation.

Malacinski, Brothers & Chung (1977), Malacinski, Chung & Woo Youn (1978*a*) and H. M. Chung & Malacinski (1980, 1981) found that the dorsal blastoporal lip of embryos developed from vegetally UV-irradiated *Rana pipiens* eggs is a poor neural inductor, in contrast to that of non-irradiated embryos. Replacement of the dorsal lip of an UV-irradiated embryo by a non-irradiated lip abolished the UV damage, whereas replacement of the ventral lip had no effect. The effect of UV was maximal at a wavelength of 280 nm (Woo Youn & Malacinski, 1980). The authors assume that UV acts on a cytoplasmic factor essential for neural induction and suggest that proteins are involved in the initial reaction. This conclusion is in agreement with the observation of Malacinski (1972, 1974) that eggs can be rescued from UV damage by injection of a protein component from oocyte germinal vesicles, but is more difficult to reconcile with recovery from UV damage as a result of rotation of the egg (Scharf & Gerhart, 1980; H. M. Chung & Malacinski, 1980). The latter observation would rather suggest a disturbance of bonds between the egg plasmalemma and the cytoskeleton by UV irradiation.

X-irradiation impairs the capacity of the ectoderm to form neural structures. Presumptive hind brain is more susceptible (3.0 Gy) than forebrain (5.5 Gy), presumptive spinal cord being least susceptible (10 Gy) (Reyss-Brion, 1963). X-irradiation does not abolish the inductive capacity of the dorsal blastoporal lip but affects the reacting ectoderm. The cells of the irradiated region of the neural plate are damaged and subsequently extruded. Heavy irradiation of the dorsal blastoporal lip with up to 1500 Gy finally affects its inductive capacity, and prevents regional differentiation (Reyss-Brion, 1964). In our opinion these results indicate that X-rays do not impair the induction process itself but interfere with the subsequent cellular differentiation of the CNS.

Neural induction in the avian embryo

Cellular and ultrastructural changes during neural plate formation

As shown in Chapter 11, p. 144, the ingression of the presumptive mesoderm starts after the presumptive embryonic endoderm has left Hensen's node and the anterior part of the primitive streak. The first ingressing mesoderm moves forward, forming the prechordal plate. As soon as the primitive streak begins to regress, notochord and somites are laid down in cranio-caudal sequence, while the presumptive neurectoderm swings towards the dorsal midline (Spratt, 1952, 1955, 1957 *a*, *b*). Neural induction by the prechordal plate and chordomesoderm therefore occurs in a cranio-caudal sequence, exactly opposite to the situation in the amphibian embryo, where it spreads caudo-cranially during gastrulation (see Nieuwkoop, 1966*; B. R. Rao, 1968). This implies that in the avian embryo the rhombencephalon is induced slightly later than the prosencephalon, and the various cranio-caudal regions of the spinal cord at successively later times. The strength of induction decreases in cranio-caudal direction, as deduced from the width of the neural anlage (see also Hara, 1961 and 1978*).

In avian development, as in the amphibians, neural induction is initiated in the mid-dorsal region, where a band of nuclei appears in the ectoderm overlying the presumptive notochordal cells (England, 1973). Whereas at the definitive streak stage (stage 4 H. & H.) the various cell organelles are still scattered throughout the cytoplasm in both ectodermal and mesodermal cells, and a distinct basal lamina lines only the ventral surface of the ectoderm, at the head-process stage (stage 5) cell organelles are lacking in the vicinity of the basement membrane in both ecto- and mesodermal cells. This region is also free of ribosomes (England, 1974). England & Cowper (1975) observed junctional cell contacts between ecto- and mesodermal cells, with tufts of fibrous basal lamina material between the two layers. A single mesodermal cell may be in contact with several ectodermal cells (England & Cowper, 1976). Induction may be synchronised by these contacts. The same phenomena occurred when a neural plate was induced by grafting a Hensen's node between the epi- and hypoblast in the border region of the area pellucida (England, 1981 *a*). At stage 5 extracellular matrix material containing fibronectin and glycosaminoglycans is found in a fan-shaped region of the ectoderm anterior to Hensen's node (England, 1981*b*; see Fig. 18 in Chapter 11, p. 147).

The temporal aspects of neural induction

In the avian as well as in the amphibian embryo both the inducing axial mesoderm and the reacting ectoderm undergo changes during development.

The development of the neural inductor

Gallera & Nicolet (1969) observed that the middle portion of a young primitive streak, when transplanted underneath competent host epiblast, induces either another primitive streak (meso-endoderm induction) or neural structures (neural induction). The anterior portion of a medium primitive streak more frequently induces neural structures than those of a definitive primitive streak. Embryonic endoblast induces either a new primitive streak or neural structures, while axial mesoderm induces only neural tissue. The paraxial mesoderm shows the same inductive power as axial notochordal mesoderm (Gallera, 1966*a*). The inductive capacity of the chordomesoderm decreases as soon as Hensen's node begins to regress (Gallera, 1966*b*). The chordomesoderm just in front of the regressing node loses its inductive capacity by the 4 somite stage (Gallera, 1966*a*). Gallera & Ivanov (1964) found that the node of the definitive primitive-streak stage, when grafted laterally into an early primitive-streak stage embryo, induces brain.

Neural competence

The frequency of neural induction rapidly decreases in medium and definitive primitive-streak hosts, with only very weak inductions in the latter (Gallera & Ivanov, 1964). No induction occurs in hosts in which the streak is beginning to regress (Gallera, 1968). The primary and secondary neural anlagen always appear at the same time (Gallera, 1969). Contrary to Gallera and coworkers, Shieh, Ning & Tsung (1965), grafting Hensen's nodes underneath the epiblast of early head-process to 4-somite stages, reported that early hosts show neural induction in the entire epiblast, the frequency of induction declining cephalo-caudally.

The neural competence of the ectoderm declines rapidly at the neural-fold stage, when only placodal structures are formed. Whereas anterior epiblast retains its placodal competence up to the 2–4-somite stage, posterior epiblast has already lost it by that time (Gallera & Ivanov, 1964; Gallera, 1966*a, b*).

Testing the competence of different regions of the epiblast of the early primitive-streak stage (stage 2, H. & H.), Gallera (1971) concluded that neural competence decreases antero-posteriorly in the area pellucida, while the area opaca has a much lower neural competence. Cuevas & Orts Llorca (1974) observed that a quail node grafted into a stage 4 chick embryo induces nervous system, notochord and pharyngeal endoderm. Eyal-Giladi (1970*b*) obtained secondary neural inductions by the primary primitive streak in blastoderms folded transversely at the definitive-streak stage, when apparently mesodermal competence is lost so that neural competence can be revealed.

Minimal induction time

Gallera (1965) states that Hensen's node of the definitive primitive-streak stage requires at least 6 h of contact with ectoderm of a medium-streak stage for neuroid structures, and 8.5 h for brain structures to appear. Alcohol-treated chick liver, acting on early primitive-streak ectoderm, induces neuroid structures after 4 h and neural structures after 6 h of contact (Leikola & McCallion, 1967). According to Sherbet & Lakshmi (1969*b*) 3–5 h of contact between a node and competent ectoderm are needed for neural induction. Gallera (1970*b*) found that 2–6 h of contact are required for neural induction by a node in the area pellucida, but 7–9 h in the area opaca.

Regionalisation of the neural anlage

Hara (1961) tested the inductive influence of various parts of the mesoderm of the definitive primitive-streak to head-process stages on explanted epiblast of slightly younger embryos, culturing the tissues as 'open sandwiches' in intracoelomic grafts. While the compact mesoderm situated in front of Hensen's node at the definitive primitive-streak stage induces prosencephalic as well as mesencephalic structures, the prechordal mesoderm of the young head-process stage induces only prosencephalon. The short head-process of the young head-process stage induces mes- and rhombencephalon and sometimes spinal cord. At the medium head-process stage the prechordal mesoderm induces prosencephalon, the anterior portion of the head process mes- and rhombencephalon, and its posterior portion rhombencephalon and spinal cord. Hara concluded that during head-process formation the prechordal plate and chordomesoderm are responsible for the regional determination of the CNS. The data could be adequately interpreted according to Nieuwkoop's activation–transformation hypothesis originally framed for the amphibian embryo (Nieuwkoop *et al.*, 1952; Nieuwkoop, 1966*).

B. R. Rao (1968) studied the appearance of neural differentiation tendencies in the presumptive neurectoderm of the chick embryo by interrupting the induction process and rearing the isolated median and more lateral portions of the neurectoderm as intracoelomic grafts. He could show that an activation wave leading to prosencephalic differentiation spreads through the ectodermal layer, followed by a transformation wave which converts prosencephalic differentiation tendencies to those for mes- and rhombencephalon and spinal cord. Both waves spread from the midline laterally, while the activation wave also extends anteriorly into the region overlying the prechordal plate. Whereas the activating influence has its maximum in the prechordal and anterior notochordal region and decreases in caudal direction, the transforming influence, which is restricted to the chordomesoderm, increases in caudal direction and has its maximum in the caudal chordomesoderm. The regionalisation of the CNS (which is evidently quantitative in nature) is closely connected with the dynamic process of primitive streak regression, and is thus clearly time-dependent. Consequently, notwithstanding the pronounced differences in the mechanism of germ layer formation in amphibians and birds, neural induction shows great similarity in the two groups (see also Nieuwkoop, 1967* and Chapter 14, p. 192).

McCallion & Shinde (1973) observed that a quail Hensen's node of stage 3 or 4 induces neural tissue in chick epiblast of the same age, whereas posterior parts of the streak have no inductive capacity. Hensen's node of stage 4 acts mainly as a head organiser, that of stages 6–8 as a trunk–tail organiser, while the stage-5 node is intermediate in regional-inductive capacity (Tsung, Ning & Shieh, 1965). Gallera (1970a) reported that the stage-3 node always induces brain structures, while the node of stage 5^+ induces only small spinal cord structures in low frequency. It is thus evident that the inductions evoked by nodes of successively older stages show a cranio-caudal shift in regional character, while moreover the inductive capacity of the node decreases. A. Clavert (1974) and Ulshafer & Clavert (1979), studying the interactions between the induction fields of twin chick embryos, concluded that forebrain is exclusively induced by the prechordal plate, that notochord inhibits forebrain induction, and that induction of rhombencephalic structures depends on the presence of both prechordal plate and chordomesoderm. These data again reveal a clear resemblance with regional neural induction in the amphibians.

Le Douarin (1974) and Rasilo & Leikola (1976) demonstrated the *homoiogenetic* character of neural induction between pieces of neuralised and non-neuralised, competent epiblast of chick and quail primitive streak blastoderms cultured in close contact.

Biochemical data relevant to the nature of the neural induction process

Niu and coworkers postulated that RNAs may exert an organ-specific inductive action (Niu, 1958; Hillman & Niu, 1963; Mansour & Niu, 1965; Sanyal & Niu, 1966). Hillman & Hillman (1967) found that epiblast of the chick responds to brain and heart RNA (but not to liver RNA) by forming neural tube. However, they could correlate the different responses with the presence or absence of Folin-positive material in the RNA preparations and concluded that this impurity is responsible for the induction of the neural structures. They therefore agree with T. Yamada (1962) and Finnegan & Briggin (1966) that there is no organ specificity in the RNA. Lee (1973) used ^{3}H-uridine-labelled Hensen's node as inductor, which formed heavily labelled notochord, mesenchyme and foregut; the neural structures induced in unlabelled epiblast were not labelled, and neither was the non-induced ectoderm, so that no RNA transport is involved in neural induction (see also p. 163).

Gallera (1969, 1970c) observed that actinomycin-D-treated Hensen's node induces normal neural structures when implanted into stage 3 blastoderms; untreated nodes implanted into actinomycin-D-treated blastoderms induced neural structures but their differentiation was strongly inhibited. This suggests that the initial steps of the neuralisation process do not require new RNA transcription.

Chloroacetophenone (which is a specific, irreversible inhibitor of SH groups) interferes with the neural-inductive capacity of Hensen's node and reduces its glutathione content, suggesting the presence of SH groups in the living organiser (Lakshmi, 1962; Lakshmi & Sherbet, 1962; Lakshmi & Mulherkar, 1963). The same inhibitor also reduces the competence of the reacting ectoderm (Lakshmi & Sherbet, 1964). L-cysteine-hydrochloride and reduced glutathione (which contain SH groups) enhance neural induction by post-nodal pieces of chick blastoderms (Waheed & Mulherkar, 1967). However, we feel that this may represent a secondary effect of these compounds resulting from their dorsalising or cephalising action on the mesodermal inductor (see below).

Sherbet (1963), Sherbet & Mulherkar (1963) and Sherbet & Lakshmi (1967, 1969a) found that mammalian anterior pituitary is a strong neural inductor and suggested that FSH represents the active principle in neural induction. In Chapter 10, p. 120, we have already seen that the observed neural inductions can be explained by a dorsalisation of the mesoderm. Orts-Llorca & Domenech Mateu (1980) reported that Testoviron is a potent neural inductor.

Sherbet (1966a, b) noted that histones have an inhibitory effect on chick development. Hensen's node treated with histones loses its inductive capacity (Sherbet & Lakshmi, 1967). The authors claim that the histones interfere with the direct action of the inductor upon the target cell DNA. The inductive capacity can be restored with FSH.

Bjerre (1974) observed a neuralising action of dibutyryl cAMP on competent chick epiblast, but no effect of 5′-AMP and sodium-butyrate (see also neural induction in amphibians, p. 165). The lectin Con A strongly inhibits the differentiation of node grafts but apparently does not affect its neural-inducing capacity (Lee, 1976).

Heterogenous neural inductors

Pasternak & McCallion (1962) found that coagulated mammalian liver and kidney can act as neural inductors on chick epiblast. As already mentioned, mammalian anterior pituitary can enhance neural induction (Sherbet, 1962, 1963) but its action is very probably due to a dorsalisation of the inducing mesoderm by FSH (see above).

Induction through permeable membranes

Gallera (1967) reported that interposing a millipore filter between the very young primitive streak and the epiblast does not interfere with the induction of neuroid and neural structures. Although the graft emits filopodia into the pores of the filter, Gallera, Nicolet & Baumann (1968) did not observe any direct contact between inductor and reacting epiblast. They therefore concluded that diffusible substances must be involved in neural induction (see also the discussion of similar experiments on amphibian neural induction, p. 154).

Extracellular matrix and neural induction

The recent observation by Duband & Thiery (1982) that during gastrulation and neurulation the presumptive notochord and neural plate are separated by the basement membrane of the epiblast, which is rich in fibronectin (FN), may be of significance. Prior to gastrulation, when only passive cell movements occur, no FN can be detected; neither is it present on the ingressing streak cells. However, it appears on the definitive endoderm and the laterally migrating mesoderm. The authors conclude that FN is involved in active cell movement. In the trunk region intense FN staining is found between the neural tube and the somites and, later, between the neural tube and the notochord.

Two cell surface glycoproteins involved in cell adhesion (Cell Adhesion Molecules, CAM) isolated from chicken brain and liver and called N CAM and L CAM, respectively, already seem to be formed during early stages of development. Whereas both CAMs are more or less evenly distributed in the extracellular coat during early development, N CAM becomes localised exclusively in the neural anlage and L CAM in the endodermal derivatives, reflecting the divergent early differentiation of these derivatives (W. A. Thomas *et al.*, 1981; Edelman, 1983; Edelman *et al.* 1983; Hoffman & Edelman, 1983). The authors suggest that these molecules may also play a role in inductive interactions because they show a pattern of distribution that strongly resembles the fate map of the early chick embryo. However, fate maps give only the localisation of presumptive organ anlagen but do not reveal anything about their state of determination.

Neural induction in uro- and cephalochordates

Vandebroek (1938) already suggested that in ascidians the nervous system may develop epigenetically under the influence of the hypo- and chordoblast. From defect experiments on *Ascidiella* embryos, Vandebroek (1961) concluded that the anterior animal endoblast of the 16–32-cell stage, which contains the chordoplasm, is a strong neural inductor, while the vegetal endoblast is only a weak inductor. Reverberi, Ortolani & Farinella-Ferruza (1960) had come to similar conclusions. In transplantation experiments Ortolani (1961) found that in *Ascidiella*, *Phallusia* and *Ascidia* the nervous system is evoked in the competent presumptive neurectoderm by the invaginating chordomesoderm, while the presumptive epidermal region does not show any neural competence. Reverberi & Farinella-Ferruzza (1961) and Farinella-Ferruzza (1961) showed that treatment of the antero-animal and antero-vegetal blastomeres of 8-cell stages of *Ascidia malaca* with lithium chloride prevents the formation of the brain, sense organs and palps.

Wu & Cai (1964) mention that removal of the notochord prevents neural tube formation in *Branchiostoma*. T. C. Tung *et al.* (1961, 1962*b*) demonstrated that grafting veg$_2$ blastomeres into the animal hemisphere of 32–64-cell stages, or implanting a dorsal blastoporal lip into the blastocoel of a blastula or early gastrula of *Branchiostoma belcheri*, leads to the induction of a secondary neural tube when the graft differentiates into notochord, but not when it differentiates into other tissues.

Von Ubisch (1963) expressed doubt about the conclusions of Reverberi and Ortolani and of Tung and coworkers (see above) because he obtained negative results in grafting experiments in *Branchiostoma lanceolatum*. Fuldner & von Ubisch (1965) implanted an upper blastoporal lip into the blastocoel of *Branchiostoma lanceolatum*; the results were negative at 17–20 °C but positive at 20–4 °C. They explained this by assuming that neural induction is an artifact and does not take place in normal development. We suspect, however, that the negative results may be due to the different fluidity of the blastocoelic fluid in *belcheri* and *lanceolatum* (the fluid being gelatinous in the latter; Conklin, 1932, 1933).

Evaluation

Neural induction is undoubtedly an epigenetic event which is caused by the spatial interaction of the invaginating archenteron roof of the amphibian embryo or the ingressing mesoderm of the avian embryo with the overlying ectoderm. The interaction is clearly reciprocal in nature.

While the spatial extension of the neural anlage and the adjacent placodal ectoderm is due to the spreading of the primary, activating action in the ectodermal layer, the subsequent regional differentiation of the neural plate into the various cranio-caudal regions of the CNS depends on a secondary, transforming or caudalising influence emanating from the chordomesoderm and acting upon the neuralised ectoderm. This holds both for the amphibian and the avian embryo.

Both actions start from, and are chiefly restricted to, the median region of the archenteron roof, i.e. the prechordal plate and the notochordal anlage, and spread in the overlying ectoderm in anterior and medio-lateral directions. Both actions show a field-like distribution in the ectodermal layer, but with different centres of activity. They seem to propagate homoiogenetically in the ectoderm, spreading from cell to cell. Propagation apparently occurs with decrement and is dependent on the local intensity of the inductive action. Recent experiments of Albers have made it seem very likely that the neural anlage becomes spatially delimited as a result of the gradual loss of neural competence in the reacting ectoderm during the anterior and medio-lateral propagation of the inductive stimulus (see also Nieuwkoop, 1973*).

Whereas the activating action can be mediated across Nuclepore filters through which no cellular processes can penetrate, and even through a dialysing membrane, the transforming action seems to require direct cellular contact. This difference may not be so fundamental as it looks, however, since extracellular matrix material produced by the interacting cells can presumably penetrate a filter in the form of soluble or colloidal precursors. Although pinocytosis of extracellular matrix material by ectodermal cells has been reported, neuralisation has not yet been achieved by extracellular matrix components, as is the case for mesodermal induction. It must be emphasised that the nature of the inductive actions *in vivo* is still unknown: diffusible ions and small molecules as well as high MW components of extracellular matrix may be involved in the activating action, while for the transforming action direct cell membrane interaction must also be taken into consideration. It is most likely that, whatever the chemical or physical nature of the inducing factors, they represent normal, possibly transient manifestations of cellular differentiation of the interacting systems as these develop along diverging developmental pathways and become temporarily or permanently apposed to one another.

POSTSCRIPT TO PAGE 180

Electrical coupling is reestablished between explanted chordo-mesoderm and gastrula ectoderm after 3–6 h contact. Pretreatment of the ectoderm with Con A prevents coupling and interferes with neural induction. The latter is restored in only one-third of the cases after 3 h contact. These data suggest that cellular communication plays an important role in neural induction (Suzuki, Hakatake & Hidaka, 1984; see also Warner, Guthrie & Gilula, 1984 and Slack, 1984 on p. 94).

13

The process of neurulation or neural tube formation

The shaping of the flat neural anlage into the tubular CNS in amphibians and fishes

The process of neurulation is the result of the preceding induction of the neural anlage, during which the cells acquire new properties (see Chapter 12, p. 153). These manifest themselves in, among other things, the autonomous folding of the initially flat anlage into a tube-like structure. In a recent interview (Mehr, 1982) Holtfreter emphasised that the induced primordium of the CNS organises itself into a complex structure with only supporting influences from surrounding tissues.

C. O. Jacobson (1962), who studied the morphogenesis of the neural plate by means of vital staining, observed similar dorsal convergence and stretching movements in the neural plate as in the underlying mesodermal mantle. The isolated neural plate has an intrinsic capacity to fold itself up. However, the elongation of the neural plate is due to the strong adhesion between the notochordal anlage and the overlying neural plate (see also C. O. Jacobson, 1964*b**).

Changes in selective cell affinity play a role in the fusion of the neural folds and in the separation of the epidermis from the neural material (C. O. Jacobson, 1968). C. O. Jacobson & Löfberg (1969) found that the paraxial mesoderm hardly plays a role in the neurulation process. The later morphogenetic movements of the neurectoderm could be effectively inhibited by β-mercaptoethanol, which does not affect the expansion of the epidermis (C. O. Jacobson, 1972). Under these circumstances well-developed neural folds are nevertheless formed, which pleads in favour of a supporting influence of the expanding epidermis in neural fold elevation. This was already suggested by Løvtrup (1965*c**) and has recently been confirmed by Brun & Garson (1983), who noted that colchicine prevents epidermal spreading. Christ (1971) assumed that the attachment of the notochord to the overlying ectoderm interferes with the expansion of the lateral ectoderm, thus causing epidermal thickening.

C. O. Jacobson & A. G. Jacobson (1973) removed longitudinal strips of epidermis, neural fold or neural plate and concluded from these and other experiments that there are three main morphogenetic forces: (1) elongation and narrowing of the neural plate by notochordal stretching and column-arisation of the neural epithelium; (2) formation of the neural folds when the neural plate becomes wedge-shaped in cross section, the competence for neural-fold formation being restricted to a relatively narrow marginal zone of the neural plate; (3) dorsad spreading of the lateral ectoderm. The three forces can operate independently but are coordinated in normal development. However, it was found later that there seems to be a tension in the epidermis at all stages of development, leading to the formation of a gaping wound after incision, which pleads against epidermal support of neurulation (A. G. Jacobson, 1978).

In newts the neural plate consists of a sheet of cells essentially one cell layer thick. Changes in surface area are inversely correlated with changes in height of the neural epithelium. In the neural plate the cells retain their contact relationships with neighbouring cells, so that displacements of groups of cells are brought about by deformation of the cell sheet resulting from regional differences in cell shape change (Burnside & A. G. Jacobson, 1968). The circular arrays of microfilaments in the apical portions of the neural plate cells are thought to be responsible for apical cell constriction during neurulation (Schroeder, 1973). A. G. Jacobson & Gordon (1976a) suggested that two forces are required for the shaping of the initially hemispherical neural anlage into the key-hole-shaped older neural plate: (1) a regionally programmed shrinkage of the neural plate surface, in which contraction of the surface area is strictly correlated with elongation of the neural plate cells due to the fact that the volume of the neural anlage remains constant; and (2) a shearing force due to the antero-posterior elongation of either the notochord or the overlying median neural plate or both. A. G. Jacobson & Gordon (1976b) stated that induction may play a role in establishing this pattern.

Gordon & A. G. Jacobson (1978) and A. G. Jacobson (1980) made a computer simulation of the shape changes of the neural plate on the basis of a grid placed over the neural anlage and a description of the changes in form of each of the initial grid squares. Burnside & A. G. Jacobson (1968) had already determined the distortion of the geometry of such a grid, which is correlated with the shrinkage of the outer surface of cells and areas, the extent of shrinkage varying over the neural plate. The computer simulation programme developed by A. G. Jacobson (1980) included two spatial coordinates, the volumes of the neural plate areas, the initial height, and a 'height programme' giving the regional increases in height. The formation of the normal key-hole shape of the older neural plate required an additional shearing force, which is represented by the elongation of the region overlying the notochord. A second region of shear exists at the boundary between neural plate and epidermis. The shearing forces may be thought of as altering the formation of gap junctions at the

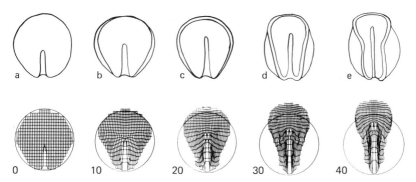

Fig. 22. Outline drawings of normal neural plate development of *Taricha torosa* from stage 13 to 15 (top row, a–e), and photographs from a computer graphics terminal showing every tenth step of a computer simulation of neural plate development (bottom row, 0–40). (After A. G. Jacobson & R. Gordon, 1976*a*, *b*).

cell boundaries. Fig. 22 shows that there is good correspondence between the simulated form changes and those observed in normal development.

Neural plate formation in the anuran amphibians is complicated by the presence of two more or less separate cell layers, a thin superficial epithelial layer, which will form the ependymal layer of the neural tube, and a massive deep sensorial layer, which will furnish the bulk of the nervous system as well as the neural crest (Nieuwkoop & Florschütz, 1950; Karfunkel, 1971 – both in *Xenopus laevis*). In the anurans 'direct' neural tube formation by folding of the neural plate and subsequent fusion of the neural folds occurs over the entire length of the neural anlage including the neurenteric canal (Nieuwkoop & Florschütz, 1950). Schroeder (1970) at first suggested that in the double-layered anuran neurectoderm the shaping forces for neurulation are chiefly generated in the deep layer, with the superficial layer sliding over it. However, later Schroeder (1971) came to the conclusion that in *Xenopus laevis* 'flask cells' of the superficial layer contribute the necessary forces that initiate folding. The notochord provides the longitudinal axis along which infolding occurs and at the same time secures the floor of the neural groove in the midline of the neural plate. In amphibians five mechanical factors would operate in neurulation: (1) microtubule-dependent cell elongation; (2) microfilament-dependent cell surface contraction; (3) swelling of the notochordal cells by intracellular vacuolisation; (4) the capacity of flask cells to realign themselves; and possibly (5) epidermal spreading.

In *Ambystoma mexicanum* direct neural tube formation is restricted to the brain and the anterior portion of the spinal cord, while 'indirect' tube formation by cavitation of the originally solid neural cord occurs in the posterior trunk and tail region.

In the teleost, *Oryzias latipes*, neural tube formation is entirely indirect:

a solid neural cord is formed from deep ectodermal cells, which divides into a left and a right column of cells, after which a primitive neural canal is formed, first ventrally and then dorsally (Miyayama & Fujimoto, 1977).

Gallera (1961) tried to block the shape changes in the neural plate of *Triturus alpestris* by embedding late gastrulae in gelified agar. Whereas morphogenetic processes were markedly inhibited, histogenesis and cyto-differentiation were not influenced. Herkovits (1977, 1978) repeated Gallera's experiments on *Bufo arenarum* embryos and observed dramatic deformations, from which however the embryo recovered to a considerable extent after being freed from the agar. Malacinski & Woo Youn (1981a) found normal morphogenesis of the neural plate with normal neural-fold closure and axial stretching in notochord-defective or notochord-less *Xenopus laevis* embryos obtained after mild UV-irradiation of the vegetal hemisphere of fertilised uncleaved eggs. They concluded that the notochord is not essential for normal morphogenesis of the CNS. In our opinion this conclusion is only partially correct. In the absence of the notochord the somitic mesoderm is responsible for the more or less normal stretching of the axial anlage, but instead of two lateral neural masses there is only a single medio-ventral neural mass overlying the fused somites (see Spemann, 1936*, reprinted in 1968*).

Bergquist (1963*) and Kallén (1968*) stated that the notochord and the prechordal plate determine the bilateral symmetry of the CNS, but the adjacent mesoderm is of great importance for the shaping of the neural tube. The mitotic activity of the neural tube is regulated by inductive influences from the chordomesoderm and the migrating neural crest, local increases in mitotic activity leading to local outbulgings of the brain wall, the 'neuromeres' (see Fig. 24, Chapter 14, p. 191). However, Cooke (1973b) observed normal brain shaping in *Xenopus laevis* during long-term mitotic inhibition from an advanced gastrula stage onwards. He therefore questions the shaping influences of differential mitotic rates (see also Cooke, 1980* and in birds, p. 186).

Cellular and ultrastructural changes during neurulation

A condensation of pigment along the dorsal midline, caused by apical cell contraction, is the first sign of the neurulation process in *Ambystoma*, *Taricha* and *Rana* (Karfunkel, 1971).

Balinsky (1961) observed that during neurulation in *Phrynobatrachus natalensis* the outer cell surface of individual cells is thrown into folds by the contraction of an underlying contractile layer. In *Ambystoma mexicanum* Voss (1965) found much extracellular glycogen in the vicinity of the neural plate at neural-fold stages, which disappeared by the time of neural tube closure. Waddington & Perry (1966) noted large numbers of microtubules inside the neural plate cells, which were oriented in the direction of cell elongation. Burnside & A. G. Jacobson (1968) concluded from cell surface measurements that the contraction of microfilaments is responsible for the shrinkage of the outer cell surface. According to Burnside (1973), both microfilaments and microtubules play an important role in changing the cell shape. These observations were confirmed by Karfunkel (1971), among others. Löfberg (1974) described the formation of large numbers of adhesive surface projections on the bottle-shaped cells of the folding neural plate.

In *Ambystoma mexicanum* Löfberg & C. O. Jacobson (1974) could inhibit neurulation with vinblastine and colchicine, which both disassemble microtubules. In contrast, guanosine triphosphate accelerates neurulation, leading to overabundance of microtubules. They concluded that the microtubular system is associated particularly with the process of cell elongation and, through that, with morphogenetic movements. According to C. O. Jacobson (1970) colchicine and β-mercaptoethanol do not affect the dorsad expansion of the epidermis, but Brun & Garson (1983) state that colchicine prevents epidermal spreading (a process which is only slightly affected by nocodazole). Hydrostatic pressure (400 psi) acting for 180 min disrupts the apical microfilaments, while 330 min of treatment is required for microtubule disintegration, leading to similar effects as cytochalasin B and colchicine respectively. While microfilaments seem to be responsible for cell constriction, microtubules are primarily involved in cell elongation (Messier & Sequin, 1978).

Papaverine, which interferes with Ca^{2+} fluxes, inhibits neural-fold formation. Ionophore A 23187 and EGTA promote rapid cellular contraction and accelerate neural-fold formation. During neurulation Ca^{2+} is released into the medium, a process which is impeded by papaverine. Ionophore A 23187 induces a Ca^{2+} influx while EGTA causes a Ca^{2+} efflux (Moran, 1976). The availability of free Ca^{2+} is crucial in controlling microfilament-generated morphogenetic movements (Moran, 1978a). Moran & Rice (1976) suggested that papaverine affects cell junctions, microfilaments and mitochondria, but not microtubules. Ionophore treatment of papaverine-inhibited embryos leads to a disruption of the neural folds. La^{3+}, which replaces Ca^{2+} in biological membranes, causes disaggregation of neural tissue (Moran, 1978a).

In neural plate and neural groove stages Waddington & Perry (1966) found an accumulation of fine granular material on the external cell surfaces, and flanges of similar material between neighbouring cells of the neural plate. Inside the cells they observed large numbers of small vesicles. Moran & Rice (1975) detected a large amount of lanthanum-nitrate-positive cell surface coat material (CSM) on, and numerous 'CSM-coated' vesicles inside the neural plate cells of *Ambystoma maculatum* by the time the neural folds appear. In contrast, the neural ridges themselves were covered with only a small amount of uniformly distributed CSM, and vesicles were sparse in their cells. During neural-fold elevation there is a progressive increase in the amount of CSM in the presumptive tube region, with a pronounced increase in CSM at the leading edges of the converging neural folds. The approaching neural folds become physically bridged by La^{3+}-positive material. Finally, during the fusion of the folds and tube formation no CSM remains in the interspaces. Mak (1978) found that during neural-fold fusion in *Hyla regilla*, *Rana pipiens* and *Xenopus laevis* the contacting surfaces become adhesive. Prior to fusion the formation of cytoplasmic vesicles, changes in surface morphology and filopodial extensions were observed. Mak noted that the glycosaminoglycan composition of adhering surfaces differs from that of non-adhesive epidermal cells.

Decker & Friend (1974) observed the presence of junctional complexes in the form of zonulae occludentes along the apical periphery of amphibian neural plate cells. However, these 'belts' are no longer clearly discernible as neurulation proceeds. As a consequence tracers like lanthanum, ruthenium red, etc. can now readily penetrate. After the closure of the neural tube all evidence of apical occluding zones has vanished. They are replaced by macular gap junctions. Gap junctions are formed independently well below the junctional complexes. Decker (1981) described how, during neurulation in *Rana pipiens*, junctional complexes quickly fragment into smaller domains, leading to the breakdown of the permeability barrier. Ultimately the junctional complexes disappear altogether, leaving randomly scattered gap junctions in their wake. The disappearance of the zonulae occludentes may express changes in membrane polarity. We believe that these changes may be related to the remodelling of the outer neural plate surface into the inner surface of the neural tube (see also Burgess & Schroeder (1979*)).

Neural tube formation in the amniote embryo

Mitosis may play a more important role in neuromorphogenesis in the amniotes than in the amphibians, since in the former extensive growth of the CNS occurs (A. G. Jacobson, 1978). Jelínek & Friebová (1966) suggested that in the chick embryo cell proliferation is the main dynamic factor in neurulation, while apical contraction of the anchored neural plate cells constitutes the second factor. In our opinion it seems more likely that apical contraction represents the primary force, while neurulation may be supported by cell proliferation.

At posterior levels of the neural tube of the chick embryo 'direct' tube formation by folding and 'indirect' tube formation by cavitation occur side by side and overlap spatially (Criley, 1969). The transition zone is constituted by a caudal proliferation centre, which regresses along with the primitive streak immediately behind the notochord-forming centre. The proliferation centre produces a solid medullary cord. Neural fold elevation and closure give rise to a medullary groove dorsally, while ventrally the solid medullary cord is penetrated by a blindly ending cavity. After closure of the medullary groove the two lumina fuse. More caudally, cavitation of the medullary cord predominates (Jelínek, Siechert & Klika, 1969). In indirect neural tube formation cell segregation is preceded by extracellular matrix deposition (Schoenwolf & Delongo, 1980). Gap junctions are distributed radially around the indirectly formed lumen (Schoenwolf & Kelley, 1980).

Lee *et al.* (1976) and Lee, Sheffield & Nagele (1978) observed a thick coat of extracellular material (ECM) between the leading edges of the approaching neural folds in the chick embryo, while the cells of the inner surface of the neural tube have folded surfaces and much less ECM. The ECM serves as a temporary adhesive bond between the leading edges of the neural folds, until junctional complexes are established. The fusion of the neural folds can be inhibited by trypsin treatment and by the lectin Con A. Lee (1976) suggests that Con A blocks neural tube formation in the chick by inhibiting interkinetic nuclear migration. Messier (1978) emphasises that both cell broadening by interkinetic nuclear migration and cell elongation play a role in neuromorphogenesis. Microtubules are involved in both processes. Microfilaments are particularly well organised in the apical portions of the neuro-epithelial cells. All microfilaments bind heavy meromyosin (Nagele & Lee, 1980). Ca^{2+} is concentrated in mitochondria and coated vesicles of neuro-epithelial cells. The coated vesicles are more numerous in uplifted neural folds than in the flat neural plate. The vesicles move along oriented microtubules towards the cell apex, accumulating Ca^{2+} during their migration (Nagele, Pietrolongo & Lee, 1981). These authors therefore suggest that the Ca^{2+}-containing vesicles play a role in the contractile activities of the apical microfilament system.

Y. K. Takeuchi & Takeuchi (1980) described features of the neurulation process in the rat embryo which seem to be similar to that in birds. Cell division is more rapid in the lateral ectoderm than in the future frontal neurectoderm in early post-implantation mouse embryos (Poelmann, 1980).

Some biochemical changes during neurulation in amphibians

During the neurula stages a rapid synthesis of rRNA and 5S RNA takes place, while mRNA synthesis continues. Whereas the mRNA is continuously degraded and resynthesised, the rRNA and 5S RNA are more stable and thus accumulate. The rRNA content of the egg only doubles between fertilisation and feeding stages, while the 5S RNA content increases 30-fold in the same period (Brown & Littna, 1964, 1966*a*, *b*). Due to the enormous stockpile of rRNA accumulated during oogenesis the embryo is able to maintain equal amounts of 28S + 18S RNA and 5S RNA, although the regulatory mechanism is not yet understood.

In RNA–DNA hybridisation experiments the mRNA of neurulae competes to a large extent with that of older embryos and even of adult tissue cells and must therefore be closely similar, but a small proportion of the mRNA no longer seems to be present in older embryos and thus probably represents stage-specific mRNA. As development proceeds the mRNA has an increasing number of sequences in common with the mRNA of differentiated tadpoles (Denis, 1968*). Embryonic development is concomitant with a progressive release of genetic information encoded in the nuclear DNA. Whereas the mRNA of gastrulae is unstable and apparently made for immediate use, tail bud stages contain both unstable and stable mRNA. The latter may have been synthesised at earlier stages by genes which are no longer active (Denis, 1968*).

Relatively little is known about regional differences in synthetic activity. This is essentially due to the fact that the currently available methods of analysis still require rather large amounts of material, which it is difficult to collect by manual dissection. Gurdon (1968*b**) stated that different RNA populations are being synthesised in different regions of the embryo; unfortunately, there is still little experimental evidence for this statement. Flickinger (1963*) found an antero-posterior gradient in protein synthesis in the neurula and interpreted it in the light of Child's physiological gradient theory. However, according to Duspiva (1962*) it is caused by a gradient in mitotic rate, more nuclei being present in the higher than in the lower ranges of the gradient. No net increase in mRNA is detectable during embryogenesis, so that the transcribed mRNA must be translated and then broken down without much delay (Gross, 1967*a**).

Evaluation

The process of neurulation is the direct consequence of neural induction, as a result of which the neurectoderm cells acquire new properties. These are expressed as changes in cell affinity and alterations at the cell surface – particularly the formation of extracellular material on both the inner and the outer cell surfaces – and the modulation of a highly flexible cytoskeleton consisting of a chiefly apically located microfilament meshwork and longitudinally oriented microtubules. While the extracellular material secreted at the outer surface of the neuro-epithelium plays an important role in the mutual adhesion of the approaching neural folds, both the microfilament and microtubular system are involved in the shape changes of the neuro-epithelium.

Neuromorphogenesis is essentially based on shape changes of individual neuro-epithelial cells, which are then transmitted to the neuro-epithelium as a whole by means of the junctional complexes, thus transforming the flat cellular sheet in a highly coordinated manner into a tube-like structure. All these activities are transient expressions of processes of cellular differentiation.

14

The development of the central nervous system (CNS)

Morphogenesis of the developing vertebrate CNS

After the initial studies of Woerdeman and Manchot in the late 1920s, C. O. Jacobson (1959) determined the presumptive cerebral regions in the early neural plate of the axolotl embryo by means of vital staining. Similar results were obtained by von Woellwarth (1960) in *Triturus alpestris* (Fig. 23).

In the vertebrates, after neurulation the tube-like brain anlage shows local distensions corresponding to the future prosencephalon or forebrain, the future mesencephalon or midbrain, and the future rhombencephalon or hindbrain. Subsequently, the prosencephalon divides into two separate vesicles: the telencephalon, which differentiates into separate hemispheres each consisting of an olfactory lobe and a cerebral cortex, and the diencephalon, which forms the evaginating optic vesicles and the hypothalamic region. The rhombencephalon divides into the metencephalon, consisting of cerebellum and pons, and the myelencephalon or medulla oblongata, which forms the brain stem and the upper spinal cord. The telencephalic hemispheres are accompanied by the olfactory placodes, the diencephalon by the hypophyseal anlage, and the myelencephalon by the auditory placodes and a series of placodes giving rise to the cephalic ganglia (V to X.) (Kallén, 1965b*). The floor of the neural plate folds down around the tip of the notochord, forming the hypothalamic protuberance. The epithelial roof and floor plates represent further morphogenetic specialisations; the roof plate forming the ependymal telae of the third and fourth ventricles and the ependymal roof tissue, and the thin median floor plate enabling the bilateral folding of the neural plate (Nieuwkoop, 1947). The thinning of the roof plate occurs where the brain anlage is in contact with the mitosis-inhibiting mesenchyme of neural crest origin (see further under eye development, p. 197; Takaya, 1956a, b).

In the metamerisation of the vertebrate CNS Bergquist (1963*) distinguished three successive types of neuromery: proneuromery, neuromery proper and postneuromery, with interneuromeric phases I and II (Fig. 24). Neuromeric outbulgings develop cranio-caudally as far back as the spinal

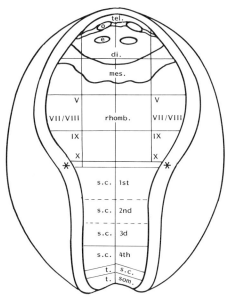

Fig. 23. Projection of presumptive brain and spinal cord regions on to open neural plate stage of urodele embryo (stage 15). di., diencephalon; e., eye; mes., mesencephalon; o., olfactory bulb; rhomb., rhombencephalon; s.c. 1st to 4th, first to fourth quarter of spinal cord; t.s.c., tail spinal cord; t.som., tail somites; tel., telencephalon; V to X, origins of fifth to tenth cephalic nerve; asterisk, boundary of presumptive brain and spinal cord regions (after C. O. Jacobson, 1959; C. von Woellwarth, 1960).

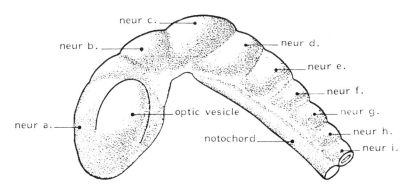

Fig. 24. Side view of the brain of an *Ambystoma punctatum* larva of *c.* 8 mm length, showing the first nine neuromeres (a–i). (After H. Bergquist, 1963*).

cord and disappear later in a caudo-cranial sequence (Kallén, 1965 b*). After the transverse neuromery has been established, longitudinal columns appear: for instance, the rhombencephalon segregates into two ventral columns forming the future basal plates and two dorsal columns forming the alar plates (see also Starck, 1982*).

In both brain and spinal cord the ventricles are lined by the ependymal or matrix layer. Proliferation of this layer and subsequent cell migration leads to the formation of the lateral mantle layer consisting of differentiating neurons. Still more laterally the cell-free marginal layer of fibre tracts develops. Neuronal migration from the matrix layer is patterned three-dimensionally. Areas of high activity are called 'migration areas' (Bergquist, 1963*). Migration may be continuous, producing a gradually thickening mantle containing brain nuclei, or discontinuous, giving rise to a stratification in concentric layers, as e.g. in the telencephalic hemispheres and the mesencephalic tectum opticum. Beside these radial migrations there are other types of migration involving whole nuclei and even groups of nuclei. Kallén (1965 a) states that degenerative processes also play a role in the morphogenesis of brain and spinal cord.

Causal factors in the morphogenesis of the CNS

In Chapter 12 we have seen that the primary patterning of the CNS into fore-, mid-, hindbrain and spinal cord is due to the two inductive actions emanating from the underlying chordomesoderm and prechordal endomesoderm; the activating one, which determines the spatial extension of the neuralised area, and the superimposed transforming one, which is responsible for the cranio-caudal organisation of the CNS. The anterior one-quarter of the neural plate, which has only been under the activating influence, forms prosencephalon, while the posterior three-quarters is subsequently transformed into mes-, rhombencephalon and spinal cord depending on the intensity of the transforming action (Nieuwkoop *et al.*, 1952; Nieuwkoop, 1966*). Kallén (1965 b*) states that the primary pattern of the CNS is based on the presence of two fields, an anterior prosencephalic field and a posterior rhombencephalic spinal field, which is essentially in accordance with the activation/transformation hypothesis.

The prosencephalon
The isolated anterior one-quarter of the neural plate of early neurulae of *Triturus alpestris* differentiates into a complete prosencephalon, which however forms only a single telencephalon and a single eye evagination. The normal bilateral configuration of telencephalon and eyes is due to an inhibiting influence from the underlying prechordal plate mesoderm (Adelmann, 1932, 1934, 1936*; Boterenbrood, 1962).

After dis- and reaggregation of the prosencephalic region of the neural plate and adjacent ectoderm, neural and epidermal cells sort out towards the centre and periphery, respectively. In the centrally located neural mass one or more cavities appear, while tel- and diencephalic neural and eye structures differentiate in the walls. The number of individual structures varies considerably, the simplest cases strongly resembling the organisation of the presumptive prosencephalon upon isolation without disaggregation. Smaller reaggregates show a higher percentage of telencephalic structures than larger ones, while inclusion of some mesencephalic material markedly interferes with telencephalic differentiation. Peripheral conditions seem to enhance telencephalic differentiation, while weak transforming influences favour diencephalic differentiation. It is evident that in the reaggregates a new prosencephalic pattern develops due to an autonomous organisation process (Boterenbrood, 1962).

Explanted presumptive ectoderm of late blastulae or early gastrulae of *Ambystoma mexicanum*, and dis- and reaggregated ectoderm of *Rana pipiens*, become partially neuralised. The explants form small prosencephalic formations consisting of tel- and diencephalic and eye structures, surrounded by placodal formations and epidermis (Nieuwkoop, 1963). These 'local activations' segregate concentrically into a central neural mass, an intermediate thickened placodal epithelium and an outer ectodermal region. This shows that in the primary activation process placodal cell material represents an intermediate step between neural and epidermal development. This is probably due to the persistence of placodal competence after the disappearance of neural competence (Nieuwkoop, 1958). Definite placodal anlagen are subsequently formed from this placodal cell material as a result of local, secondary inductive actions (see Chapter 15, p. 203). Rollhäuser-ter Horst (1981) describes a similar organisation of artificial neural inductions in three *Triturus* species.

The most anterior part of the presumptive prosencephalon of early neurulae of *Triturus alpestris* forms predominantly telencephalic, and the most posterior part predominantly diencephalic neural structures. Portions of increasing size, starting from the *anterior* boundary, form a decreasing percentage of telencephalic and eye structures and an increasing percentage of diencephalic ones, while portions of increasing size starting from the *posterior* boundary show the opposite sequence of differentiation (Nieuwkoop, Niermeyer & Jansen, 1964).

Corner (1966) studied eye differentiation in small grafts of the presumptive prosencephalic region of the early neural plate of *Xenopus laevis* and concluded that the prosencephalic region behaves as a morphogenetic field. Although this field is no longer equipotential at the early neurula stage, field properties still persist throughout the neurulation process.

Successive, small antero-posterior grafts taken from the median strip of the presumptive prosencephalic region of the neural plate of *Triturus alpestris* demonstrate the existence of an essentially concentric, field-like

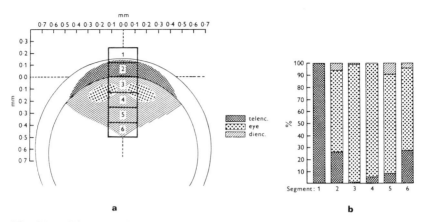

Fig. 25. a, Diagram of presumptive tel-, diencephalic and eye areas of *Triturus alpestris* neurula of stage 14½, with outlines of segments 1–6 used as grafts into the cranio-ventral mesoderm-free region of host embryos of stages 13½–14½. b, Average proportions of tel-, diencephalic and eye structures formed in the differentiated grafts of segments 1–6, expressed in volume percentages (after E. C. Boterenbrood, 1970).

organisation of the prosencephalon. While both anterior and posterior fragments show a relatively high percentage of telencephalic differentiation, the middle portions form eye structures almost exclusively (Fig. 25). Telencephalic differentiation tendencies are apparently suppressed in the more caudal region of the prosencephalon by weak transforming influences emanating from the adjacent mesencephalic anlage (Boterenbrood, 1970). Choudhury & Khare (1978) and Khare & Choudhury (1984) did similar experiments on the chick embryo at stages 4 and 5 (H. & H.) and concluded that there is an essentially concentric organisation in the presumptive prosencephalon, comparable to that in amphibians.

C. O. Jacobson (1964a) concluded from rotation experiments that the neural plate of the axolotl has become a mosaic around the stage of the raised neural folds. The antero-posterior axis is apparently already firmly determined at an early neurula stage (Stage 13, Normal Table of *Ambystoma punctatum*, Harrison, 1969) (Roach, 1945), while the medio-lateral axis becomes fixed at stage 15 (Sládeček, 1952). Shimada (1965) found that in *Bufo bufo formosus* the early neural plate no longer shows any antero-posterior but still a pronounced medio-lateral regulation, except for the di- and mesencephalic regions. Kallén (1965b) essentially confirmed these findings. In later stages only reparative processes occur without true regulation of brain structures.

Whereas the neural tube of the 24 h *Xenopus laevis* embryo (± stage 22, N. & F.) consists of a mosaic of regions with their own developmental tendencies, as shown in 180° rotation and transplantation experiments of the neural tube (S. H. Chung & Cooke, 1978), the polarity for selective nerve connections remains labile. For instance, optic axons invading a reversed optic tectum form a normal retinotectal projection. However, the axial polarity of the latter changes whenever the position of the diencephalon is altered relative to the mesencephalic tectum.

The development of the vertebrate eye

The vertebrate eye is a very complex structure with a long and intricate developmental history. Its structure was extensively described by Reyer (1977*) and its developmental history extensively discussed by Lopashov (1961*, 1965*); Lopashov & Stroeva (1964*); A. J. Coulombre (1965*); Lopashov & Hoperskaya (1967*b*); Keating & Kennard (1976*) and Reyer (1977*). In this chapter we shall only give the main features of the complex process, but certain tissue interactions will be discussed in somewhat more detail.

The principal eye structures. Morphologically the eye consists of three concentric sheaths, the *tunica interna, tunica vasculosa* and *tunica fibrosa*. The *tunica interna* comprises the innermost retinal layer or neural retina and the tightly adhering pigment epithelium or tapetum. The *tunica vasculosa* is composed of the chorioid or vascular layer and the stroma of the ciliary body and iris, both of which are provided with muscle fibres. Finally, the *tunica fibrosa* is represented by the outer, rather thin fibrous scleral layer and the slightly thicker cornea.

The pigmented ciliary body and iris differentiate along the rim of the eye cup. The lens, which is situated between cornea and eye cup, consists of an outer epithelial layer and an inner layer of highly elongated cells, the lens fibres (Fig. 26). The neural retina, which is built up of several layers, shows two axes (Goldberg, 1976) and is connected in a very precise manner with the optic tectum of the mesencephalon. Growth of the retina occurs by continuous cell proliferation in the marginal growth zone, followed by active migration of newly formed cells into the nuclear layers (Hollyfield, 1968, 1971).

Causal development of the eye anlage. In the amphibians, the eye anlagen arise during the beginning of neurulation from the anteriormost portion of the neural plate. The capacity for eye formation appears simultaneously with the capacity for the formation of anterior brain structures (von Woellwarth, 1952, 1960; Masui, 1960*c*). Holtfreter (1939) pointed to the emergence of a negative tissue affinity of the eye anlage vis-à-vis the prosencephalic brain structures as one of the guiding forces in the evagination of the primary eye vesicles from the closing neural tube. As already mentioned (p. 194) the capacity for eye formation is initially highest in the centre of the prosencephalic portion of the neural plate. The eye field becomes split into two by an inhibiting influence from the

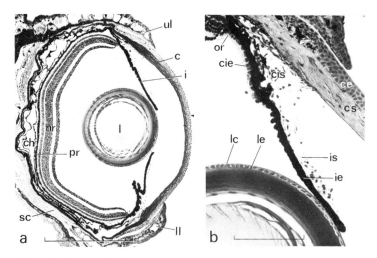

Fig. 26. a, Median sagittal section through the eye of the newt *Notophthalmus viridescens*; bar = 1 mm. b, Detail of cornea, lens, dorsal iris and ciliary body as indicated with arrows in part a; c, cornea; ce, corneal epithelium; ch, chorioid; cie, ciliary epithelium; cis, stroma of ciliary body; cs, corneal stroma; i, iris; ie, iris epithelium; is, iris stroma; l, lens; lc, lens capsule; le, lens epithelium; ll, lower eye lid; nr, neural retina; or, ora serrata; pr, pigmented epithelium; sc, sclera; ul, upper eye lid; bar = 0.25 mm (courtesy R. W. Reyer).

underlying prechordal mesoderm, so that two separate eye evaginations develop (Adelmann, 1937). Kaufman (1979) described the evagination of the optic cups in the rat embryo.

The primary eye vesicles make contact with the overlying ectoderm, to which they then firmly adhere, preventing the penetration of mesenchyme between the two tissues. In the homozygous eyeless mutant (*e/e*) of *Ambystoma mexicanum* the basal lamina covering the external surface of the optic vesicle forms prematurely; subsequently the eye vesicle is separated from the overlying ectoderm by the invasion of mesenchyme (Ulshafer & Hibbard, 1979). The intimate contact between the eye vesicle and the ectoderm is responsible for (1) the development of the lens placode in the placodal ectoderm situated peripherally to the neural anlage; and (2) for the invagination of the outer wall of the primary eye vesicle leading to the formation of the double-walled eye cup. This interaction is therefore clearly reciprocal.

The development of the double-walled eye cup into the inner retinal layer and the outer pigment epithelium is again a complex process. Extensive morphogenetic movements are involved in the evagination of the primary optic vesicle and in its subsequent invagination to form the optic cup (Holtfreter, 1939; Townes & Holtfreter, 1955; P. Weiss & Taylor, 1960). Typical eye cup formation requires continuous firm adherence between the lens placode and the outer border of the eye vesicle. This on the one hand conditions a thickening of the inner presumptive retinal layer, which shows increased cell proliferation, and on the other hand leads to a simultaneous thinning of the outer presumptive pigment epithelium. Thinning of the outer layer entails more extensive contact with the surrounding mesenchyme, which is for the greater part of neural crest origin. Absence of neural crest may be responsible for the incomplete evagination of the eye anlagen in artificially induced prosencephalic formations (Nieuwkoop, 1963). In the complete absence of ectoderm and mesenchyme the inner sensorial layer of the anterior part of the neural plate *senso strictu* of *Rana temporaria* and *Rana esculenta* differentiates into forebrain and retina tissue, with only iris tissue and lentoids in between (Hoperskaya, 1972).

Mesenchyme enhances the thinning of the outer layer of the eye cup (Lopashov, 1961*) and suppresses mitotic activity (Takaya, 1956*a*, *b*). Artificial thinning of eye anlage material by compression favours pigment epithelium development and prevents retinal differentiation, whereas thickening of eye vesicle material by culturing it in the body or in the brain cavity stimulates retinal differentiation (Lopashov & Hoperskaya, 1967*a**, *b**). While retinal condensation seems to be largely autonomous, pigment epithelium development is strongly influenced by the inductive action of the surrounding mesenchyme. The pigmented epithelium is stretched out into a thin sheet of flattened cells, while the retina develops into a massive multilayered structure (Lopashov, 1961*). Neural retina is characterised by glutamine synthetase activity, providing a quantitative measure for its differentiation (Moscona, Saenz & Moscona, 1967; Moscona, Moscona & Saenz, 1968). The regularly observed extrusion of pigment granules from the retinal layer points to an intracellular antagonism between retinal and pigment epithelium differentiation. The surrounding mesenchyme guarantees an abundant blood supply, which is indispensable for normal eye morphogenesis.

Twitty (1955*) found that the two layers of the optic cup are initially interchangeable, but when the neural retina begins to differentiate, its capacity for transformation into pigment epithelium decreases. In the urodeles the pigment epithelium remains capable of differentiation into retina throughout life (Stone, 1950; Keefe, 1973*a*, *b*, *c*, *d*), but in the anurans this capacity is only maintained until the onset of metamorphosis (Sato, 1953). In rats it lasts until the 15th day of embryonic life and in the fish *Acipenser* only until the onset of pigment formation (see Lopashov, 1961*). In frogs the transformation of pigment epithelium into retina is prevented by the presence of Bruch's membrane, to which the pigment epithelium is tightly attached (Lopashov, 1977). In homozygotes of the periodic albinism mutant (a^p/a^p) of *Xenopus laevis* the pigment epithelium and iris, as well as the skin melanophores, lose their pigment at an early tadpole stage. In contrast to the wild-type, the pigment-free tapetum of mutant frogs is capable of transformation into retina (O. A. Hoperskaya, unpublished results, see in Lopashov, 1977). Under tissue culture conditions chick retinal tissue can transdifferentiate into pigment epithelium up to day 15 and into lentoids up to day 18 (Nomura & T. S. Okada, 1979; T. S. Okada *et al.*, 1979*a*; T. S. Okada, Yasuda & Nomura, 1979*b*). Dissociated neural retina of 9 day chick embryos cultured *in vitro* forms a monolayer. Addition of amphotericin B leads to precocious pigment formation and tyrosinase activity in foci of epithelial cells, which will then develop into pigment epithelium (Itoh, Ide & Hama, 1980).

The ciliary body and iris constitute the transition zone between the pigment epithelium and the retina. While strongly flattened amphibian eye vesicle material tends to differentiate into pigment epithelium, less strongly flattened material forms iris tissue (Lopashov & Hoperskaya, 1967*b*). The development of the ciliary body and iris depends on the presence of the epithelial layer of the lens, as observed in the rat embryo (Stroeva, 1963). Implantation of a second lens primordium into the optic cup of the chick embryo leads to the formation of another pupillary border and a greatly enlarged pupil. Another anterior eye chamber with corneal epithelium develops in front of the new pupil (Genis-Galvez, 1966).

In amphibians the inner layer of the iris has a strong capacity to regenerate retina, a capacity that is weaker in the outer layer. Transformation of pigment epithelium into retina and lens requires extirpation of both retina and lens, or devascularisation with subsequent degeneration of retina and lens (see Reyer, 1977*). The iris stroma plays a repressive role in the transformation of iris into retina and lens. In *Triturus viridescens* Wolffian lens regeneration from the dorsal iris is prevented by the existing lens as long as the lens is contiguous with the iris (T. Yamada, 1972*; see also Lopashov, 1977). Glücksmann (1965) describes cell death as a regular phenomenon in the differentiation of the retina of *Rana temporaria*, as well as in the ectoderm during lens induction; this would be due to an overproduction of cells or to their inability to adapt to changes in the cellular environment.

The formation of the cornea. The cornea represents the transparent, non-keratinising area of the skin and underlying stroma, situated above the iris of the eye cup.

Genis-Galvez, Santos-Gutierrez & Rios-Gonzalez (1967) concluded from various types of experiment that in the chick the lens in conjunction with the optic cup plays an active role in the induction and morphogenesis of the cornea. This tallies with the fact that the development of the iris depends on that of the lens. The lens is also required for the persistence of the cornea. It is particularly the lens epithelium that is responsible for the development of the acellular corneal stroma.

Y. Katoh (1975) found that outer corneal epithelium of a young chick embryo can still form keratinised epidermis when cultured with $6\frac{1}{2}$ day dorsal dermis, even when a millipore filter is inserted between them. In older embryos the corneal epithelium loses the capacity for keratinisation. Corneal stroma does not inhibit the keratinisation of $6\frac{1}{2}$ day dorsal skin.

In the chick the acellular primary stroma of the cornea consists of collagen fibrils embedded in glycosaminoglycans. Embryonic corneal epithelium produces chondroitin and heparan sulphate and collagen. Their synthesis is greatly enhanced when isolated corneal epithelium is grown on a collagenous substratum (Meier & Hay, 1974a). The corneal epithelium forms primary stroma under the inductive influence of the lens; collagen of the lens basement membrane stimulates extracellular matrix production by the corneal epithelium and the polymerisation of this matrix (Meier & Hay, 1974b). When isolated corneal epithelium is grown in transfilter culture on a collagenous substratum, epithelial collagen synthesis is directly proportional to the pore size of the filter and inversely proportional to the filter thickness. These correlations are therefore based on the growth of cell processes traversing the filter and making contact with the collagenous substratum. Epithelial collagen production thus depends on an interaction of the epithelial cell surface with extracellular matrix as 'inductor' (Meier & Hay, 1975). The contact-mediated collagen/cell-surface interaction requires the continuous presence of collagen for the maintenance of epithelial stroma synthesis. The collagen acts upon the epithelial cell surface without entering the cells (Hay & Meier, 1976). The interaction is an autocatalytic process. The primary stroma between lens and corneal epithelium is later invaded by mesenchymal cells which form the fibroblasts of the cellular secondary stroma (Hay, 1977a*).

The formation of the lens (see Chapter 15, p. 206).

Melanogenesis in the pigmented epithelium, ciliary body and iris. In homozygotes of the periodic albinism mutant (a^p/a^p) of *Xenopus laevis*, albinism is expressed in the skin and in the chorioid coat and pigment epithelium: all melanocytes disappear following a short initial phase of pigmentation. Confrontation of mutant eye vesicles with wild-type endomesoderm and vice versa shows that the absence of melanogenesis in the mutant is not related to a deficiency in tyrosinase synthesis, to hormonal influences or to the expression of genes coding for the melanoprotein complex, but to a deficiency in the inducer of melanogenesis (Hoperskaya, 1978).

Melanogenic induction starts at a late gastrula stage. A melanogenic factor (MgF), probably proteinaceous in nature, is produced by the endomesoderm (Hoperskaya, 1981). The competence for melanogenic induction lasts until stage 30 (N. & F.) (Hoperskaya & Golubeva, 1980). Embryonic extract of $+/+$ *Xenopus* embryos stimulates melanin synthesis in a^p/a^p pigment epithelium and melanophores (Hoperskaya *et al.*, 1984). Melanogenic induction is not species-specific (Hoperskaya & Golubeva, 1982). In normal development the prechordal plate endomesoderm acts as the primary source of MgF at the late gastrula stage, but at tail bud stages the presumptive heart mesoderm situated just posterior to the eye vesicles is a secondary source of MgF. 2 h of contact between inductor tissues and ectoderm is sufficient for melanophore induction (Hoperskaya, Zaitzev & Golubeva, 1982; Hoperskaya & Golubeva, 1982). In the mutant a^p only the prechordal endomesoderm is active (Hoperskaya, 1981).

Sclera formation. Removal of the retina does not affect the normal formation of the sclerotic layer. Implantation of pigment epithelium with associated mesenchyme into cephalic mesenchyme of the 15–22 somites chick embryo *in ovo* leads to additional sclera formation; pigment epithelium alone can also induce a sclerotic layer in head mesenchyme (Reinbold, 1968). Stewart & McCallion (1975) found that scleral cartilage can be induced in periorbital mesenchyme by pigmented epithelium, confirming Reinbold's earlier results. M. C. Johnston *et al.* (1979) demonstrated that, in the chick, most cells of the scleral cartilage originate from the cranial neural crest. The neural crest cells arrive in the neighbourhood of the pigmented epithelium by day 14 (H. & H.).

The rhombencephalon and spinal cord
In the vertebrate CNS Bergquist (1963*) observed the appearance and disappearance of neuromeric outbulgings. Kallén (1965*b**) emphasised that normal morphogenesis of the CNS requires a continuous interaction of the segregating neural tube with the surrounding meso- and ectodermal tissues: for instance, neuromerisation and longitudinal column formation do not occur in the absence of the underlying chordomesoderm.

Jacob, Christ & Jacob (1975) and Jacob *et al.* (1976) concluded from cranio-caudal and dorso-ventral rotations of the neural tube in the chick that both axes are firmly fixed at stages 12–14 (H. & H., 1951). However, after 180° dorso-ventral rotation of the rhombencephalon in 48 h chick embryos ($\pm$ stage 12) A. H. Martin (1977) observed normal ventral motor neuron formation in the original basal plates, while additional motor neurons were induced in the now ventrally located alar plates: obviously dorso-ventral polarity is only partially determined at this stage.

In *Ambystoma gracile* somite tissue enhances hindbrain differentiation more than notochord does, whereas notochord favours spinal cord histogenesis more than somites do (Landzman, 1967). Whereas A. H. Martin (1971) found that in the 48–60 h chick embryo the differentiation and proliferation of the neural tube is not affected by the removal of adjacent somites, Strudel (1970) reported that extirpation of several pairs of somites affects the number of vertebrae as well as that of ganglia and nerve roots of the spinal cord, so that an interaction between the skeletal and neural axes must occur.

Mitolo, Ferrannini & Franchini (1968) noted that enlargement or reduction of peripheral organs (e.g. as a result of implantation or extirpation of wing buds) leads to ipsilateral increase or decrease in mitotic rate in the neural tube. Using tail or postumbilical grafts, which are self-innervating, Mitolo, Jirillo & Neri (1970) could show that in early stages the morphogenetic influences of the periphery on the spinal cord centres do not travel along nerve paths.

The development of some early cell types in the rhombencephalon and spinal cord. The presumptive *Mauthner cells* (which furnish the primary sensory neurons for larval swimming movements), are localised at the extreme lateral edge of the presumptive motor neuron area of the medulla oblongata (C. O. Jacobson, 1964*a*). Antero-posterior reversal of the presumptive hindbrain including the Mauthner cells leads initially to outgrowth of Mauthner fibres in accordance with the polarity of the tissues through which they are growing. This pleads in favour of Weiss' (1934, 1941*, 1969*) contact guidance hypothesis. However, further outgrowth follows the normal cranio-caudal course. Model (1978) and Model & Wurzelmann (1982) found complete positional and functional regulation of Mauthner cell development following unilateral rotation of the presumptive hindbrain. Extirpation of the primordium of the VIIIth cranial ganglion in *Ambystoma* causes the absence of Mauthner cells in about one-third of the cases, demonstrating the importance of the VIIIth root fibres for the differentiation of the Mauthner cells (Piatt, 1969).

In the amphibian spinal cord the *Rohon-Beard cells* are large neuroblasts which arise from the inner side of the neural folds and become localised mid-dorsally in the spinal cord. They function as temporary sensory elements of the early larva until the spinal ganglia become functional. They can be forced to leave the spinal cord, just as neural crest cells do, by application of hypertonic salt solutions (Holtfreter, 1947*b*).

Evaluation

The primary patterning of the CNS into fore-, mid- and hindbrain and spinal cord is due to the primary inductive actions emanating from the underlying substratum, the activating action being responsible for the spatial extension of the neuralised area and the superimposed transforming influence for the cranio-caudal regional differentiation of the CNS. Whereas the former influence initiates prosencephalic differentiation tendencies, the latter transforms these into mes-, rhombencephalic and spinal cord tendencies.

The spatial organisation of the prosencephalon is largely an autonomous segregation process, leading to tel- and diencephalon and eye formation. In normal development the pattern is partially duplicated by an inhibiting influence of the underlying prechordal plate mesoderm. A continuous, or at least sequentially repeated interaction between chordomesoderm and neural plate is responsible for the further organisation of the rhombencephalic and spinal cord regions of the CNS, while adjacent placodal anlagen exert additional local influences.

The complicated development of the vertebrate eye is based upon a series of interactions between the evaginating primary eye vesicle, the overlying ectoderm and the surrounding mesenchyme, the latter being mainly of neural crest origin. The interaction between the primary eye vesicle and the overlying ectoderm leads to the formation of the lens (see Chapter 15, p. 206). The neural retina, which develops from the inner layer of the secondary eye cup (or *tunica interna*), is essential for the polar differentiation of the lens (see Chapter 15, p. 209). The lens in its turn is responsible for the invagination of the primary eye vesicle to form the secondary eye cup, and for the development of the retina, the iris and the overlying cornea. The surrounding mesenchyme induces the outer layer of the future eye cup to differentiate into the thin pigmented epithelium, thus supporting eye cup formation. Under the influence of the pigment epithelium the surrounding mesenchyme forms the *tunica vasculosa* and *tunica fibrosa* of the eye. The formation of the cornea is caused by a cell-surface/intercellular-matrix interaction in which matrix components stimulate their own production by the corneal epithelium. The extracellular matrix of the lens capsule acts as the primary inducer for corneal stroma production.

In the majority of these interactions, if not in all, extracellular matrix components representing transient products of cellular differentiation seem to play an important if not a leading role.

15

The development of the cephalic placodes

The cephalic placodes comprise the anlagen of adenohypophysis, nose, lens and ear as well as those of the cranial ganglia.

Nieuwkoop (1963) observed that in 'artificial activations' appearing in early gastrula ectoderm of *Rana pipiens* after dis- and reaggregation of the ectoderm, or in ectoderm of *Ambystoma mexicanum* after simple explantation, a region of thickened placodal ectoderm forms around each prosencephalic neural anlage. In the placodal ectoderm definite placodes are formed under the influence of additional inductive actions from parts of the developing prosencephalon. The same holds for the rhombencephalon. Folds made of ectoderm taken from successively older *Ambystoma* donors and attached to the mid-rhombencephalic region of host neurulae showed loss of neural competence of the ectoderm between stages 11 and 12, and the appearance of placodes at least till stage 14 (Nieuwkoop, 1958). It therefore seems likely that a competence for placodal development remains after the disappearance of neural competence. This view was recently supported by B. Albers (unpublished results), who found that the medio-lateral propagation of neural induction in the ectodermal layer becomes spatially limited by the loss of neural competence. Competent ectoderm implanted beyond the neural plate boundary at the open neural plate stage, and even some distance away from this boundary, can still be neuralised. This shows that the inductive principle, whatever it may be, spreads beyond the boundary of the neural plate and the adjacent placodal ectoderm. Therefore the spreading of the inductive principle as such cannot be responsible for the size of the neural and placodal regions; it must be based on the emergence and loss of competence.

A. G. Jacobson (1963a) found that in the West Coast newt (*Taricha torosa*) the anlagen of nose, lens and ear are not yet sufficiently determined for autonomous development at an early neurula stage, but require additional inductive influences from adjacent organ anlagen, either neural, mesodermal or endodermal. This also holds for the adenohypophysis (Ferrand, 1972). Jacobson's conclusions were in general confirmed by Michael & Nieuwkoop (1967) and Michael (1968). Removal of the endoderm at the neurula stage had its greatest effect on nasal development,

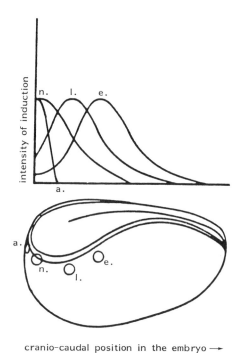

Fig. 27. Cranio-caudal distribution of intensity of induction for the formation of the (unpaired) adenohypophysis (a), the nasal placode (n), the lens placode (l) and the ear placode (e) in the placodal ectoderm of the amphibian neurula (after C. O. Jacobson & R. Gordon, 1976).

less on lens and still less on ear development. Antero-posterior reversal of the endoderm had no effect. In the absence of the endoderm, neural tissue is the principal inductor of nose, lens and ear. Mesoderm participates in lens and ear induction, but not in nose induction (N.B. the nasal anlage has no mesodermal substrate). The neural fold plays no obvious role in lens induction (A. G. Jacobson, 1963*b*). Antero-posterior reversal of the placodal ectoderm with or without increasingly larger parts of the neural folds and neural plate demonstrated that the positioning of the nasal, lens and ear placodes depends on instructions received from all neighbouring tissues. When instructions are conflicting – as e.g. after antero-posterior reversal of the placodal anlagen – two different organs form side by side without competitive interaction (A. G. Jacobson, 1963*c*). Jacobson concluded that the differentiation of a given group of placodal cells is determined by its entire history of interaction with all tissues in its environment, and not by one or another specific inductor tissue alone. Different portions of the endoderm, mesoderm and neural tissue act successively as primary, secondary and tertiary inductors for nose, lens and ear (A. G. Jacobson, 1966*; see Fig. 27). Michael & Nieuwkoop (1967) found that the sensory placodes are less easily induced in medio-laterally reversed placodal ectoderm. In normal orientation all ectodermal placodes can form in the absence of the neural inductor. Removal of the entire neural plate and folds usually prevents placode formation, but this may be due to a dislocation of the placodal ectoderm with respect to the underlying archenteron roof consequent upon the closure of the large wound (Michael, 1968).

The adenohypophysis

The adenohypophysis develops from the mid-dorsal region of the ecto-dermal stomodaeal invagination which is situated directly in front of the anterior neural fold (see Fig. 27), in close association with the infundibulum of the diencephalon.

Isolation of the presumptive anlage in the nurse frog (*Pelobates fuscus*) and *Rana esculenta* at successive stages of development revealed that the determination of the adenohypophysis is a stepwise process (Pehlemann, 1962). Hammond (1974) claimed that in the chick an initial inductive action for the development of the adenohypophysis emanates from the prechordal mesoderm. In our opinion this primary action may give rise to the appearance of competence. Ferrand (1972) demonstrated that in the chick before stage 25 (H. & H.) the anlage of the adenohypophysis requires an inductive action from the floor of the prosencephalon to become more firmly determined. Prechordal mesenchyme remains indispensable for final differentiation, but there is no particular preference for the type of mesenchyme.

Takor Takor & Pearce (1975) concluded from a histological study that in birds the adenohypophysis is of neuro-ectodermal rather than ectodermal origin. We think this is a misinterpretation possibly due to the firm attachment of the stomodaeal invagination to the floor of the CNS at the stages investigated.

The nasal placode

Nieuwkoop (1963) found a highly significant correlation between the volume of the telencephalic structures and that of the olfactory placodes in prosencephalic formations developing upon artificial activation of *Rana pipiens* and *Ambystoma mexicanum* gastrula ectoderm. This correlation apparently expresses a reciprocal interaction. Removal of the presumptive telencephalon often prevents olfactory placode formation (A. G. Jacobson, 1966*; Michael, 1968), while removal of the olfactory placodes interferes with the formation of the telencephalic hemispheres, particularly the bulbus olfactorius (Clairambault, 1968, 1971). Removal of the nasal placode of the just-hatched trout larva (*Salmo irideus*) leads to complete blockage of the development of the bulbus olfactorius (Chanconic & Clairambault, 1975). The naso-lacrimal duct grows out from the olfactory organ during metamorphosis (Bijtel, 1958), the dermis being indispensable for the differentiation of the epithelial duct (Yvroud, 1971, 1974).

The development of the lens

The development of the lens in the chick, as described by McKeehan (1951), is very similar to early lens development in the urodeles and only slightly different from that in the double-layered ectoderm of the anurans (compare neurulation in urodeles and anurans, Chapter 13, p. 182–3).

In the chick the optic vesicle makes contact with the ectoderm from the 9-somite stage onwards. Optic vesicle and ectoderm adhere strongly between the 10- and 21-somite stages. Lens placode formation takes place between the 12- and 19-somite stages (44–50 h of incubation). From the 26-somite stage onwards lens and optic cup are again readily separable. Lens invagination occurs between the 20- and 21-somite stages by a contraction of the microfilament meshwork present in the apical ends of the columnar epithelial cells (McKeehan, 1951), while oriented microtubules play a role in cell elongation (Byers & Porter, 1964). During early lens formation interkinetic nuclear migration occurs towards the apical side of the epithelium where mitosis takes place; the nuclei then return to a more basal position for DNA synthesis (Zwaan, Bryan & Pearse, 1969). Zwaan & Hendrix (1973) suggested that lens placode formation is partially mechanically determined by the resistance to expansion of the ectoderm, firmly attached as it is to the margin of the optic cup while mitosis and growth continue in the placode ectoderm.

Woerdeman (1962*) noted that in some amphibian species the entire gastrula ectoderm has lens competence, which then becomes restricted to the head region with the onset of neurulation, while in other amphibian species only the head ectoderm shows lens competence. Becker (1960) found that eye cup implantation in *Triturus vulgaris* results in lens induction in all body regions, but lenses are smaller in the trunk than in the head and heart regions. In most amphibians any part of the early ectoderm can form a lens if sufficiently exposed to lens-inducing tissue(s). Early mesoderm is far less likely to form a lens and early endoderm is not at all likely to do so (A. G. Jacobson, 1966*; see also Reyer, 1977*). A. G. Jacobson summarises the whole problem as follows. In the gastrula the presumptive lens ectoderm is underlain by the presumptive pharyngeal endoderm, which acts as primary lens inductor. Later the presumptive lens ectoderm comes under the influence of the presumptive heart mesoderm, i.e. the anterio-lateral edges of the mesodermal mantle; this acts as secondary lens inductor. After neurulation the evaginating optic vesicles

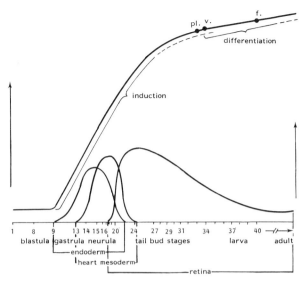

Fig. 28. Graphic representation of lens induction and determination in the axolotl. The ordinate represents the relative time of development in Harrison stages. Left part of figure (in a logarithmic scale) shows the response curve of the epidermis, with levels of determination as lens placode (pl.), as lens vesicle (v.), and for lens fibre formation (f.). Right part of figure (in a linear scale) illustrates the inducing capacity of the three successive inductors, viz. endoderm, heart mesoderm and retina (after R. W. Reyer, 1977*).

come into close and prolonged contact with the presumptive lens ectoderm, acting as the tertiary and final lens inductor (see Fig. 28).

Upon prevention of the primary contact with foregut endoderm the ectoderm responds far less readily to the eye vesicle as inductor. A. G. Jacobson (1966*) interprets his results in terms of a simple additive or synergistic action of primary, secondary and tertiary inductors (see also Reyer, 1977*). Hoperskaya (1968) suggested that lens induction occurs directly by the cranial endomesoderm and indirectly via the neural plate and eye anlagen, the same lens-inducing agent being involved; the agent would originate from the cranial endomesoderm. Under experimental conditions the presence of this agent in the eye anlage may also be responsible for lentoid formation from various parts of the adult eye. Hoperskaya (1976) observed lentoid induction in competent gastrula and neurula ectoderm of *Rana temporaria* by lens epithelium, but not by lens fibres (see further under lens regeneration, p. 211). It must however be realised that the various inductive actions do not occur simultaneously but act sequentially on an ageing reaction system, and may therefore be different in nature. Wolff (1973*) agrees with Mizuno (1970) that the first contact of the cephalic ectoderm with the endomesoderm is specific for the formation of pre-lens ectoderm, or in other words, for the emergence of lens competence, while the subsequent inductive actions are far less specific. In this context Karkinen-Jääskeläinen (1978 *b*) speaks of 'lens-forming bias'. Mizuno (1972) noted that while cephalic chick epiblast alone cannot be induced to form free lentoids or lenses by $6\frac{1}{2}$ day trunk dermis or $13\frac{1}{2}$ day tarso-metatarsal dermis, cephalic epi- and hypoblast together can respond to these inductors. Other mesenchymes of various origins can do the same but are less effective.

S. Takeuchi (1963) found that in *Triturus pyrrhogaster* the eye vesicle shows maximal inductive capacity at an early tail bud stage (stages 26–27, O. & I.). Van der Starre (1977) observed maximal inductive capacity of the optic cup in the 72 h chick embryo, when the head ectoderm is still fully competent.

Early electron microscopic studies by Grobstein (1955*), Eakin & Lehmann (1957) and Ferris & Bagnara (1960) have shown the presence of an intercellular matrix in the interface between eye vesicle and ectoderm. McKeehan (1956) and Hunt (1961) noted that in the chick at stages 11–14 (H. & H.) the interepithelial space between the optic vesicle and the lens anlage contains filaments, particles, basal lamina and PAS-positive interepithelial 'clouds'. Hendrix & Zwaan (1974) demonstrated the presence of glycoproteins in the interspace between the optic vesicle and the lens anlage. They believe that these may play their main role in lens differentiation. The matrix material may however not only play a role in the adhesion between the two layers but may also be involved in the induction process itself, as suggested by the fact that the eye vesicle can be replaced by various adult mesenchymes (Mizuno, 1972).

It is a well-known fact that penetration of head mesenchyme – which is chiefly of neural crest origin – between the optic vesicle and the ectoderm prevents lens formation in amphibians (S. Takeuchi, 1963 and others). Bose & Medda (1965) found that chloramphenicol, an inhibitor of protein synthesis, does not affect optic vesicle formation but prevents lens induction by interfering with adhesion, allowing head mesenchyme to penetrate between the optic vesicle and the ectoderm.

Von Woellwarth (1961) studied the formation of so-called free lenses in *Triturus alpestris*. After removal of the anterior neural plate and cephalic neural folds, free lenses were formed in 95% of the cases, while this occurred in only 26.5% of the cases after removal of the anterior neural plate alone. He concluded that the migration of the cephalic neural crest restricts lens formation to the area of contact with the eye anlage. The frequency of free lens formation is moreover dependent on the temperature of culturing, being 95% at 14 °C, 64% at 16 °C and 53% at 23 °C. A. G. Jacobson (1966*) ascribed the temperature effect to the shorter or longer duration of exposure to the primary lens inductor.

In *Triturus vulgaris* Becker (1960) observed a cranio-caudal gradient in the occurrence and size of free lenses and lentoids after eye cup extirpation. In *Xenopus laevis*, under similar conditions, free lenses develop by aggregation of epidermal cells instead of regular placode formation (Deuchar, 1972*). Free lenses and lentoids represent true lens formations since they contain specific lens proteins, as could be demonstrated with specific lens antigens (Mizuno & Katoh, 1972 and others; see also Reyer, 1977*).

Oppenheimer (1966) noted that in the teleost *Fundulus* lenses never develop in association with pigment epithelium. Lenses may be induced by neural retina but can also develop in its absence.

Balinsky (1957) found that in the amphibian eye the size of the lens depends on the size of the eye cup, the reactivity of the ectoderm, and a species-specific size-limiting factor. De Graaff (1960) observed that treatment of presumptive lens ectoderm of *Xenopus* late gastrulae/early neurulae with thiomalic acid leads to a marked increase in the upper size limit of the lens formed. Reyer (1962, 1966) emphasised that the eye cup is important not only for the final phase of lens induction but also for the further growth and differentiation of the lens. In studies in the mouse *in vitro* Muthakkaruppan (1965) reported that an isolated lens primordium dedifferentiates. In the presence of mesenchyme it forms a lens vesicle, while lens fibre formation occurs in the presence of the optic cup, particularly the neural retina. Continued action of the retina is required for differentiation of the lens fibres. This was confirmed for the chick by Gunia & Tumanishvili (1972), who removed the retinal anlage from the eye cup and found that the lens vesicle showed no polar differentiation. The formation of the fibrillar portion of the lens thus depends on the presence of the neural retina and is always oriented towards the latter. In the chick embryo outward rotation of the lens through 90° or 180° leads to partial or complete reversal of its polar differentiation by degeneration of the original lens fibres and new fibre formation in the now inwardly oriented lens epithelium (J. L. Coulombre & Coulombre, 1963; Genis-Galvez, 1964).

Ribonucleoprotein (RNP) particles are abundant in the chick optic vesicle at stages 11 through 13 (H. & H.) and decrease in amount at stage 14, when they become much more abundant in the overlying lens ectoderm. This led McKeehan (1956) and Hunt (1961) to suggest that RNP particles are transferred from the inductor to the reaction system. Eisenberg & van Alten (1964), however, found that lens RNA, which seems to promote the subsistence of chick cephalic ectoderm, does not induce lens formation; the same holds for liver and yeast RNA. The increase in RNA content of first the inductor and then the reacting tissue may only reflect a successive increase in the rate of RNA synthesis in the two tissues, and may have nothing to do with RNA transfer (see also Chapter 12, p. 163). Mezger-Freed & Oppenheimer (1965) observed an increase in the amount of RNA per milligram dry weight and a change in RNA base composition in the eye-cup/lens system of *Rana pipiens* during lens induction between stages 15 and 20 (Shumway, 1940, 1942).

In the chick, antibodies against adult lens interfere with lens induction (Langman, 1956; Maisel & Langman, 1961; Flickinger & Stone, 1960). Lens antigens were found in the lens, the iris and the pigment epithelium (Maisel & Langman, 1961). They were detected in the lens primordium before any differentiation of the lens anlage occurred, and also in several surrounding embryonic tissues (Langman, 1959; Clarke & Fowler, 1960; Perlmann & De Vincentiis, 1961; J. C. Campbell, 1965; Vyasov, Averkina & Petrosjan, 1965).

The presence of lens antigens outside the lens was contested by Ikeda & Zwaan (1966), who attributed earlier results of others to rapid postmortem diffusion of antigens from the lens, but more recent investigations by McDevitt & Clayton (1979) and R. M. Clayton (1982*) and others confirmed the presence of lens antigens in various embryonic and adult eye tissues (see p. 213). Kirzon, Averkina & Vyasov (1969) mentioned that 'common' lens antigens (antigens also present in other tissues) are detectable in all developmental stages of *Rana temporaria*, including unfertilised eggs, whereas 'organ-specific' lens antigens can only be demonstrated in the lens placode at tail bud stages.

By molecular hybridisation with cDNA, Jackson *et al.* (1978) showed that in the chick neural retina and pigment epithelium contain in their cytoplasmic mRNA a low concentration of the most abundant lens mRNAs, and that these are absent in the mRNAs of headless animals. Shinohara & Piatigorsky (1976) found that in the chick the initiation of δ-crystallin mRNA accumulation in the presumptive lens ectoderm coincides with the initiation of lens placode formation which is 8–9 h after the beginning of lens induction. They suggest that δ-crystallin synthesis is regulated by the accumulation of δ-crystallin mRNA during lens induction.

Clarke & Fowler (1960) reported that fluorescent sera against adult chick lens react with the optic vesicle of the chick prior to lens induction, thus interfering with lens induction. According to these authors this suggests that substances reacting with lens antibodies are essential in the induction process. In our opinion this conclusion needs further confirmation.

In *Rana tigrina*, lens induction can be prevented by heat shock (37 °C) given at an early neurula stage (Bose & Chatterjee, 1964). In *Rana pipiens* β-mercaptoethanol does not affect lens induction by the optic cup but inhibits further lens development (Wolsky & de Issekutz-Wolsky, 1968).

McKeehan (1958) observed lens induction in the chick through 20 μm thick agar strips inserted between the eye vesicle and the presumptive lens ectoderm prior to their normal contact. No cellular processes penetrated the agar. He therefore attributed lens induction to diffusible substances rather than to cell–cell interaction. In the mouse Muthakkaruppan (1965) found that inductive influences emanating from the eye vesicle can pass through a 25 μm thick millipore filter of 0.45 μm average pore size. Karkinen-Jääskaläinen (1978 *a*) obtained lentoid formation, with lens-specific protein synthesis in 2 day chick optic vesicle/quail trunk ectoderm recombinates, and vice versa, using 100 μm thick millipore filters (0.80 and 0.45 μm average pore size), nucleopore filters with 0.6, 0.2 and 0.1 μm pores, as well as dialysing membrane that permits the passage of molecules of MW up to 12000 daltons (see also Saxén, 1980 *b**). In the chick, Van der Starre (1978) achieved induction through a millipore filter of 0.65 μm average pore size in 36% of the cases. He found that 0.5 mm thick agar slices which have been inserted between the optic vesicle and the ectoderm of 72 h chick embryos for 3–4 h (6 h gave still better results) induced lenses in competent ectoderm in 43% of the cases. Ultrafiltration of material extracted from the agar slices yielded an active protein of MW between 5000 and 10000 daltons.

We have seen that in transfilter experiments extracellular matrix components may penetrate the filters, so that the ectodermal cells may interact with such components without the actual penetration of cytoplasmic processes. In our opinion extracellular matrix components may likewise accumulate in agar slices and cause lens induction when these are subsequently brought into contact with competent ectoderm.

Lens regeneration

A few words may be said about the transformation of non-lens tissues into lens, a process called *lens regeneration* and bound up with what is now called cellular transdifferentiation. Although these processes usually take place in later development, which falls outside the scope of this book, the great similarity between lens regeneration and normal lens development justifies a brief discussion.

Lens regeneration in amphibians may occur from the corneal epithelium, the neural retina or the iris (see Deuchar, 1973* and Reyer, 1977*). When corneal epithelium of *Xenopus laevis* larvae is brought into direct contact with a lentectomised eye, the dorsal cornea and the adjacent pericorneal epidermis are transformed into lens tissue. No lens formation takes place with intact cornea (Bosco, Filoni & Connata, 1979). The basal lamina of the corneal epithelium encloses the regenerating lens and contributes to the lens capsule (see also Reyer, 1977*). Denatured tissue from a lentectomised eye, or protein pellets made from it, can induce lens formation in the outer cornea when implanted between the outer and inner cornea in *Xenopus laevis* (Filoni *et al.*, 1983). Lens regeneration from the corneal epithelium represents a direct transformation of corneal cells into lens cells (transdifferentiation, see below).

After extirpation of the retina and lens in the chick, or after devascularisation and subsequent degeneration of the retina and lens, metaplasia occurs in the pigment epithelium as well as in the ciliary body and iris, resulting in the regeneration of retina and lens (Gaze & Watson, 1968).

Dissociated neural retina cells of 8 day chick embryos *in vitro* form a sheet-like monolayer. After a culture period of more than 30 days, allowing for several cell division cycles, pigment epithelium and α-crystallin-antiserum-positive lentoids develop (T. S. Okada *et al.*, 1975). Okada compared these phenomena to transdetermination in insects. Gross synthesis of crystallin begins after 12–16 days of culture, which is at least 10 days before any visible lentoid formation takes place (de Pomerai, Pritchard & Clayton, 1977). Whereas 8 day neural retina culture does not contain any mRNA for crystallins, a 42 day neural retina culture that is undergoing transdifferentiation into lens cells contains abundant crystallin mRNA (Thomson *et al.*, 1978; R. M. Clayton, Thomson & de Pomerai, 1979). The transdifferentiation into lenses and pigment epithelium takes place in 'foci' in the epithelial cell sheet. Extensive contact of the cell sheet with the culture dish and culture medium is favourable for transformation into pigment epithelium. Moreover, the area of pigment epithelium formed is proportional to the sodium bicarbonate concentration in the medium. Lentoid bodies arise where cells are crowding and multilayering occurs. Their survival is inversely proportional to the sodium bicarbonate concentration in the medium. Mitosis may also be essential for pigment epithelium differentiation, since crowding, which inhibits mitosis, interferes with it (Pritchard, Clayton & de Pomerai, 1978). This seems to be at variance with embryonic pigment epithelium development, in which the surrounding mesenchyme inhibits mitosis (see Chapter 14, p. 197). Araki, Yanageda & Okada (1979) studied crystallin synthesis in lentoid bodies formed in long-term cultures of $3\frac{1}{2}$ day chick embryo neural retina. Moscona & Degenstein (1982*) suggest that lentoids formed from neural retina cells are of glial origin.

Lens regeneration from the dorsal iris of the adult eye, known as 'Wolffian lens regeneration', occurs in *Triturus* and *Ambystoma* after removal of the lens. The pupillary margin of the dorsal iris begins to swell at 4 days post-lentectomy, and depigmentation of the epithelial cells starts at 8 days. Subsequently a lens vesicle is formed by delamination of the double-layered epithelium, and normal polar lens differentiation follows (T. Yamada, 1967*). Competence for lens development is highest in the dorsal iris and decreases circumferentially. Lens regeneration depends on a stimulus from the neural retina. The dorsal iris epithelium goes through a process of dedifferentiation, with autophagy and exocytosis of cell organelles, and subsequent differentiation into lens cells representing a direct transformation of differentiated iris cells into lens cells (transdifferentiation) (T. Yamada, 1972*; T. Yamada & McDevitt, 1974; T. Yamada *et al.*, 1978). The presence of a lens interferes with lens regeneration from the dorsal iris, but for this the lens must be in contact with the iris (S. D. Smith, 1965). In the mouse, implantation of embryonic lens vesicles into an adult eye leads to the formation of small additional lenses, but normal growth and differentiation occurs only in the absence of the host lens. A 6 day mouse embryo lens implanted into a lentectomised adult mouse eye grows normally and attaches itself to the ciliary body (Reyer, 1966). In the adult mouse reversed lenses show elongation of epithelial cells facing the retina and formation of a new lens epithelium on the corneal side. No growth of normal or reversed lenses occurs after removal of the retina (Y. Yamamoto, 1976).

Implantation of an adult pituitary graft into the anterior eye chamber of the newt *Notophthalmus* can lead to secondary lens formation from the dorsal iris; the lens fibres are oriented towards the graft, so that the graft not only counterbalances the inhibitory influence of the host lens but also mimics the stimulating influence of the neural retina (Powell & Segil, 1976).

R. M. Clayton (1982*) correlates lens regeneration from iris, neural retina, pigment epithelium and cornea with the presence of low levels of lens crystallins in these organs and advances the hypothesis that transdifferentiation may be caused by an elevation of the level of expression of gene products (mRNAs).

For further details about lens induction and lens regeneration the reader is referred to the reviews by Woerdeman (1962*); Croisille (1963*); A. G. Jacobson (1966*); Saxén & Kohonen (1968*); Kratochwil (1972*); Wolff (1973*); Saxén *et al.* (1976*); Hay (1977*b**); Reyer (1977*) and R. M. Clayton, Truman & Bird (1982*).

The cephalic ganglia

The cranial or cephalic sensory ganglia originate from ectodermal placodes and not from neural crest derivatives (Chibon, 1966). Weston (1971*) proposed that the cephalic ganglia, though developing from ectodermal placodes, may be induced by the neural crest. In our opinion their particular location near certain regions of the CNS rather pleads in favour of local influences from the segregating neural tube.

The development of the auditory complex

From transplantation experiments A. G. Jacobson (1963*a*) concluded that in the amphibians the placodal ectoderm is determined to form the otic vesicle under the influences of the underlying archenteron roof, the rhombencephalon and the rhombencephalic neural crest. He then extended this to a determinative role for the entire history of inductive interactions with all surrounding tissues (1963 *c*) (see Fig. 27, p. 204). Chulitskaia (1962) emphasised the significance of the inductive influence of the underlying mesoderm. Its effect is greater in anurans than in urodeles and still smaller in sturgeons. In extirpation experiments in the chick, Orts-Llorca & Murillo-Ferrol (1965) found that auditory vesicle formation can take place independent of the rhombencephalon as soon as somite segmentation has begun. Whereas Benoit (1964) found that removal of the cephalic neural crest from the $1\frac{1}{2}$ day chick embryo ($\pm$ stage 10, H. & H.) or removal of the hyoid mesectoderm from the $2\frac{1}{2}$ day embryo ($\pm$ stage 17) did not affect the development of the otic vesicle, Cuevas (1977) could prevent otic vesicle formation in the stage 4 chick embryo by removal of the presumptive head mesenchyme.

Removal of the cephalic neural crest in the $1\frac{1}{2}$ day chick embryo prevented the formation of the columella, which is of mesectodermal origin, but removal of the otic vesicle did not. In contrast, the presence of the ear vesicle is necessary for the differentiation of the stapes (Benoit, 1964).

The analysis of several mouse mutants showing abnormal neural tube formation suggests the existence of an inductive influence of the neural tube on the differentiation of the inner ear in mammals, similar to that demonstrated for amphibians and birds (Deol, 1966).

Evaluation

The development of the cephalic placodes of adenohypophysis, nose, lens and ear from the adneural placodal ectoderm is the consequence of a series of successive inductive interactions with surrounding tissues. The cephalic endoderm constitutes the primary inductor, the mesoderm or mesectoderm may act as secondary inductor, while different regions of the central nervous system act as tertiary and final inductors. What is here called the action of the primary inductor is considered to represent the evocation of competence for further inductive interactions. The intercellular matrix seems to play an important role in the firm adhesion between the optic vesicle and the presumptive lens ectoderm. Although the role of humoral influences in the inductive interactions cannot be excluded (as exemplified by the influence of the retina on the polar differentiation of the lens, these tissues being far apart in the developing eye) it seems likely that intercellular matrix components are involved in the induction processes themselves, as clearly demonstrated for corneal development (see Chapter 14, p. 198).

The signals acting in the successive inductive interactions probably represent transient or definitive products of cellular differentiation of the various inducing tissues. Lens regeneration and the tissue interactions involved in it show great similarity to embryonic lens development.

Introduction to Chapters 16–18 (organ formation)

As stated in the Preface (p. viii) this book deals with inductive interactions in early chordate development. Therefore we cannot extensively discuss organogenesis but must restrict ourselves to the initial steps of organ formation, leaving morphogenesis and differentiation of the various organ systems out of consideration. It is however not always easy to draw the borderline between the early determinative events and the phase of organogenesis. We already met this difficulty in the chapter on the development of the CNS, where we discussed the origin of its regional organisation and the self-organisation of the prosencephalon. Evidently, we have to draw a line between early determinative events and actual organogenesis for each organ system separately, in order to provide the proper perspective for the development of its separate parts, which often represent distinct subunits.

Any classification is artificial; for instance, the development of the CNS might just as well have been classified under organogenesis. It has however been treated separately and taken as the direct consequence of the early events of induction, its further development being characterised by self-organisation and additional inductive interactions. Placodal development has also been considered as belonging to a later phase of the neural induction process, with local inductive actions superimposed on it. One might also object to the classification of neural crest development under organ formation, since the neural crest does not represent a particular organ, not even a particular tissue, but is only a transient embryonic anlage. Yet it gives rise to many different structures of the embryo and contributes to many different organ systems. For this reason it has been treated at the beginning of the chapters on organogenesis.

Wessels (1973*) states that virtually all organ systems arise as a result of embryonic tissue interactions between dissimilar cell populations. Hay (1977b*) classifies organ formation into epithelial–epithelial interactions (e.g. lens induction), epithelial–mesenchymal actions (e.g. ureteric bud on metanephric mesenchyme) and mesenchymal–epithelial actions (e.g. enamel formation by epithelial cells under the influence of tooth mesenchyme). This classification is essentially based on the different adhesive properties of epithelial and mesenchymal cell populations but says nothing about the character of the interaction, which may be largely identical, so we did not adopt this distinction.

Since the literature on organ formation often deals with early as well as later phases of organ development, we have had to select the data on early determinative events from among a much more extensive literature. For a proper understanding of organ formation the reader is referred to a number of review articles at the beginning of each chapter.

16

Development and fate of the neural crest

For detailed information the reader is referred to the following review articles and books: Yntema & Hammond (1947*); Holtfreter (1968*); Weston (1971*); Chibon (1974*); Leikola (1976b*); Le Douarin, Teillet & Le Lièvre (1977*); Noden (1978c*); Le Douarin (1979*, 1980a*, b*); Le Douarin, Le Lièvre, Schweizer & Zeller (1979*); Le Douarin et al. (1980*); Black & Patterson (1980*); Bronner-Fraser & Cohen (1980*) and Cochard & Le Douarin (1982*).

Origin of the neural crest

In the urodele amphibians the neural crest material occupies the most lateral portion of the neural plate and subsequently part of the elevating neural folds. The anterior, transverse neural fold, which forms part of the future prosencephalon, does not contain any neural crest material (Nieuwkoop, Oikawa & Boddingius, 1958). During neurulation the neural crest separates from the neural plate *senso strictu* and from the lateral epidermis. After closure of the neural tube the two masses of neural crest cells fuse into one dorso-medially (Raven, 1936). In the anuran amphibians neural crest formation is restricted to the inner, sensorial layer of the double-layered neuro-ectoderm. The neural crest material occupies the most lateral region of the inner layer of the neural plate. During closure of the neural tube, which primarily involves the outer epithelial layer, the left and right neural crest cell masses stay apart. They may partially fuse later, e.g. during dorsal fin formation (Nieuwkoop & Florschütz, 1950). Neural crest development in birds and mammals markedly resembles that in urodele amphibians, but shows a much more pronounced cranio-caudal sequential development. Vermeij-Keers & Poelmann (1980) describe neural crest formation in the mouse embryo and suggest differentiation *in situ* without migration in the facial region. In the human embryo neural crest cells have been recorded to emigrate also from the outer wall of the primary eye vesicle (Bartelmez, 1954). It is not known whether this is a more general feature of mammalian development and whether it also occurs in other amniote groups.

Recently Hirano & Shirai (1982) studied neural crest formation in urodeles with TEM and SEM techniques. In the early neural plate stage the neural crest primordium consists of pseudostratified columnar epithelial cells, in contrast to the dorsal neural plate and the lateral epidermis, which both consist of a single-layered cuboidal epithelium. At the neural-fold stage the neural crest cells lose their epithelial arrangement and become polygonal. At the time of the fusion of the neural folds the neural crest forms a single mass of cells situated medially between the epidermis and the neural tube. The neural crest cells begin to migrate at an early tail bud stage. Hirano & Shirai conclude that already during neurulation the cells make preparations for their future migration by changing their cellular affinities, shape and cellular activities. Christ (1971) suggests that in the chick the growth of the peripheral ectoderm may contribute mechanically to the thickening of the neural crest anlage.

The study of the migration routes

Raven (1931) followed the early development of the neural crest in the urodeles on the basis of differences in shape of neural crest cells as compared to neural, epidermal and mesodermal cells. After hetero- or xenoplastic exchange of homotopic regions of the neural crest Raven (1937) could follow their migration routes further, using species-specific differences in pigment granules and yolk platelets and in nuclear size as criteria. The next improvement was the technique of labelling the nuclei of donor embryos with ^{3}H-thymidine and subsequently exchanging homotopic regions of the neural crest between labelled and unlabelled chick embryos (Weston, 1963, 1971, 1980*; M. C. Johnston, 1966; Chibon, 1966, 1967, 1970; Le Douarin, 1973; Noden, 1975). However, all these methods are only applicable during the early phases of migration, since the cytological markers gradually disappear and the nuclear marker becomes diluted by cell division. The most successful marking method to date is the homotopic exchange of quail and chick neural crest (Le Douarin, 1973). Quail cells can be distinguished from chick cells by the presence of a compact mass of heterochromatin in their nuclei, as against dispersed heterochromatin in chick nuclei. This criterion lasts at least for the whole of embryonic development. The method may only be limited by incompatibility reactions, which are however weak or nearly absent between these rather closely related species (Le Douarin, 1974). Bellairs *et al.* (1981) caution against the general validity of the quail/chick chimaera technique, since they observed important differences in behaviour between certain types of chick and quail cells grown *in vitro*. Thiébaud (1983) recently developed a staining technique with the fluorescent dye quinacrine for differential nuclear marking of heteroplastic grafts between *Xenopus laevis* and *X. borealis*.

Table 1. *Derivatives of the neural crest*

Neuronal elements (sensory)
Spinal ganglia
Some neurons of the cephalic ganglia V, VII, IX and X
Rohon Beard cells (amphibians)
Autonomic ganglia
Supportive elements of the nervous system
Schwann and sheath cells
Supportive cells of the cephalic ganglia VII, VIII, IX and X
Supportive cells of spinal and autonomic ganglia
Pigment cells
Melanocytes of skin, mesenteries and internal organs and melanophores of the iris
Endocrine and para-endocrine cells
Adrenal medulla
Calcitonin-producing cells
Type I and II cells of the carotid body
Mesectodermal derivatives
Bones and cartilages of the facial and visceral skeleton
Dermis of the face and ventral neck region
Some striated muscles in the facial and visceral regions
Connective tissue of buccal and pharyngeal glands
Fibroblasts and 'endothelium' of the cornea
Connective tissue of the thymus
Musculo-connective tissue of the large arteries derived from the aortic arches
Ciliary muscles

After le Douarin (1980).

These various techniques have demonstrated that there are two main migration routes: a dorso-lateral route between the epidermis and the dorso-lateral face of the somitic mesoderm, and a ventral route between the neural tube and the median face of the somitic mesoderm.

It has been shown that the neural crest gives rise to a broad range of different cell types: (1) mesenchymal cells (called mesectoderm, as distinct from mesenchyme of mesodermal origin), which can subsequently develop into connective tissue, cartilage and bone; (2) neuronal elements of the sensory and autonomous peripheral nervous system; (3) neuron-associated cells such as Schwann cells and glial cells; (4) different types of endocrine cells belonging to the amino-precursor-uptake-and-decarboxylation (APUD) series, and (5) pigment cells. These different elements participate in the development of many organ systems (see Table 1 and Chapter 18, pp. 250 and 262).

The various derivatives of the neural crest arise from different cranio-caudal levels of the neural tube. In the chick cranial or cephalic neural crest forms abundant mesectoderm, which gives rise to head mesenchyme and

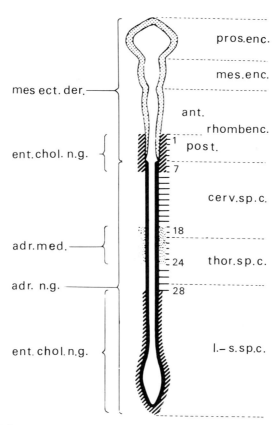

Fig. 29. Diagram showing the segments of the CNS of the chick embryo (right) and the presumptive areas of the neural crest (left), as inferred from chick/quail chimaeras. adr.med., adrenal medulla; adr.n.g., adrenergic ganglia; ant., anterior; cerv.sp.c., cervical spinal cord; ent.chol.n.g., enteric cholinergic ganglia; l.-s.sp.c., lumbo-sacral spinal cord; mesect.der., mesectodermal derivatives; mesenc., mesencephalon; post., posterior; prosenc., prosencephalon; rhombenc., rhomben-cephalon; thor.sp.c., thoracic spinal cord; 1–28, somite numbers (after N. Le Douarin *et al.*, 1977).

to cartilage and bone of the neurocranium and the visceral skeleton (Le Lièvre & Le Douarin, 1975). In the chick, the migration of the most cranial cephalic neural crest seems to be controlled by the mesodermal metamery in the region anterior to the first differentiating somites (Anderson & Meier, 1981). Mesencephalic neural crest also furnishes the ciliary ganglia, and rhombencephalic neural crest components of the cephalic ganglia V–X. In the amphibians the cephalic neural crest forms similar structures (Hörstadius & Selman, 1946). In the chick the parasympathic intramural ganglia and the ganglion of Remak arise from the vagal neural crest, corresponding in level to somites 1–7, as well as from the lumbo-sacral neural crest located behind the level of somite 28. The entire trunk neural crest, corresponding in level to somites 8–28, furnishes sensory and orthosympathic ganglia, while the adrenomedullary cells arise from the neural crest region at the level of somites 18–24 (Andrew, 1964, 1969, 1970; M. C. Johnston, 1966; Chevallier, 1972; Teillet & Le Douarin, 1974; Fontaine & Le Douarin, 1977; Le Douarin *et al.*, 1977; Bronner-Fraser, Sieber-Blum & Cohen, 1980 and others; see Fig. 29).

The nature of neural crest migration

Townes & Holtfreter (1955) and Holtfreter (1966*) already suggested that differences in cell affinity, described as differential cell adhesion by Steinberg (1964*), and chemotaxis play a role in neural crest cell migration. Abercrombie (1970) proposed contact inhibition as one of the mechanisms guiding cell migration (see also Martz & Steinberg, 1973). P. Weiss (1961 *a*, *b**) and Hay (1968*) emphasised the importance of the extracellular matrix as providing the necessary adhesive substrate for cell migration. Bronner-Fraser *et al.* (1980) hold the opinion that contact inhibition and differential adhesion are likely to be the key mechanisms in neural crest migration, but that chemotaxis cannot be ruled out.

Davis & Trinkaus (1981) observed neural crest migration in hydrated collagen lattices. It occurs in the form of tongues or cell streams and apparently does not depend on contact inhibition of movement. The rate of translocation is dependent on the collagen concentration, being about $1 \ \mu m \ min^{-1}$ at low and only $0.5 \ \mu m \ min^{-1}$ at high concentrations. Keller & Spieth (1984) found that melanocytes of dark (D/d) axolotl larvae migrate dorso-ventrally at a rate of $0.7 \ \mu m \ min^{-1}$ (see also Steinberg & Poole, 1981, 1982).

In quail/chick chimaeras Teillet (1971) reported that different derivatives of the neural crest move in different directions: presumptive spinal ganglion cells migrate predominantly dorso-ventrally, future sympathetic ganglion cells displace more antero-posteriorly as they become located more ventrally, while prospective pigment cells move dorso-ventrally as well as anteriorly and posteriorly, thus showing the existence of specific pathways of migration. Weston (1971*) and Le Douarin *et al.* (1977*) found that the migratory behaviour of neural crest cells depends on the cranio-caudal level of the embryo, each level offering a different route. These routes show preferential sites of arrest according to a precise pattern, each site being able to accommodate only a certain number of cells (Le Douarin *et al.*, 1977*). When a site of arrest is saturated, other arriving cells continue to migrate past, thus gradually filling up all the successive sites along a migration route. Le Lièvre *et al.* (1980) reported that cephalic neural crest cells grafted into the trunk region can penetrate into the dorsal mesentery and can occupy sites in the ganglion of Remak and the enteric plexuses, while trunk neural crest cells cannot pass the dorsal mesentery. Both cephalic and vagal neural crest cells are apparently able to invade the splanchnopleuric mesoderm. Therefore, the ultimate behaviour of neural crest cells is determined by their special migration routes as well as by their origin.

Keller & Spieth (1984), among others, observed the presence of long filopodia along the leading edge of migrating neural crest cells (the leading edge being oriented parallel to the long axis). M. S. Cooper & Keller (1984) noted that neural crest cells of *Ambystoma* and *Xenopus* orient themselves perpendicular to a DC electric field, thus migrating towards the cathode. This behaviour is probably due to ion fluxes through the anode- and cathode-facing cell surfaces.

In the axolotl Löfberg & Ahlfors (1978) followed neural crest cell migration into an already present highly organised extracellular matrix containing a prominent fibrillar network of collagen and proteoglycans. The neural crest cells seem to adhere firmly to the fibrillar network, which may serve for their guidance. Pratt, Larsen & Johnston (1975) found that in the chick cephalic neural crest cells migrate into the cell-free space between the head ectoderm and underlying mesoderm between stages 9 and 10 (H. & H.). At that time the cell-free space becomes heavily labelled after ^{3}H-glucosamine administration. The main component of the intercellular matrix seems to be hyaluronic acid, its appearance being correlated with the size of the cell-free space and the migration of the neural crest cells. Hyaluronic acid is apparently essential for neural crest cell migration since treatment with hyaluronidase prevents migration (Pratt *et al.*, 1975). Bolender, Seliger & Markwald (1980) noted that in the chick at stages 7 and 8 (H. & H.) the intercellular space through which the cranio-facial neural crest cells migrate contains fine filamentous strands of non-sulphated, carboxyl-rich glycosaminoglycans (GAG) representing mainly hyaluronate, with lesser amounts of chondroitin and only some sulphated GAG. This is replaced at stage 9–10 by coarse fibrillar strands and amorphous material coating the surface of surrounding mesenchymal cells as well as that of the migrating neural crest cells, and consisting chiefly of strongly sulphated polyanionic chondroitin sulphate. These authors believe that GAG chiefly stimulates motility rather than providing specific morphogenetic cues. Future dorsal root ganglion cells seem to accumulate where the hyaluronic acid concentration is locally reduced (Pratt *et al.*, 1975; see also Le Douarin, 1980*b**).

Moran (1974) found that the lectin Con A, which binds to embryonic membranes, inhibits the migration of neural crest cells. Its effect is both time- and concentration-dependent. Moran (1978*b*) observed that amphibian trunk neural crest cells cultured *in vitro* continuously synthesise extracellular material when labelled with ^{3}H-glycosamine. The cells leave behind radioactive 'images' of extracellular material during their migration.

In the chick Newgreen & Thiery (1980), Newgreen *et al.* (1982) and Thiery *et al.* (1982) demonstrated the presence of fibronectin (FN) in the basal lamina of the epidermis, somites, notochord and neural tube that surround the neural crest cell population during migration, while FN is absent on migrating neural crest cells. They consider FN the possible first expression of mesenchymal differentiation, since it is only synthesised by cranial and not by cervico-lumbar neural crest cells. The absence of FN production in migrating neural crest cells may magnify their response to FN provided by neighbouring tissues. Newgreen & Gibbins (1982) studied the factors controlling the time of initiation of neural crest migration in the chick.

The stepwise determination of the neural crest cells

Many observations, some of them to be discussed below, plead in favour of a stepwise determination of the neural crest cells, the majority of the cell types being definitively determined only late in the process. We will try to summarise the present insight into the successive steps.

While Leikola (1976*b**) stated that the mode of primary determination of the neural crest is still completely unknown, Rollhäuser-ter Horst (1977*a*, *b*, 1979) claimed that neural crest cells develop where neural and epidermal influences meet. As mentioned earlier, artificial activation of amphibian gastrula ectoderm leads to prosencephalic neural formations surrounded by thickened placodal ectoderm, but without any neural crest formation (Nieuwkoop, 1963). The latter is restricted to that region of the neural plate which comes under subsequent transforming influences emanating from the notochordal portion of the archenteron roof. In analogy with the formation of the neural/placodal ectoderm boundary, which seems to be due to the persistence of placodal competence when neural competence has disappeared, we want to suggest that the competence for neural crest arises (or persists) when the competence for neural transformation begins to subside. Since the transforming action, as the preceding activating one, spreads mainly from the dorsal midline laterally, as visualised in attached ectodermal folds (Nieuwkoop *et al.*, 1952; Nieuwkoop, 1958), neural crest formation will preferentially occur near the periphery of the neural plate (see Fig. 30).

Heterotopic grafting of various neural crest regions in the avian embryo has shown that the various regions of the trunk neural crest are fully interchangeable, which pleads in favour of late determination. Cephalic and trunk neural crest are not fully interchangeable, however; cephalic neural crest grafted in the trunk region partially differentiates into

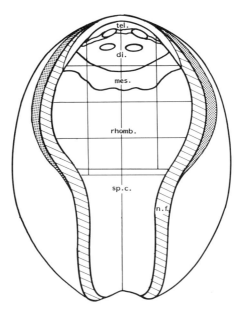

Fig. 30. Localisation of the cephalic placodal ectoderm (stippled) outside the anterior neural plate, and localisation of cephalic and trunk neural crest (hatched) in the neural folds (n.f.; see text for further explanation). di., presumptive diencephalon; mes., presumptive mesencephalon; rhomb., presumptive rhombencephalon; sp.c., presumptive spinal cord; tel., presumptive telencephalon – see, for further details, Fig. 23 on p. 191 (Redrawn from C. O. Jacobson, 1959 and adapted to the views of P. D. Nieuwkoop, 1963 and B. Albers (unpublished).)

mesectoderm, while trunk neural crest grafted in the head region cannot form mesectoderm. As far as neural derivatives are concerned, the two regions are, however, still isopotential. The formation of mesectoderm from the rather cell-rich cephalic neural crest must already be determined prior to migration (Chibon, 1974*; Le Douarin & Teillet, 1974 and others). Weston & Butler (1966) hold the opinion that except for the mesectodermal bias of the cephalic neural crest, neural crest cells are pluripotent, their ultimate differentiation into different cell types depending upon environmental cues encountered during migration or at their ultimate destinations, or both. This was corroborated by Le Douarin & Teillet (1973); Bronner-Fraser & Cohen (1980*); Black & Patterson (1980*) and others. Black & Patterson (1980*) state that neural crest cells possess some unique properties prior to migration, e.g. the ability to concentrate and decarboxylate L-dihydroxyphenylalanine (L-DOPA).

We want to suggest that the segregation of cephalic from trunk neural crest may be due to the close proximity of the pharyngeal endoderm to the cephalic portion of the neural plate, the two tissues only being separated by the thin paramedian prechordal and anterior chordomesoderm. Newly condensed avian cephalic neural crest can differentiate into cartilage when cultured *in vitro* (Hall & Tremaine, 1979), while amphibian cephalic neural crest still needs an additional direct confrontation with the pharyngeal endoderm for cartilage formation. The capacity of avian neural crest to form cartilage is strongly inhibited by excess vitamin A (Hassell, Greenberg & Johnston, 1977).

There are indications that in the peripheral nervous system adrenergic neuronal differentiation is favoured by influences emanating from the ventral spinal cord, the notochord and the somites (A. M. Cohen, 1972; Weston, 1970*; Norr, 1973). Teillet, Cochard & Le Douarin (1978) noted that this type of differentiation is supported by somitic mesenchyme but also by gut mesenchyme. Teillet & Le Douarin (1983) state that spinal and sympathetic ganglia require the presence of notochord and particularly ventral spinal cord via their influence on the formation of somitic mesenchyme. Sieber-Blum & Cohen (1980) found that extracellular matrix from both quail somites and chick skin fibroblasts stimulates adrenergic neuronal differentiation. It cannot, therefore, be a very specific influence.

Although adrenergic neuronal differentiation may be favoured during passage of the neural crest cells along the axial organs, the majority of neural crest cells still seem to be pluripotent and capable of differentiating into various specific cell types, depending upon the environment in which they ultimately find themselves. For instance, diencephalic neural crest, which normally does not form neurons, forms sensory and autonomous neurons and Schwann cells when grafted to the metencephalic region (Noden, 1978 *a*, *b*). Parasympathetic cells of the ciliary and Remak ganglia, which already synthesise acetylcholine, reinitiate migration, populate sympathetic ganglia and adrenal medulla and start synthesising catecholamines when placed in the back of early embryos (Le Douarin *et al.*, 1978). These observations demonstrate the possibility that neural crest cells become reprogrammed under the influence of new environmental conditions. Le Douarin *et al.* (1979*) propose the following cellular determinative steps in the development of the peripheral nervous system in the chick: (1) determination for either sensory or autonomic precursor cells; then determination for (2) either neuronal or glial precursor cells; the autonomic neuronal cells ultimately become determined (3) for either cholinergic or adrenergic neurons. This last determinative step remains labile for a long time. This scheme is in accordance with the fact that ventrally migrating neural crest cells first meet the sensory dorsal root ganglion sites, then the sympathetic ganglion and adrenal medullary sites, while neural crest cells from the vagal and lumbo-sacral regions may finally reach the enteric ganglion sites, where they develop into cholinergic neurons (Cochard & Le Douarin, 1982*).

It seems rather likely that the extracellular matrix which fills up the intercellular spaces along the migration routes and at the ultimate arresting sites of the neural crest cells determines their ultimate differentiation in a cell-surface–matrix type, or perhaps even in a matrix–matrix type of interaction.

Clonal cultures of quail neural crest cells may differentiate into both pigmented and unpigmented cells but may also form either of these two cell types alone (A. M. Cohen & Konigsberg, 1975). The former case clearly shows the pluripotency of neural crest cells, while the latter case pleads in favour of heterogeneity in the cell population.

Chibon (1974*) noted that cell aggregation favours neuronal differentiation and ganglion formation, whereas cell dispersion stimulates pigment cell differentiation. Nerve growth factor (NGF) enhances neurogenic differentiation by retarding cell dispersion. The presence of NGF in somitic mesoderm might support the aggregation of migrating neural crest cells into ganglia (Weston, 1971; Glimelius & Weston, 1981).

The autonomous differentiation of pigment cells has been clearly demonstrated by Twitty (1945) in hanging drop cultures of amphibian neural crest. Koecke (1960) studied the regional differentiation of melanocytes in the duck embryo. He showed that the head neural crest forms fewer melanoblasts than the trunk neural crest, and that it colonises a smaller area.

In the axolotl larva phenylalanine does not seem to play a role in pigment cell differentiation but is involved in protein synthesis in differentiated cells (Moran, Palmer & Model, 1973). Matsudo & Kajishima (1978, 1980) found that melanophore differentiation *in vitro* from trunk neural crest of *Cynops pyrrhogaster* is suppressed by ectoderm-conditioned medium but not by neural-fold/neural-crest-conditioned medium. Whereas melanocytes of the dark (D/d) axolotl larva migrate actively, those of the white (d/d) genotype show hardly any migration and remain confined to the dorsal midline of the trunk region (Keller & Spieth, 1984). Their migration seems to be inhibited by an altered matrix structure in the subepidermal space (Spieth & Keller, 1984). K. Ito & Takeuchi (1982) studied mouse melanocyte differentiation *in vitro* (see also Chapter 14, p. 199).

Finally it should be mentioned that reciprocally, neural crest derivatives exert inductive influences upon other tissues during embryonic development; e.g. in the formation of the enamel organ from the buccal epithelium and in balancer, external gill and dorsal fin development (see Chibon, 1974*).

Evaluation

The neural crest is a transient embryonic structure, which gives rise to a large number of different cell types and contributes to the formation of many different organs.

The mode of primary induction of the neural crest is practically unknown. We suggest that a separate competence for neural crest formation appears or persists after the subsidence of the competence for transformation of the neural anlage into mes- and rhombencephalon and spinal cord under the influence of the notochordal portion of the archenteron roof, in analogy with the emergence of a separate competence for placodal ectoderm formation after the subsidence of neural competence. Consequently, neural crest formation will preferentially occur along the lateral margin of the neural plate.

While mesectodermal competence (in amphibians) and mesectodermal differentiation tendencies (in birds) already appear in the cephalic neural crest prior to migration, possibly under the influence of the underlying pharyngeal endoderm, the various trunk neural crest regions are still fully interchangeable at that stage. The migrating neural crest cells, which follow different routes at different cranio-caudal levels and fill up successive sites of arrest, may become partially predetermined during migration but are only fully determined after reaching their ultimate sites of arrest.

There is good evidence that neural crest cell migration is guided by the composition of the extracellular matrix of the tissues through or along which migration takes place, while the same or other components of the extracellular matrix may be responsible for the predetermination and ultimate determination of the neural crest cells during or after cessation of migration. Neural crest development and differentiation must be considered as one of the most typical examples of sequential induction and determination of pluripotent cells by transient cellular products of surrounding tissues in a cell-surface–matrix type or even in a matrix–matrix type of interaction.

17

The development of the mesodermal organ systems

In the amphibian embryo the mesoderm and cephalic endoderm – together called the meso-endoderm – are induced in the totipotent animal moiety of the blastula by the peripheral region of the vegetal endodermal moiety, the central portion of which is spatially separated from the roof by the blastocoelic cavity (see Chapter 10, p. 95). From the boundary between the vegetal yolk mass and the blastocoelic roof meso-endoderm induction spreads in the direction of the animal pole, leading to the emergence of a ring-shaped, tube-like meso- and endodermal anlage. The spreading of the inductive action is a slow process, starting at an early blastula stage and covering only the lower half of the presumptive marginal zone at the early gastrula stage (Kaneda & Hama, 1979). Meso-endoderm induction continues during gastrulation and neurulation, ultimately to comprise the entire mesodermal mantle, which thus achieves its definitive pattern (see Chapter 10, p. 101).

In the tube-like marginal zone two polar axes can be distinguished: a cranio-caudal one extending from the boundary with the vegetal yolk mass animally, and a dorso-ventral one oriented perpendicular to the former. Quantitative differences in the intensity of the inductive action, due to the presence of a meso-endoderm-inducing centre in the (by definition) 'dorsal' region of the yolk mass, are responsible for the dorso-ventral differentiation of the meso-endoderm as well as for its cranio-caudal extension and segregation. The inductive action extends markedly further in the dorsal than in the lateral and particularly the ventral blastocoelic wall. The spreading of the inductive action occurs with decrement, which manifests itself in the partial endodermisation of the animal moiety near the boundary with the inducing yolk mass, and in the decrease of notochordal differentiation tendencies in the dorsal marginal zone with increasing distance from this boundary.

During gastrulation the cranio-caudal axis of the entire meso-endoderm becomes inverted, thus acquiring its definitive orientation in what is now called the 'mesodermal mantle', situated between an outer, ectodermal and an inner, endodermal layer (see Nieuwkoop, 1977*). During gastrulation, when just invaginated and non-invaginated portions of the marginal zone interact, the anterior portion of the converging dorsal chordomesoderm changes into the expanding prechordal meso- and endoderm, while the posterior portion acquires notochordal differentiation tendencies (see p. 104). The dorsal mesoderm forms the axial notochord and somites, the more lateral mesoderm forms the nephrogenic tissues, and the ventral mesoderm forms the lateral plates and blood islands (in the urodeles, the primordial germ cells also; see Chapter 10, p. 102).

In the amniotes, of which only the birds have been extensively studied, the embryonic meso- and endoderm as well as the extraembryonic mesoderm are induced by the endodermal hypoblast in the overlying epiblast of the flat blastoderm in more or less concentric configuration: the embryonic endoderm in the centre and the mesoderm around it (see Chapter 10, p. 116). During primitive streak formation the embryonic endoderm is the first to ingress through Hensen's node and the anterior portion of the streak. It is followed by the mesoderm, the axial mesoderm ingressing through Hensen's node and the anterior portion of the regressing streak, the extraembryonic mesoderm ingressing mainly through its posterior portion. In this manner a flat mesodermal layer with cranio-caudal and dorso-ventral (medio-lateral) polar axes is formed in between the outer epidermal and the inner endodermal layer. During the elevation of the embryo from the yolk sac the flat embryonic anlage is transformed into a tube-like structure.

The most characteristic aspect of the differentiation of the mesoderm is its bilaterally symmetrical organisation. It forms a medio-dorsal notochordal anlage, flanked on both sides by a row of somites, followed more laterally by nephrogenic tissue and lateral plate. This bilateral symmetry is also established under experimental conditions; for instance, grafting a half dorsal blastoporal lip into the ventral marginal zone of another early *Triturus* gastrula leads to homoiogenetic induction of the missing structures (Spemann & H. Mangold, 1924; Holtfreter, 1968*; Cooke, 1980 and others). Isolated halves of dorsal blastoporal lips of *Hynobius nebulosus* early gastrulae regulate internally into a bilaterally symmetrical structure by the formation of a median notochordal anlage (Ikushima, 1961). After dis- and reaggregation of the vegetal yolk mass, which destroys its polarity, and recombination with either a dis- and reaggregated or an intact animal cap of an axolotl blastula, mesoderm is arbitrarily induced along the entire periphery of the recombinate. This mesoderm organises itself into a single bilaterally symmetrical axial system or into two opposing systems, and sometimes even into more than two systems (see Nieuwkoop, 1973*). These results demonstrate that the development of bilateral symmetry is a self-organising capacity of the mesoderm.

In an extensive series of transplantation and recombination experiments T. Yamada (1937, 1939*a*, *b*, *c*, 1940) demonstrated that the dorso-ventral sequence of mesodermal differentiations in the mesodermal mantle depends on a morphogenetic factor emanating from the notochord and spreading dorso-ventrally through the mesodermal mantle, declining in intensity with distance. Consequently the lateral plate and blood islands (as well as the primordial germ cells in urodeles) represent the lowest level, nephrogenic tissues the next higher level, somites the next and notochord the highest level of mesodermal differentiation. Although Muchmore (1951) could confirm Yamada's main results, he nevertheless concluded that a single morphogenetic factor, such as initially proposed by Yamada, cannot account for the characteristic cranio-caudal localisation of e.g. heart, pronephros and blood islands. Yamada (1950b) extended his theory by postulating an additional cephalo-caudal factor. Muchmore (1951) found that the notochord is not indispensable for muscle and pronephros development *in vivo*. It has been observed by several authors that small parts of a dorsal blastoporal lip are not always able to form notochord, but may form a single median row of somites as their highest level of differentiation. (This is also achieved after local removal of the notochordal anlage (Christ, 1970).) Muchmore (1951) therefore concluded that the differentiation of each anlage in the mesodermal mantle depends on influences from the entire complex of surrounding ecto-, meso- and endodermal organ anlagen. Muchmore (1957a) also observed an effect of tissue mass on muscle differentiation. Takaya (1961) deduced from grafting experiments on *Triturus pyrrhogaster*, *Hynobius nebulosus* and *Rana nigromaculata* neurulae that notochord enhances somite development both qualitatively and quantitatively.

The development of the notochord

Woo Youn, Keller & Malacinski (1980) studied mesodermal differentiation in the anurans *Xenopus laevis* and *Rana pipiens* and the urodeles *Ambystoma mexicanum* and *Pleurodeles waltlii* with SEM and observed that the notochordal cells become distinguishable from the paraxial mesoderm by their cell shape and closeness of packing. Notochordal elongation is accompanied by a decrease in cross section and by a specific cell arrangement. During neurulation, vacuolisation occurs. Takaya (1973) noted that in *Cynops pyrrhogaster* vacuolisation occurs in almost all embryonic cell types. Whereas vacuolisation is limited and transient in most tissues it is extensive and persistent in notochordal cells. According to Takaya & Kayahara (1978) the presence of surrounding tissues is required for notochord vacuolisation. The notochordal anlage forms a thick extracellular matrix, the so-called notochordal sheath (Vasan, 1981, 1983). Bancroft & Bellairs (1976) studied notochordal differentiation in the chick with SEM and TEM. They describe three successive stages, designated as bilaminar, rod-like-unvacuolated and rod-like-vacuolated, and the accompanying changes in cell shape, orientation and position.

Notochordal differentiation tendencies are only present in the presumptive 'anterior' half of the dorsal blastoporal lip of the early gastrula of *Hynobius nebulosus* (Ikushima, 1961; Ban-Holtfreter, 1965), the presumptive 'posterior' half acquiring them only when approaching the dorsal blastoporal lip (Kaneda & Hama, 1979). Notochordal differentiation tendencies are further strengthened during neural plate induction (Nieuwkoop & Weijer, 1978, see Chapter 10, p. 105). Notochord formation is restricted to the invaginating archenteron roof and extends only secondarily into the tail bud (see also p. 236). Schoenwolf (1977) reported that ³H-thymidine-labelled chick tail bud grafts form neural tube, somites and mesenchyme, but neither notochord nor hind and tail gut. Therefore tail bud formation in birds resembles that in amphibians.

Ranzi, Vailati & Vitali (1972) found that sodiumthiocyanide (NaSCN) treatment of chick primitive streak stages enhances notochord formation, both in volume and in cell number, which is probably due to a transformation of presumptive somitic mesoderm cells into notochordal cells. Hydroxyurea, which inhibits DNA synthesis, prevents both somite and neural tube formation in *Pleurodeles waltlii* but did not interfere with notochordal differentiation according to Beetschen & Buisan (1977). We think that notochordal differentiation may already have been too far advanced at the time of treatment.

L. G. Barth & Barth (1974) and L. J. Barth & Barth (1974) observed that small aggregates of *Rana pipiens* gastrula ectoderm treated with lithium chloride showed notochordal differentiation when cultured in a medium containing 2.3–5.3 mM K^+ but formed nerve and pigment cells at lower or higher K^+ concentrations. They explained this by the effects of K^+ on membrane depolarisation, protein and DNA synthesis and intracellular Ca^{2+} concentration. In our opinion the phenomenon should primarily be viewed as a strengthening of either neural or mesodermal competence of the pluripotent ectoderm under the influence of different ionic balances.

Somitogenesis

One of the most intriguing but still little understood problems in embryogenesis is that of the segmentation of the somitic mesoderm. Somitogenesis has been studied extensively in birds and also in amphibians (see the reviews by Nicolet (1970b*), Bellairs (1974a*, b*) and Deuchar (1973*)).

SEM analysis shows that in the anurans *Xenopus laevis* and *Rana pipiens* and the urodeles *Ambystoma mexicanum* and *Pleurodeles waltlii* the somitic mesoderm becomes distinguishable from the lateral plate mesoderm by changes in cell shape and by antero-posterior cell orientation, followed by segmentation (Woo Youn et al., 1980). In *Bufo bufo* segmental differentiation starts in the middle region of each segment and progresses towards the dorsal and ventral surfaces (Brustis, 1979), which may contribute to intrasomitic 'virtual slit' formation (see below under birds, p. 232).

As already mentioned, somite formation in amphibians is largely dependent on the notochord, although it can also occur autonomously. Somitogenesis in *Xenopus laevis* and *Rana temporaria* is characterised by a wave of cell change – first a rounding up and then elongation – propagating antero-posteriorly and preceding segmentation (Pearson & Elsdale, 1979). There are two heat-shock-sensitive periods: late blastula to middle gastrula and neurula to tail bud stages. While early heat shock affects the wave of rapid cell change, late heat shock disturbs cellular coordination; the period around the late gastrula stage is refractory to heat shock (Elsdale & Pearson, 1979). The authors conclude that there are two physiologically independent temporal patterns of cellular processes which interact, specifying the segmental somite pattern: (1) the coordination of pre-somitic cells; and (2) their recruitment into a segmental prepattern, the latter being established before visible segmentation.

In *Xenopus laevis* Cooke (1975b, 1981) observed constancy of somite number in embryos of reduced size obtained by removal of the ventral quarter of the blastula, the number of cells per somite adjusting to the overall size of the embryo.

Deuchar & Burgess (1967) concluded that in *Xenopus laevis* embryos the orientation of the V-shaped intersomitic grooves must be governed by some global effect, since reversed grafts of somitic mesoderm reorientate according to their new surroundings.

In the *avian* embryo the presumptive somitic mesoderm ingresses through the regressing primitive streak just caudal to Hensen's node. Somewhat later the somitic cells begin to condense into segmental aggregates, a process which progresses cranio-caudally (Nicolet, 1970b*). In the chick embryo at stages 2–3 (H. & H.) the paraxial mesoderm adheres firmly to the basal laminae of both epi- and hypoblast. At stages 4–6 the mesodermal cells have multiplied to form a layer 4–6 cells thick. Now the presomitic mesoderm adheres preferentially to the basal lamina of the neurectoderm. At stages 6–7, when the neural plate contracts, the presomitic mesoderm becomes more compact by elimination of intercellular spaces and by the formation of cell junctions at the apical ends of the cells. During neurulation at stages 8–9 the somitic mesoderm remains attached to the neural plate, but it later separates from the neural tube and the notochord as a result of accumulation of extracellular matrix material produced by these two tissues. By stage 12 bundles of collagen fibrils connect the somitic mesoderm with the notochord and spinal cord (Lipton & Jacobson, 1974).

Bellairs & Portch (1977) and Bellairs, Curtis & Sanders (1978) studied the role of cell adhesion in somite segmentation in the chick, while Bellairs (1979) followed somite segmentation between stages 8 and 14 (H. & H.) with TEM and SEM. The segmental plate mesoderm, which is the paraxial mesoderm situated posterior to the last-formed somites, consists of loosely arranged mesenchymal cells, whereas the newly formed somites are composed of elongated, spindle-shaped cells arranged radially around a

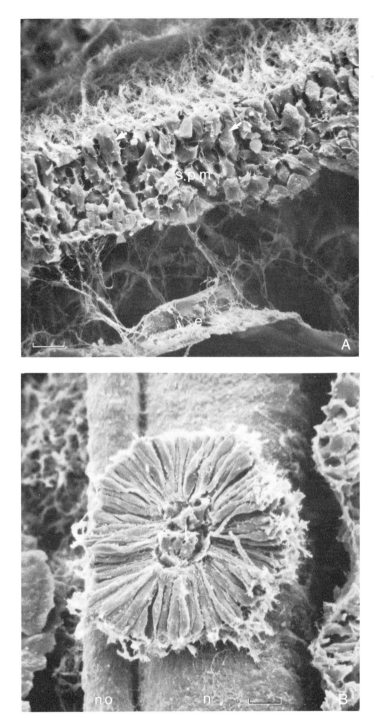

Fig. 31. For legend see p. 234.

myocoel in the form of a single cell layer (see Fig. 31). Two factors seem to be responsible for cell-shaping during segmentation: (1) the presumptive somite cells become anchored with their outer surfaces to the surrounding neural tube, notochord, epidermis, endoderm and aorta through collagen fibrils; and (2) a change in cell adhesiveness occurs, causing the free ends of the cells to adhere to each other at the inner myocoel surface. In more differentiated somites of stages 17–18, the somitic cells undergo a rotation and become oriented antero-posteriorly, as in the amphibians.

Many hypotheses have been advanced to explain somitogenesis in the avian embryo. We only briefly mention here the 'somite-forming centre' hypothesis of Spratt (1955, 1957*a*, *b*), the 'primitive streak regression' hypothesis of Bellairs (1963), the 'Hensen's node and notochord' hypothesis of Nicolet (1970*b**, *c*) and the 'neural plate' hypothesis of Fraser (1960) and Butros (1967). Thus Fraser (1960) attributed importance to the regressing node in the laying-down of axial tissues but considered the node not to be mandatory for somitogenesis. The neural plate would be chiefly responsible for the induction of the paraxial somites, contact between it and the somitic mesoderm being required for segmentation. Nicolet (1968) found that Hensen's node can induce somite formation when grafted into the posterior primitive streak, which by itself cannot form somites. Implanted notochord can also induce somite formation in post-nodal regions (Nicolet, 1970*b**). Hornbruch, Summerbell & Wolpert (1979) concluded from grafts of quail Hensen's node into chick blastoderms that the median region of the area pellucida, but not its margin, possesses competence for somite induction. Lipton & Jacobson (1974) emphasised that no single factor can be made responsible for somitogenesis. Although the neural plate may be the principal somite inductor, the notochord plays an important role in shearing the somitic mesoderm into left and right halves, and in the initial release of somite-forming capacities.

Lanot (1971) observed that in the chick the segmental plate mesoderm can form somites autonomously but will always do so in continuity with newly formed somites. The segmental plate comprises about 11 presumptive somites in embryos ranging from 5 to 21 pairs of somites. Lanot found that removal of the apical end of the segmental plate prevents somitogenesis or at least leads to incomplete development. However, Packard & Jacobson

---

Fig. 31. A, SEM micrograph of a stage 10 (H. & H.) chick embryo broken transversely caudal to the somites, with ectoderm removed. A dense meshwork of fibrils can be seen on the dorsal side of the segmental plate mesoderm (s.p.m.). Many fibrils are also present between the segmental plate and the endoderm (en.). Some of the most dorsally situated cells are already elongated (arrows). Bar = 10 μm. B, SEM micrograph of a longitudinal section through a somite of a stage 11 chick embryo. Spindle-shaped cells are arranged around a lumen (l) which contains other cells. Extracellular material covers the surface of the somites. n., neural plate; no., notochord. Bar = 10 μm. (Courtesy R. Bellairs).

(1976) and Packard (1978, 1980) noted in experiments *in vitro* that up to eight presumptive somites can be removed from the anterior end of the segmental plate without interfering with somitogenesis in the posterior portion. They concluded that the segmental plate mesoderm is fully programmed. Lanot (1971) stated that the neural tube and possibly also the notochord play a leading role in the propagation of segmentation. Packard & Jacobson (1976) believe that the neural plate imposes a segmentation prepattern on the presomitic mesoderm. Continued intimate contact with the axial structures would be required for further somitogenesis and somite maintenance (see also Christ, Jacob & Jacob, 1972). Reversed segmental mesoderm shows a reversed cranio-caudal segmentation sequence, so that this sequence must be firmly determined in the presomitic mesoderm (Christ, Jacob & Jacob, 1974). Longitudinal splitting of the segmental plate mesoderm leads to the formation of two rows of somitic vesicles, of which only the medial row bordering on the neural tube and notochord undergoes differentiation (Menkes, Sandor & Elias, 1968; Menkes & Sandor, 1969; Sandor & Amels, 1970, 1971; Sandor, 1972).

Bellairs, Sanders & Portch (1980) showed that lateral plate cells, segmental plate cells and cells of newly formed somites, in that order, decrease in adhesiveness to artificial substrates such as glass and plastic but increase in cell-to-cell adhesiveness. Treatment with analogues which interfere with collagen formation causes a reduction in number and size of the somites and leads to retardation of somite formation. Bisected unincubated quail blastoderms from twin embryos of reduced size which show a spatially adjusted somite shape (Veini & Bellaris, 1983). Unilateral incision along the neural tube and notochord causes the formation of fewer somites and slows down the segmentation process on the operated side (Bellairs & Veini, 1980). Treatment of young embryos with UV leads to the absence of the endoderm; the fact that normal mesoderm segmentation occurs shows that the endoderm is not essential for somitogenesis (Bellairs & Veini, 1980).

De Bernardi-Laria *et al.* (1977), Bolzern *et al.* (1979), Cigada-Leonardi *et al.* (1980) and Ranzi (1981) claim that myosin is the natural inductor of the somites. Puromycin, which blocks the translation of myosin mRNA (isolated from skeletal muscle) into myosin, prevents somite formation in chick mesoderm *in vitro*. It is ineffective in the presence of myosin, though only in 33% of the cases, which makes their conclusion doubtful. Moreover, these observations have not been confirmed.

Although there is hardly any experimental evidence, normal development suggests that somitogenesis in reptiles and mammals occurs in much the same manner as in birds (P. D. Nieuwkoop, unpublished observations).

Somatopleure mesoderm grafted ortho- or heterotopically between quail and chick embryos differentiates into cartilage, smooth muscle, tendons and connective tissue but not skeletal muscle. The skeletal muscles developing from the somatopleure are of somitic origin (Christ, Jacob & Jacob, 1979).

A few words must be said about the origin of the tail somites. In amphibians these do not originate from the archenteron roof but from the caudal portion of the neural plate (Bijtel, 1931, 1936). Spofford (1945, 1948) found that the tail somitic mesoderm acquires its final determination when the neural folds are closing, as a result of an inductive action emanating from the posterior part of the archenteron roof. Niazi (1969) showed that presumptive tail somite material isolated from the hind part of the neural plate of young axolotl neurulae differentiates into neural tissue, but forms tail somites when combined with caudal chordomesoderm. This demonstrates the complete transformation of neuralised ectoderm into mesoderm in the case of the tail somites (see also Chapter 12, p. 159). These observations have been confirmed by Beetschen & Buisan (1977) in *Pleurodeles*.

The further development of the somitic mesoderm

In the chick somitic differentiation begins on the dorso-lateral side with the development of the dermal plate, which represents the combined anlagen of dermatome and myotome. At 3 days of incubation (36 pairs of somites) the dermal plate separates into the pseudostratified dermatome, situated directly underneath the trunk epidermis, and the deeper myotome. The dermatome later forms the dermal mesenchyme while the myotome forms the segmental muscles. The medio-ventral portion of the somite disperses and forms the sclerotomic mesenchyme (see Nicolet, 1970b* and Fig. 32).

Whereas DNA synthesis is still active in the dermatome of the 3 day chick embryo, it is completely inhibited in the myotome, where synthesis of specific proteins has started (Langman & Nelson, 1968). The sclerotome shows incorporation of $^{35}SO_4^-$ into mucopolysaccharides that enter the collagenous matrix (Lash, Holtzer & Whitehouse, 1960).

Gallera & Ivanov (1964) and Gallera (1966c) found that in the chick embryo the dermal plate is always oriented towards the epidermis. Somites inserted between two layers of epidermis *in situ* show double dermatome and myotome formation, while somites inserted between two layers of endoderm show cellular dispersion and loss of metamery. However, dispersion of somites into sclerotomic mesenchyme occurs also in the absence of the endoderm. The epidermis inhibits this spontaneous dispersion. Both embryonic and extraembryonic epidermis can induce dermal plate formation in early somites.

Brustis (1976, 1978) claims that in *Rana dalmatina* and *Bufo bufo*, dermatome, myotome and sclerotome segregate autonomously from

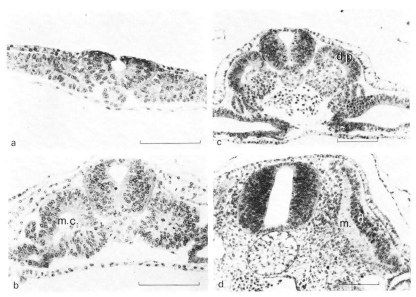

Fig. 32. Successive stages in the morphological development of the anterior somites in the embryo of the marine turtle, *Chelonia mydas*. a., Embryo of 6 somites, with condensation of somitic cells into segmental aggregates showing cell elongation towards the centre. b, Embryo of 12 somites, with formation of myocoel (m.c.) in centre of somite. c, Embryo of 20 somites; the dorsal wall of the somite becoming the dermal plate (d.p.), while the ventral wall disperses into mesenchymal cells occupying the space between dermal plate, spinal cord, notochord and primitive aorta. d, Embryo of 31 somites. The dermal plate has segregated into the pseudostratified epithelium of the dermatome (d), situated directly underneath the trunk epidermis, and the myotome (m), forming a layer of cells with distinct nuclei and large nucleoli (presumptive myoblasts) and apposed to the lower surface of the dermatome (original). Bars each represent 0.1 mm.

explanted somite mesoderm, but the isolation may have been performed at too late a stage. Jacob, Christ & Jacob (1974) corroborated Gallera's conclusion concerning the determinative role of the epidermis in chick myotome development but deny such a role for the neural tube. Muchmore (1964*a*, *b*) found that in amphibians embryonic spinal cord enhances, whereas embryonic brain suppresses muscle differentiation from somitic mesoderm. The spinal cord is indispensable for the maintenance of the myotomes.

According to Packard & Jacobson (1976) the notochord and the ventral part of the spinal cord are essential for the development of the median portions of the somites into sclerotomic mesenchyme. Gonzalo-Sanz (1972) reported that ablation of the neural crest causes delayed and incomplete development of the sclerotome and suggests that the neural crest is also necessary for normal sclerotome development. Moreover, neural crest cells can only migrate through normal sclerotomic mesenchyme.

Chondrogenesis

There are two main sources of chondrocytes in the vertebrate embryo, the mesectodermal cells of the cephalic neural crest (see below) and the sclerotomic cells of the medio-ventral portions of the somites (see p. 239). Study of cartilage formation from the mesectoderm has been chiefly carried out on amphibian material, while sclerotomic chondrification has primarily been studied in birds. For more detailed information the reader is referred to the reviews by Holtzer & Matheson (1970*), Holtzer & Mayne (1973), Hall (1977*) and Lash & Vasan (1977*).

Mesectodermal chondrification. The formation of the branchial arches in amphibians requires the presence of the pharyngeal endoderm, which gives rise to the gill pouches that develop between the branchial arches (Balinsky, 1948). The various contributions of the cephalic neural crest to the visceral skeleton were carefully analysed by Hörstadius & Selman (1946). They confirmed that the pharyngeal endoderm is the leading structure in the formation of the visceral skeleton. T. S. Okada & Sirlin (1960) reported that in *Xenopus* $^{35}SO_4^-$, which is incorporated into mucopolysaccharides, is first taken up by the cytoplasm of precartilage cells and is later found in the intercellular matrix. Petriconi (1964) observed that extract of pharynx endoderm is capable of inducing cartilage formation in cephalic neural crest cells in *Bombina*.

The cephalic neural crest requires direct contact with the pharyngeal endoderm to form cartilage. Neural crest cells of *Triturus alpestris*, which show fibroblast-like motility and contact inhibition *in vitro*, lose their motility in direct contact with pharyngeal endoderm but maintain it some distance away from the endoderm (Drews, Kocher-Becker & Drews, 1972). This was confirmed by Epperlein (1974, 1978 a, b) and Epperlein & Lehmann (1975), who observed increased mutual adhesiveness of neural crest cells in contact with the pharyngeal endoderm. They believe that mitotic divisions are responsible for the spreading of the inductive message within the responding ectomesenchyme.

Corsin (1975) found that in *Pleurodeles waltlii* chondrification of cephalic neural crest cells *in vitro* is also possible under the influence of dorsal mesoderm, so that the influence of the pharyngeal endoderm is not specific.

In birds Le Lièvre (1971) studied the contributions of the cranial neural crest to the formation of the visceral cartilages and bones, using quail/chick chimaeras. The nearly complete absence of branchial mesenchyme did not prevent the formation of the visceral pouches and clefts in the foregut endoderm but severely affected the visceral skeleton. Also in birds, the foregut endoderm seems to be the leading structure in visceral skeleton formation (Le Lièvre, 1974), although, as already discussed under neural crest development, the cephalic neural crest of the chick embryo is capable of autonomous differentiation into cartilage.

The pigment epithelium of the chick eye causes chondrification of periocular cephalic neural crest cells, leading to *sclera* formation (Wedlock & McCallion, 1969; Newsome, 1972). The inductive action can pass through a Nuclepore filter of 0.8 μm pore size but not through 0.2 μm pores, so that the induction requires either contact of neural crest cells with non-diffusible matrix components of the epithelium, e.g. basal lamina, or direct cell–cell interaction (L. Smith & Thorogood, 1983).

Sclerotome chondrification. The sclerotomic mesoderm breaks up into nests of cells which fill up the intersomitic spaces, leading to vertebral arch formation, and accumulate around the notochord, giving rise to the vertebral centres, these two structures being the main components of the vertebral column. Holtzer & Detwiler (1953) in amphibians, and Strudel (1953) in the chick, demonstrated that both notochord and ventral spinal cord promote cartilage formation in sclerotome cells. The processes are so much alike in the two groups that a common treatment seems justified.

Bancroft & Bellairs (1976) observed the presence of glucosaminoglycan granules and collagen fibrils around the chick notochord and ventral spinal cord at stage 10 (H. & H.) and extensive extracellular matrix and basal lamina formation at stage 17, shortly before sclerotome cell migration starts at stage 18. Strudel (1953) noted absence of neural arches after removal of spinal cord and absence of vertebral centres after notochord extirpation. While the notochord induces a single layer of cartilage directly adjacent to it, the spinal cord induces cartilage formation at some distance, with interjacent mesenchyme (Strudel, 1953; Watterson, Fowler & Fowler, 1954; Lash, Holtzer & Holtzer, 1957). In the chick, Lash *et al.* (1957) obtained cartilage induction through a millipore filter. Flower & Grobstein (1967) observed cartilage formation in mouse somitic mesoderm in direct contact with the millipore filter with notochord as inductor, but at some distance from the filter when spinal cord was used as inductor. Moreover, notochord seems to be more active than spinal cord. Only the ventral half of the spinal cord has the capacity to evoke cartilage formation.

Cartilage can also be induced by cartilage in sclerotomic mesenchyme, a case of homoiogenetic induction (G. W. Cooper, 1965; Holtzer & Matheson, 1970*). Holtzer (1968*) found that spinal cord and notochord can only induce chondrogenesis in sclerotomic mesenchyme, not in other kinds of mesenchyme, which apparently are not competent. The noto-chordal sheath plays an important role in sclerotome formation (see p. 230). Large molecular aggregates of proteoglycans synthesised by the notochord induce somitic chondrogenesis (Vasan, 1981, 1983). Treatment of notochord with trypsin as well as with purified chondroitinase and hyaluronidase removes extracellular matrix material and reduces its chondrogenic effect on somitic mesoderm (Kosher & Lash, 1975).

In an extensive review Holtzer & Abbott (1968*) stated that cell division is essential in the reprogramming of cells, and proposed the hypothesis that determination occurs during particular cell divisions, called 'asymmetrical' or 'quantal' mitoses, which are interspersed with 'symmetrical' or 'proliferative' mitoses. Holtzer & Mayne (1973*) and Holtzer *et al.* (1975*) give further arguments for this hypothesis. However, Lash (1968*a**, *b*) emphasised that the notochord and spinal cord only enhance and accelerate the chondrogenic bias (competence) of somitic mesoderm. Green *et al.* (1968) found that in *Xenopus laevis* collagen synthesis begins during gastrulation and strongly increases during neurulation, while Kosher & Searls (1973) observed that in *Rana pipiens* $^{35}SO_4^-$ incorporation into heparan, heparan sulphate and chondroitin sulphate already starts during cleavage but is strongly enhanced, first in the invaginating chordomesoderm and later also in neural tissue. Extracellular matrix components seem to be the most plausible candidates as agents of chondrogenic induction by notochord and spinal cord, the induction having the character of a non-specific stimulation of pre-existing potentials (Strudel, 1971; Hall, 1977*; see also Kratochwil, 1972*; Hay & Meier, 1974; Lash & Vasan, 1977*; Lash, Ovadia & Vasan, 1978). Although it is very likely that mitosis plays an important role in determinative events by facilitating, or perhaps even making possible certain nucleocytoplasmic interactions after the disintegration of the nuclear membrane, in our opinion there is no need to postulate the existence of a particular mitosis for each determinative step. Determinative events also occur in the absence of mitosis (Cooke, 1973*b*).

From an extensive biochemical analysis Zilliken (1966*) concluded that a multitude of stimuli, working in well-synchronised and concerted fashion, are required for chondrogenesis. Holtzer & Matheson (1970*) confirmed this conclusion and emphasised the significance of enzymes involved in chondroitin sulphate synthesis. Lash, Glick & Madden (1964) had already found that stimulation of chondrogenesis in somitic mesoderm cells is accompanied by the appearance of sulphate-activating enzymes.

A few words must be said about 'spontaneous' cartilage formation in cultured somites. This phenomenon was first mentioned by Avery, Chow & Holtzer (1956). Strudel (1963) reported that chondrification only occurred in cultured somites when taken from 27–30-somite chick embryos. Lash (1968*b*) observed 'spontaneous' cartilage formation in somitic mesenchyme cultured in a nutrient-enriched medium. Kosher (1976) suggests that 'spontaneous' chondrification of somitic mesoderm occurs by a feedback mechanism involving components of the extracellular matrix. He found that the cAMP analogues dibutyryl cAMP and 8-bromo-cAMP severely impair 'spontaneous' chondrogenesis in somites by inhibiting the formation of small amounts of cartilaginous matrix normally formed by embryonic somites *in vitro*. The inhibitory action of cAMP derivatives is mimicked by the cAMP phosphodiesterase inhibitor theophylline. Kosher concluded that extracellular matrix components interact with the cell surface of scleroblasts, in which intracellular cAMP acts as 'second messenger', a high intracellular cAMP concentration being antagonistic to somite chondrogenesis. The interaction of notochord, spinal cord and collagen with somitic mesoderm is also cAMP sensitive.

Grobstein & Holtzer (1955) and G. W. Cooper (1965) reported that the notochord retains its inductive capacity up to $11\frac{1}{2}$ days of incubation, while ventral spinal cord loses it between days $7\frac{1}{2}$ and 9 in the chick and between days 12 and 15 in the mouse embryo. However, Tremaine & Hall (1979) obtained chondrogenic induction by ventral spinal cord irrespective of the age of the donor (up to 18 days of incubation) which, according to them, reflects the continuous character of extracellular matrix synthesis.

In the chick O'Hare (1972) observed heterologous somite chondrification under the influence of 3–4 day embryonic epiblast, which is probably due to the formation at that time of its basal lamina containing sulphated glycosaminoglycans and collagen.

Pugin (1973) noted that cartilage induction is not even class specific, since chick somitic mesoderm also responds to mouse and rat notochord, spinal cord and otic vesicle (see also Leibel, 1976).

Development of other skeletal components. Different regions of the CNS are responsible for the development of certain dermal bones of the cranium in the chick embryo (Schowing, 1968 *a*, *c*), while the notochord influences the development of the cranial floor (Schowing, 1968 *b*). In the absence of notochord and brain the skull fails to develop (Schowing, 1974). The nasal and otic placodes induce the development of the nasal and otic capsules, respectively, while the eye anlage is needed for the normal development of the orbital wall (Corsin, 1972).

Rib formation is a developmental potency of the thoracic, not of the cervical somitic mesoderm (Kieny, Mauger & Sengel, 1972).

Heart development

In the amphibians heart-forming potencies arise in the cranio-ventral mesoderm in separate left and right anlagen, which subsequently fuse into a single medio-ventral rudiment. Preventing fusion of the two anlagen by removal of medio-ventral ectoderm (leading to fistula formation) leads to the formation of two separate hearts, of which the left one starts pulsating 10–14 h earlier than the right one; while the left heart has a normal *situs*, the right one shows *situs inversus*. Vital staining experiments have shown that the left heart anlage contributes more to the ventricle than does the right one (Zwirner & Kuhlo, 1964).

A. G. Jacobson & Duncan (1968) concluded from recombination experiments *in vitro* that heart formation in the salamander is a gradual, cumulative process of inductive and suppressive interactions of the presumptive heart mesoderm with the following nearby embryonic tissues: while the anterior endoderm acts as a specific heart inductor, a general stimulating influence is exerted by the overlying epidermis and an inhibitory influence by the cranial neural fold and plate.

In the newt the capacity to induce a heart is strongest in the anterior endoderm, which retains its inductive capacity until tail bud stages. The presumptive heart mesoderm has the highest competence for heart formation, but the mesoderm lying immediately posterior to it can also be induced to form a heart, although the action requires much more time. The existence and extent of the so-called heart field is the result of the spatial distribution of the heart-inducing capacity of the endoderm (Fullilove, 1970).

Lemansky, Marx & Hill (1977) found that in the *c* mutant of *Ambystoma mexicanum*, which in homozygous condition shows no beating-heart formation, the anterior endoderm is more advanced morphologically and loses its inductive capacity prematurely. Transplantation of mutant presumptive heart mesoderm into a normal embryo leads to normal development (Lemansky, 1978).

Orts-Llorca & Ruano-Gil (1965) concluded from extirpation experiments that in the chick the endoderm underlying the presumptive heart anlagen is indispensable for heart development. Inversion of mesoderm plus endoderm of one or both pre-cardiac anlagen up to stage 7⁻ (H. & H.) leads to regulation of the endoderm and normal heart development. This regulative capacity is rapidly lost at the 2-somite stage, which corresponds to the time when heart mesoderm migration ceases and the pre-cardiac mesoderm adheres to the underlying endoderm (Orts-Llorca & Jimenez Collado, 1967, 1969). While the antero-posterior axis of the mesodermal heart anlage is determined at stage 7⁺, the heart mesoderm isolated together with the underlying endoderm can already form pulsating vesicles at stage 5 (Orts-Llorca & Jimenez Collado, 1970).

Clusters of chick pre-cardiac mesoderm cells migrate actively on anterior endoderm. Migration is random at stages 4–6 (H. & H.) but becomes oriented antero-posteriorly at stage 6⁺, when the so-called cardiogenic crescent is formed mid-ventrally (de Haan, 1964). De Haan suggests that migration becomes directional as a result of the elongation of the endodermal cells, which thus serve as a substrate exerting contact guidance. The pre-cardiac mesodermal cells continuously send out and withdraw filopodia. They adhere maximally to the endoderm in the anterior intestinal portal region. Embryonic heart cells synthesise glycoproteins as part of their basal lamina and extracellular matrix. The pre-cardiac mesoderm cells make contact with glycoproteins newly synthesised by the endoderm, enabling cell recognition and migration (Manasek, 1976).

Blood island formation

We have already seen that in the amphibians the blood islands represent one of the differentiations of the ventral mesoderm, but that they can also be formed from isolated lateral or dorsal mesoderm (T. Yamada, 1940). Ventro-caudal endoderm seems to stimulate blood island formation (Tseng, 1958). Latzis & Saraeva (1978) observed differentiation of blood cells from presumptive pronephros and liver material of late neurula to early tail bud stages of *Rana temporaria* cultured *in vitro* and assumed that they originate from local precursor cells.

Battikh (1971) found that in the chick embryo ecto- plus mesoderm of medium- or definitive-streak and 3-somite stages is capable of forming blood islands in the absence of endoderm. Ecto- plus mesoderm of the area vasculosa of the definitive-streak stage shows much better development of blood islands in millipore transfilter experiments with underlying endoderm. Posterior area opaca in contact with endoderm anterior to the head fold is also capable of forming blood islands, but not if the endoderm is first denatured (Miura & Wilt, 1969).

During later development erythropoeisis occurs in the liver and spleen in birds and, in mammals, in the bone marrow as well. However, this falls outside the scope of this book.

The development of the excretory organs

In the anamnia the excretory function is first performed by the pronephros (early larval stages) and then taken over by the definitive mesonephros. In the amniota the pronephros remains rudimentary and the mesonephros functions as the embryonic excretory organ, and is then replaced by the definitive metanephros. Pro-, meso- and metanephros arise from segmentally arranged nephrogenic mesodermal anlagen.

Pronephros and Wolffian duct

In the amphibians the nephrogenic mesoderm is situated between the dorso-lateral somitic mesoderm and the lateral and ventrol-lateral lateral plate mesoderm, and represents an intermediate level of differentiation of the mesodermal mantle (T. Yamada, 1940). The pronephros develops in the anterior region of the trunk while the mesonephros is formed in the posterior trunk. Here we meet again the problem of how the cranio-caudal and dorso-ventral localisation of a particular organ system is specified.

According to Muchmore (1951) influences from surrounding tissues must be responsible for the localisation of the pronephros; here both the axial organs and the endoderm may play a role.

The most caudal section of the pronephric anlage forms the pronephric or Wolffian duct, which grows out along the ventral edge of the somites all the way down to the cloaca. The cell number and volume of the Wolffian duct increase by proliferation before and during the differentiation of the mesonephros. A fragment of the Wolffian duct anlage can form a tubular structure by autodifferentiation. The outgrowth of the tip of the Wolffian duct is apparently conditioned by the environment. In a foreign environment, e.g. *Rana* Wolffian duct anlage grafted into *Bufo*, the anlage does not elongate (Cambar & Gipouloux, 1971 *a*). Gipouloux & Cambar (1961) demonstrated that the outgrowth of the Wolffian duct depends on particular surface properties of the surrounding tissues, conditions which are apparently not met in the *Rana/Bufo* combination. Steinberg & Poole (1982*) and Poole & Steinberg (1981, 1982) have recently presented experimental evidence for the notion that neither chemotaxis nor contact guidance plays a role in the outgrowth of the pronephric duct in *Ambystoma*, but that a cranio-caudally travelling gradient of flank-mesoderm-cell adhesiveness (which is in register with the wave-like propagation of somite segmentation) guides the Wolffian duct along the ventral edge of the somitic mesoderm towards the cloaca.

Mesonephros

Experimental reduction of the size of the Wolffian duct anlage in *Bufo* leads to developmental retardation or absence of the mesonephros, while experimental enlargement leads to precocious appearance of the mesonephric anlagen (Gipouloux & Cambar, 1974).

Bishop-Calame (1966*) summarises the factors involved in the development of the mesonephros in the chick as follows: the Wolffian duct is the inducer of the mesonephros, which extends along many somites in the caudal half of the trunk. Blockage of the outgrowing duct by insertion of an obstacle prevents the differentiation of the more caudally located mesonephric anlagen, which begin development but then regress. A separated terminal end of the Wolffian duct can 'migrate' independently and fuse with the cloaca. Such an isolated portion of the duct has inductive capacity, which manifests itself when the 'migration' of the duct fragment is blocked by an inserted obstacle, leading to local induction of mesonephric tubules (see also Steinberg & Poole, above).

Etheridge (1968, 1972) demonstrated in *Taricha torosa* that endoderm, early somites and early lateral plate stimulate mesonephric development, whereas premigratory neural crest, neural tube, older somites and older lateral plate suppress it. The majority of these interactions occur before the Wolffian duct makes contact with the mesonephric mesenchyme and are apparently responsible for the 'mesonephric bias' and initial differentiation tendencies of the mesonephric anlagen. The mesonephros is definitively determined at a late tail bud stage.

Although we have seen that *Rana* Wolffian duct cannot grow out in *Bufo* mesoderm, *Bufo* mesonephric anlagen form normal mesonephric excretory units in contact with *Rana dalmatina* or *Discoglossus pictus* Wolffian duct, so that the inductive action is not species-specific (Cambar & Gipouloux, 1970b). From antero-posterior or dorso-ventral inversions of lateral mesoderm in *Bufo bufo* it was concluded that the antero-posterior polarity of the lateral mesoderm is firmly established at the early neurula stage, but the dorso-ventral polarity not yet (Gipouloux & Hakim, 1978). Inversion of both axes of the lateral mesoderm at an early neurula stage results in total agenesis of the urogenital system, comprising both pro- and mesonephros. The following factors are required for urogenital anlage formation: the mesoderm must be competent, and it must receive stimulating influences from the chordomesoderm and the dorso-caudal endoderm (Hakim & Gipouloux, 1978). In our opinion the latter influences are responsible for the appearance of the nephrogenic competence or 'bias'.

During the interaction between the Wolffian duct and the mesonephric anlagen, numerous pseudopodial protrusions are formed by both the duct and the mesonephric cells, leading to close apposition of the two structures. At the interface a squamous material, possibly of mucopolysaccharide nature, and a network of collagen fibrils is present. Neither close cellular contacts nor membrane fusion have been observed (Gipouloux & Delbos, 1977). These authors conclude that the extracellular matrix may play an important role in the induction process.

Croisille, Gumpel-Pinot & Martin (1976) found that also in the chick the mesonephric anlagen already contain a nephrogenic 'imprint' before contact with the Wolffian duct. Both Wolffian duct and ureter 'permit' anlage cells to express their nephrogenic potencies. In the chick the Wolffian duct directly induces anlage cells to differentiate into excretory tubules, while very short collecting segments are secondarily formed by the Wolffian duct. The latter phenomenon is more pronounced in the quail than in the chick (C. Martin, 1976).

Mulnard, Creteur & Verbruggen (1977) showed that in the mouse the lateral mesoderm directly caudal to the mesonephric anlagen, a region which normally never forms tubules, actually has no competence for tubule formation, the metanephros anlage being situated more posteriorly.

Metanephros
Bishop-Calame (1966*) summarises the factors involved in metanephric development in the chick as follows: the young metanephric primordium of the 5 day chick embryo, which is situated in the most caudal trunk region, can initiate tubule formation *in vitro* and as a chorioallantoic graft.

The ureter, which acts as the normal inductor of the metanephros grows out from the cloaca some time after the Wolffian duct has reached it and develops at its tip primary, secondary, tertiary and quaternary branches, while secondary excretory tubules differentiate in the adjacent mesenchyme. Ureter and mesenchyme cannot differentiate separately, but do so after reassociation. Only metanephric and mesonephric mesenchyme allows the typical branching of the ureter. Mesonephric mesenchyme also forms nephric tubules in association with ureter, and metanephric mesenchyme does so in association with Wolffian duct. This brings out the close resemblance of the two systems. However, the inductor, not the competent mesenchyme determines the meso- or metanephric quality of the induction. Croisille *et al.* (1976*) emphasise that the metanephric anlage first induces the ureteric bud to branch into a complex collecting system, after which the tips of the branches reciprocally induce the adjacent anlage cells to differentiate into excretory tubules.

Kratochwil (1972*) states that not only the ureteric bud and the spinal cord but also a variety of other epithelial and mesenchymal tissues can induce metanephric tubule formation in metanephric mesenchyme, indicating the low specificity of the inductive action.

Saxén (1970) found evidence for a beginning differentiation of metanephrogenic mesenchyme prior to induction by the ureteric bud, as in the case of the mesonephros. Ureteric bud and spinal cord fail to induce tubule formation in pulmonary, salivary and gastric mesenchyme (Grobstein, 1962*). Combining induced and uninduced metanephric mesenchyme and using a chromosomal marker, Saxén & Saksela (1971) could demonstrate that no homoiogenetic induction occurs either in transfilter cultures or in direct contact, in contrast to e.g. neural and mesodermal induction. According to Saxén *et al.* (1965*) metanephric tubulogenesis follows the principle of 'selective aggregation' (Steinberg, 1964*) based on random cell motility and differential adhesiveness (see also Saxén, 1980*).

Koch & Grobstein (1963), who studied spinal-cord/metanephric-mesenchyme transfilter interaction in the mouse with ³H-leucine labelling, found correlations between the mass of the inductive tissue and (1) the number of tubules induced, and (2) the maximal distance over which induction could occur. Nordling *et al.* (1971), working on the same system, noticed a lengthening of the minimum induction time by about 12 h when induction took place through two instead of one 'TA' millipore filters of 25 µm thickness and 0.8 µm average pore size, and concluded that diffusion alone can hardly account for the longer transmission time. Wartiovaara *et al.* (1972) observed that cytoplasmic processes from both spinal cord and metanephric mesenchyme extend into Nuclepore filters of 0.1, 0.2 and 0.5 µm pore size and make contact. Saxén & Lehtonen (1977*) found that the minimum induction time is a function of Nuclepore filter thickness as well as pore size, being 18 h through a 25 µm thick filter and 30 h through a 50 μm thick filter and increasing by about 12 h when the pore size decreased from 0.5 to 0.2 μm. A pore size of 0.1 μm is the lower limit for the penetration of cell processes; no penetration and induction occur through filters of 0.05 μm pore size. The authors conclude that there is no need for long-range transmission because the interacting cell surfaces are very close. This closeness is essential for the transmission of the inductive signal(s). They propose that inductive interaction either takes place by interaction of complementary surface molecules or by the exchange of small signal molecules through specialised membrane junctions. L. Weiss & Nir (1979) suggest short-range diffusion as a possible mechanism in addition to the long-range diffusion-mediated and contact-mediated induction mechanisms already mentioned. In our opinion the dual origin of the glomerular basal lamina in chick/mouse chimaera's pleads in favour of a matrix–matrix type of interaction (Sariola *et al.*, 1983).

Transfilter experiments of varying duration involving metanephric mesenchyme and spinal cord show a stepwise induction of first proximal tubules, then glomerular epithelium, and finally distal tubules (Lehtonen *et al.*, 1983).

Gonadal development

For primordial germ cell (PGC) origin and gonad development in the chordates the reader is referred to Nieuwkoop & Sutasurya (1979*) and McLaren & Wylie (1983*). We may add that Eyal-Giladi, Ginsburg & Farbarov (1981), using quail/chick chimaeras of epi- and hypoblast, recently found that in birds the PGCs originate from the epiblast and not from the hypoblast, as previously assumed.

Evaluation

The overall spatial arrangement of the mesodermal organs seems to be based on a field-like action extending from the notochordal anlage, which is locally intensified or inhibited by surrounding tissues to yield particular organs.

The development of the mesodermal organs is the consequence of multiple, sequential inductive interactions resulting from transient or permanent contact with surrounding tissues. Among other things, these interactions lead to changes in the cell surface properties of the cells concerned, such as changes in cellular affinity and cell polarity, as a result of which new cellular arrangements emerge: examples are notochordal column formation, somitogenesis and the formation of tubular structures (heart and kidney).

In the majority of interactions extracellular matrix components seem to play an important role. The interactions may have the character of a positive feedback mechanism at the matrix level, e.g. in chondrogenesis, or may represent cell-surface/matrix interactions. Direct cell-to-cell interactions may also be involved, e.g. in nephrogenic tubulogenesis. Morphogenesis and differentiation of the various mesodermal organ anlagen in our opinion depend on transient or permanent confrontation (or both) of different cell populations engaged in divergent differentiation pathways.

18

Epitheliomesenchymal interactions during early organogenesis

Saxén & Toivonen (1962*) state that both the ectoderm and the endoderm acquire their regional determination by inductive actions emanating from the mesoderm during gastrulation and neurulation. Although in general this is correct, we think the interrelations between the three germ layers are more complex. It seems likely that a weak gradient in ectodermal differentiation tendencies and neural competence exists in the amphibian blastula, declining from the animal pole towards the equator, as the counterpart of the observed gradient in mesodermal competence, which declines in the opposite direction (Sutasurya & Nieuwkoop, 1974). In the triple-layered embryo, the former gradient gives rise to a weak cranio-caudal polarity in the ectoderm. During mesoderm induction a partial endodermisation of the animal, ectodermal moiety of the blastula occurs. The induced endoderm constitutes the most anterior portion and the most dorsal edge of the future gut (Nieuwkoop & Ubbels, 1972; Koebke, 1977; see Chapter 10, p. 96). The anterior portion shows rather strong differentiation tendencies for pharynx formation (Balinsky, 1948). The presence of this antero-dorsal endoderm results in a cranio-caudal and dorso-ventral polarity in the endoderm as a whole. Thus the mesodermal layer, which shows very pronounced cranio-caudal and dorso-ventral polarities in its regional differentiation, interacts with a weakly polarised ectodermal and a more strongly polarised endodermal layer. This inter-action is responsible for the regional organisation of the triple-layered amphibian embryo. It is likely that similar relationships hold for the avian embryo (Azar & Eyal-Giladi, 1981).

The importance of ecto-mesodermal interaction is made clear by the phenomenon of exogastrulation (see Chapter 11, p. 136). in the total exogastrula, where the ectoderm segregates completely from the endo- and mesoderm, the ectoderm only forms a mass of atypical ectodermal cells which subsequently degenerate, while the endo- and mesoderm develop into an almost complete mesodermal axis system enveloped by regionally differentiated endoderm (Holtfreter, 1933b). In the exogastrula endo-mesodermal interactions take place normally but ecto-mesodermal interactions are entirely absent, so that no ecto-mesenchymal organ systems develop.

For more detailed information on epitheliomesenchymal interactions the reader is referred to the reviews by Grobstein (1967*b**); Saxén & Kohonen (1968*); Wolff (1968*); Wessels (1970*a**, 1973*); Kratochwil (1972*); Saxén & Wartiovaara (1976*); Saxén *et al.* (1976*); Hay (1977*); and Saxén (1980*b**).

In this chapter we shall first discuss the *ecto-mesenchymal* and then the *endo-mesenchymal* interactions.

The ecto-mesenchymal interactions

A large number of special structures – such as light organs in fishes; adhesive glands and balancers in larval amphibia; scales and claws in reptiles and birds; feathers in birds; and hairs, nails and mammary glands in mammals – are formed as a result of ecto-mesenchymal interactions. These are also responsible for mouth formation in all the vertebrates and fin and limb development in fish and tetrapods respectively. A number of these structures will be treated separately in this chapter (see the reviews by Koecke, 1964*; and Kollar, 1972*).

The development of the tetrapod limb

Limb development has mainly been studied in the amphibians, but since the 1950s the emphasis has shifted to birds. The conclusions reached from avian development also seem to hold for the amphibians (see J. M. W. Slack, 1977*a*, *b*).

In this book we can only discuss the initial steps in limb development. The reader is referred to the following reviews for further information: Amprino (1965*); Saunders & Gasseling (1968*); Zwilling (1968*); Pinot (1970*); Saunders (1972*) and J. M. W. Slack (1979*).

The limbs develop from the somatopleural mesoderm of the flank. Lateral plate mesoderm experimentally covered on both sides by ectoderm develops double somatopleure layers and forms limbs in mirror-image configuration. Flank mesoderm covered by endoderm on both sides forms double splanchnopleure layers (Nieuwkoop, 1946). Whereas during normal development cell division slows down in the flank region between the fore and hind limbs, the presumptive limb bud mesoderm continues to proliferate (see Saunders, 1972*; Slack, 1979*).

The vertebrate limb is an asymmetrical structure with antero-posterior, dorso-ventral and proximo-distal polarities. The limb bud grows out in proximo-distal direction. The proximal parts are laid down first and the successively more distal parts appear later. Kieny (1967) drew up a

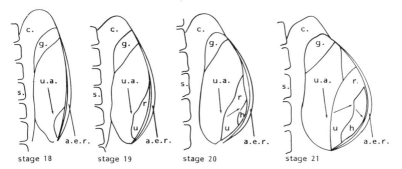

Fig. 33. Fate map of the right wing bud of the chick embryo at stages 18–21 (H. & H.). a.e.r., Apical ectodermal ridge; c., presumptive coracoid; g., presumptive glenoid region; h., presumptive hand; r., presumptive radial region of forearm; s., somites; u., presumptive ulnar region of forearm; u.a., presumptive upper arm. Arrows indicate future proximo-distal axis (after J. W. Saunders, 1972).

presumptive anlage map for the chick limb bud and Stark & Searls (1974) for the chick wing bud (see Fig. 33).

Balinsky (1933) and Rudnick (1945, 1948) showed in earlier investigations on amphibians (e.g. using implantation of otic vesicles to induce supernumerary limbs) that in the early tail bud stage the entire flank has limb-forming potencies, with separate centres of higher activity in the presumptive fore and hind limb regions. Long before any morphological signs of limb formation manifest themselves the limb-forming potencies have become restricted to fore and hind limb areas, which are however still considerably larger than the actual limb rudiments.

In the chick the territories for wing and leg formation are precociously determined in the somatopleural mesoderm, the interjacent flank region between somites 21 and 26 having no limb-forming potencies. The axial somitic mesoderm seems to exert a stimulating influence on limb development at the wing and leg levels (Pinot, 1970a, b*; Kieny, 1971). The origin of the tetrapod body pattern and the more variable pattern of appendages in fish are among the basic unsolved problems in vertebrate development.

Amprino (1964, 1975a, b, 1976, 1977, 1978), Amprino & Camosso (1966) and Camosso & Roncali (1968) emphasise the self-organising capacity of the mesoderm in the segregation, determination, growth and differentiation of the limb bud and attribute little or no role to the epidermis except that of forming the outer hull for the developing mesoderm. However, the great majority of investigators consider the limb anlage to be a typical epitheliomesenchymal interaction system (see Pinot, 1970b*; Saunders, 1972*).

After the establishment of the limb-forming territories the first inductive action in limb development emanates from the presumptive limb bud mesoderm, inducing a roughly antero-posteriorly oriented thickening in the overlying epidermis, the so-called apical ectodermal ridge (AER). Kieny (1960) was the first to show that presumptive limb bud mesoderm grafted to the flank can induce an AER in the overlying epidermis, resulting in normal limb outgrowth. Saunders & Reuss (1974) later found that AER induction occurs only in wound epidermis, not in epidermis with an intact basement membrane.

The AER shows increased mitotic activity leading to the formation of a multilayered structure in the essentially double-layered flank epidermis (see Saunders, 1972*). It is moreover characterised by the presence of many gap junctions between the cells, which are present as long as the AER remains functional (Kelley & Fallon, 1976). The numerous gap junctions apparently guarantee the electrical and metabolic coupling between the AER cells essential for the integral function of the AER during limb morphogenesis (Fallon & Kelley, 1977).

The capacity of the presumptive limb bud mesoderm to induce a supernumerary wing first appears at the 13-somite stage, that for supernumerary-leg induction at the 19-somite stage (Kieny, 1971). Prior to these stages the presumptive limb mesoderm can only express limb potentialities (wing or leg) in the presence of adjacent somitic mesoderm (Kieny, 1971). The competence for AER induction is weakest in ventral epidermis, intermediate in axial and paraxial epidermis, and strongest in flank epidermis (Kieny & Brugal, 1977). Dhouailly & Kieny (1972) observed that non-limb cells of the flank mesoderm participate in the outgrowth and organogenesis of induced supernumerary limbs. The competence of flank epidermis for AER formation under the influence of presumptive wing mesoderm, as well as the inducing capacity of the mesoderm, is lost at stage 17 (H. & H.) (Reuss & Saunders, 1965). The AER remains functional until the basic limb pattern has been laid down and growth has ceased. The initially continuous AER splits into separate digital plates during digit formation (Pautou, 1978; Jacob, Jacob & Christ, 1981). Zwilling (1956a, b, c, 1961*) and Zwilling & Hansborough (1956) observed that removal of the AER leads to cessation of further proximo-distal outgrowth of the limb, an observation that was disputed by Amprino (1965*). Kaprio & Tähkä (1978) concluded from AER removal experiments that the cessation of distal outgrowth of the wing bud is not due to mesenchymal cell death, as suggested by Amprino, but to the absence of the AER as such.

The main function of the AER is to maintain mesenchymal cells adjacent to it in a labile, undifferentiated state. As soon as cells are freed from the influence of the AER they begin to differentiate (Saunders, Gasseling & Errick, 1976; Kosher, Savage & Chan, 1979). In contrast, Amprino (1978) ascribes this role to the apical superficial blood vessel network.

Kosher *et al.* (1979) found that agents that raise the intracellular cAMP level stimulate chondrogenesis in limb bud mesoderm, which seems contrary to the situation in sclerotome chondrification (see p. 241).

Zwilling (1961*) postulated that a mesodermal maintenance factor is responsible for the persistence and continued activity of the AER. This maintenance factor can pass through a millipore filter of 0.45 μm average pore size and 25 μm thickness. Gumpel-Pinot (1980) found a correlation between the mass and density of epidermal cell processes traversing the millipore filter and the degree of cartilage induction in the underlying wing mesoderm.

Murillo-Ferrol (1965) and Reuss & Saunders (1965) hold conflicting opinions on the possible identity of the mesodermal factor involved in AER induction with the maintenance factor. Rubin & Saunders (1972) observed that any functional AER induces normal proximo-distal outgrowth of presumptive limb bud mesoderm regardless of the age of both epidermis and mesoderm, from which they concluded that the inductive signal remains qualitatively unchanged during limb development. The specific patterning of the limb anlage must therefore be programmed in the mesoderm. Whereas the AER cells of the limb bud of *Lacerta viridis* look very active, those of *Anguis fragilis*, which has transient limb anlagen, show only a moderate activity (Raynaud & Brachet, 1979). The antero-posterior polarity of the chick limb bud, which is expressed in a markedly thicker AER in the posterior region, already seems to be determined before $3\frac{1}{2}$ days, while the dorso-ventral polarity is only determined around $5\frac{1}{2}$ days of incubation (Pautou, 1977*).

Saunders & Gasseling (1968*) found that a posterior mesodermal region of the limb bud showing programmed cell death, later called the 'zone of polarising activity' (ZPA), is responsible for the antero-posterior polarisation of the limb bud. This was confirmed by, among others, MacCabe & Parker (1975). MacCabe & Parker (1976) noticed a postero-anterior gradient in ZPA activity. The ZPA region is initially diffuse but becomes more concentrated during early limb development (Summerbell & Honig, 1982). Insertion of an impermeable barrier between the anterior and posterior parts of an early limb bud restricts polarisation to the posterior part. Grafting of a ZPA into the anterior region of a limb bud leads to an anteriorly thickened AER and to subsequent limb duplication (Summerbell, 1979). The polarising activity seems to act only on the mesoderm, which in turn influences the epidermis (MacCabe & Parker, 1979). It is particularly sensitive to inhibitors of RNA synthesis (Summerbell & Honig, 1982). J. M. W. Slack (1977a) concluded from grafting experiments in the axolotl that an interaction between the flank and the limb region is responsible for the antero-posterior polarisation of the limb bud. In his 1979 article he points out that ZPA activity has been demonstrated in mammals, reptiles and amphibians as well as in birds, and therefore is a general feature of tetrapods.

Recombinates of mesodermal cores and epidermal hulls in reversed antero-posterior or dorso-ventral orientation, or both, have shown that the ectoderm plays a significant role in the dorso-ventral organisation of the limb (MacCabe, Errick & Saunders, 1974). A geometric study of experimentally modified limb development led Steding (1967) to suggest that the embryonic epidermis plays an active part in limb development, AER formation being the result rather than the cause of limb bud morphogenesis. Vardy, Stokes & McBride (1982) collected evidence for an important role of the CNS in normal limb development. MacCabe *et al.* (1975) found that genetic lesions primarily affect the mesoderm and only secondarily the epidermis of the limb bud. Recently Tomasck, Mazurkiewicz & Newman (1982) described a non-uniform distribution of fibronectin in chick limb development.

Using quail/chick chimaeras of somitic mesoderm Chevallier, Kieny & Mauger (1976) and Wachtler, Christ & Jacob (1982) could demonstrate that the musculature of the leg and wing bud is of somitic origin, the precursor cells migrating into the growing bud.

The development of the skin

Using vital staining von Woellwarth (1960) established a map of presumptive body surface regions in the neurula of *Triturus alpestris* (see Fig. 34).

The ectoderm of *Rana pipiens* begins to form cilia at the open neural-plate stage (stage 13 of Shumway (1940, 1942)). The number of cilia-bearing cells increases markedly during neurulation, while a further growth in length of the cilia occurs at early tail bud stages. Cilia are still present during hatching but begin to regress at stage 24. Some of the non-ciliated cells release mucus granules from the neurula stage till stage 25. Cilia are not formed in the neural plate, neural folds or in the cement gland epithelium, which consists of secretory cells (Kessel, Beams & Shih, 1974). About one out of three flank epidermal cells of *Ambystoma mexicanum* is ciliated. Adjacent cells are never ciliated, which implies that cilium formation is prevented in cells that are in direct contact with a cilium-bearing cell (Grunz, 1975*, 1976*; Grunz *et al.*, 1975; Landström, 1977*b*; Løvtrup, 1983*). Haploid embryos of *Bombina orientalis* show a different ratio of ciliated to non-ciliated cells and a different epidermal surface texture than diploid embryos (Ellinger & Murphy, 1979*a, b*). Cilium formation is favoured by hypotonic media and suppressed by hypertonicity. Epidermal differentiation depends on RNA and glucosaminoglycan synthesis (Landström, 1977*b*).

Vertebrate skin is composed of dermis and epidermis. The dermis consists of connective tissue cells with a large amount of intercellular matrix material formed by their own secretory activity. Bell (1960)

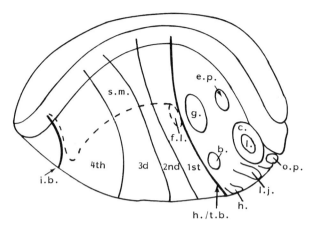

Fig. 34. Fate map of presumptive epidermal areas at the open neural plate stage of *Triturus alpestris*. h./t.b., head/trunk boundary; i.b., ultimate boundary of invagination; s.m., boundary of epidermis overlying the somitic mesoderm; 1st–4th, first to fourth quarter of trunk epidermis; b., balancer; c., cornea; e.p., ear placode; f.l., forelimb epidermis; g., gill region; h., heart region; l., lens; l.j., lower jaw region; o.p., olfactory placode (after C. von Woellwarth, 1960).

observed that in *Rana pipiens* embryos the ectoderm forms an outer surface coat and an inner extracellular matrix, both of mucopolysaccharide nature. The presence of collagen fibres plays an important role in the structure of the dermis. In later development corium, which develops from the dermis, acquires a capillary network and nerve supply and can form smooth and striated muscle fibres. The epidermis consists of a basal layer or 'stratum germinativum' of pluripotent cells and several layers of differentiating cells.

Morphogenesis of the skin and its derivatives requires a complex series of interactions between the epidermis and dermis during the entire embryonic period as well as during adult life (Wessels, 1973*). The skin shows a three-dimensional regional-specific pattern. Its antero-posterior axis is determined first, the dorso-ventral axis slightly later. Sengel (1958*, 1964*); Rawles (1963) and Reyss-Brion (1973) emphasise that, in the chick, dermis/epidermis interaction depends upon the age of both components. Depending on the stage of development both the epidermis and the dermis alternatingly act as inductor and as competent reacting system (see also Kollar, 1972*).

Sengel (1958*) showed that the dermis causes the ectoderm to form a typical multilayered epidermis. The dermis is also responsible for the maintenance of the basal, germinative layer of the epidermis (MacLoughlin, 1961*a, b, c*; Wessels, 1962, 1964).

In the chick keratinisation of epidermal cells can be suppressed by excess vitamin A, leading to their transformation into mucus-producing cells (now called 'transdifferentiation') (Fell & Mellanby, 1953). Reyss-Brion (1976) found that the dermis is particularly sensitive to actinomycin D treatment.

Moscona (1959) observed that chorionic epithelium, the outer, ecto-dermal component of the chorioallantoic membrane, keratinises upon open exposure to air. Since exposure to air under a sealed window prevents keratinisation he considers the phenomenon as dependent upon changes in the oxygen/carbon-dioxide ratio. Chorionic epithelium is not only capable of keratogenic but also of secretory and endothelial metaplasia (Moscona, 1960), which demonstrates its still embryonic, highly pluripotent state.

Mouth formation

The stomodaeal endoderm as part of the foregut induces the overlying ectoderm to form the oral invagination by direct contact between the two layers. Extirpation of the anterior archenteron at the mid-gastrula stage prevents stomodaeum formation. When gastrulation is completed the competence for mouth formation is already restricted to the presumptive mouth area (Balinsky, 1948).

Mouth formation takes place in two successive phases: formation of the mouth anlage and opening of the mouth. From studies *in vitro* and transplantations in *Discoglossus* and *Pleurodeles*, respectively, Cusimano-Carollo (1972) and Cassin & Capuron (1979) concluded that complete mouth formation only occurs in the stomodaeal ecto- and mesoderm in the presence of the neural folds, prechordal mesoderm and lateral cephalic mesoderm. Opening of the mouth requires the interaction of the stomo-daeum with the lateral cephalic mesoderm, which is of neural crest origin. In the urodeles the formation of buccal structures in the form of bones, cartilages and teeth depends on the interactions of the ectodermal stomo-daeum with neural crest derivatives, which furnish predetermined chon-droblasts and odontoblasts. In the anurans the supra- and infra-rostral cartilages induce the formation of the beak and the horny teeth (Cusimano-Carollo, 1972).

Tooth formation

In the vertebrates tooth formation occurs chiefly along the maxillary and mandibular cartilages and bones of the visceral skeleton, where they border the epithelium of the ectodermal oral invagination (Moss, 1969). The individual tooth anlage consists of an evagination of the ectodermal epithelium into the mesectodermal mesenchyme surrounding the cartilage anlage. The tooth anlage becomes cone-shaped by reinvagination of the epithelium under the influence of condensing mesenchyme forming the tooth papilla. The epithelial–mesenchymal interaction inside the cone-shaped tooth anlage gives rise to odontoblast formation in the mesenchymal papilla and ameloblast formation in the epithelium, leading to the production of the dentine and enamel of the tooth, respectively.

Koch (1967) demonstrated that the enzymatically separated epithelium and mesenchymal papilla of the 16 day mouse mandibular incisor fail to differentiate; when recombined they form ameloblasts and odontoblasts and their respective extracellular matrices. In transfilter culture both components also show overt differentiation and form their characteristic extracellular matrices on either side of the filter.

Slavkin, Beierle & Bavetta (1968) emphasised that the morphogenetic potential of the tooth anlage is closely related to the properties of both interacting cell types. The epithelial ameloblasts differentiate earlier than do the mesenchymal odontoblasts.

Using Nuclepore filters of different porosity Thesleff *et al.* (1977) found that direct contact between cellular protrusions from both epithelial and mesenchymal cells occurs inside filters of $0.2 \mu m$ pore size and larger; $0.1 \mu m$ pores prevent tooth development by blocking the penetration and thus precluding the differentiation of mesenchymal odontoblasts, which requires close association of mesenchyme and epithelium (see also Saxén, 1980 b*).

Kallenbach & Piesco (1979*) observed that during the early phase of epithelial–mesenchymal interaction in tooth development an extensive and close contact exists between preodontoblasts and the basal lamina of the epithelium, which ends when increasing amounts of extracellular matrix material are deposited between the odontoblasts and the basal lamina. The next phase is characterised either by a renewed direct contact between odontoblasts and preameloblasts (e.g. in *Hyla*), or by the extension of preameloblast processes through the basal lamina into the predentine space (in various vertebrate groups), after which the ameloblasts differentiate further.

The dental epithelium is characterised by the presence of hyaluronate and of minute quantities of heparan sulphate, while the dental papilla contains chondroitin sulphate. Large quantities of hyaluronate and sulphated glycosaminoglycans accumulate at the epitheliomesenchymal interface (Lau & Ruch, 1983).

Hata & Slavkin (1978) showed that stage 22–3 avian limb bud epithelium can induce 17 day mouse tooth mesenchyme to form cartilage, which produces type-II collagen in addition to type-I collagen, the former being typical for odontoblast differentiation. Diazo-oxo-norleucine (DON) inhibits odontoblast differentiation in the mouse by affecting the cell–matrix interaction (Hurmerinta & Thesleff, 1982).

Recombinates *in vitro* of epithelium of the first and second pharyngeal arches of the 5 day chick embryo with first mandibular molar mesenchyme of 16–18 day mouse embryos showed invasions of the chick epithelium into the mouse mesenchyme, leading to more or less typical tooth formation with molar-like dentine (Kollar & Fischer, 1980.) They concluded that the loss of teeth in birds is not due to a loss of genetic coding for enamel synthesis in the oral epithelium but to an alteration in the tissue interaction required for odontoblast formation.

The development of cutaneous appendages

The development of scales, feathers and hairs shows much resemblance. Although hairs and feathers or scales are not homologous we feel they can be discussed together. For more detailed information the reader is referred to the reviews by Sengel (1964*, 1975*) and Dhouailly (1977*).

Feather germ and hair formation

While Sengel (1964*) recognised four successive stages in feather germ formation, Dhouailly (1977*) already distinguishes six successive steps in the dermis/epidermis interaction: (1) the dermis causes the ectoderm to differentiate into epidermis with a germinative basal layer; (2) the epidermis exerts a global morphogenetic action upon the underlying dermis, enabling it to participate in appendage morphogenesis; (3) the dermis initiates the formation of epidermal placodes in precise locations and sizes; (4) under each epidermal placode the dermis forms a dermal condensation; (5) the dermis causes the transformation of the epidermal placode into an appendage rudiment (the special quality of scale, feather or hair being determined by the specific origin of the epidermis); and (6) a transfer of specific dermal information leads to the proximo-distal outgrowth of the appendage.

Normal feather formation is disturbed in at least three instances: (1) in normally featherless skin, e.g. mid-ventral skin, the dermis fails to initiate epidermal placode formation (Sengel, 1964*; Sengel, Dhouailly & Kieny, 1969); (2) in the 'scaleless' mutant the epidermis fails to exert a morphogenetic action upon the underlying dermis (Sengel, 1964*; Sawyer, 1975), and (3) in chick-epidermis/lizard-dermis combinates the dermis is unable to form dermal condensations (Dhouailly, 1977*).

The regional quality of the cutaneous appendages, taking the form of different types of feathers or scales, is determined by the dermis (Sengel, 1964*). Chick dermis can cooperate with mouse epidermis in hair formation, and mouse dermis with chick epidermis in feather formation. Whereas the specification of a hair or feather (and probably also of a reptilian scale) is under genetic control of the epidermis (Reyss-Brion, 1973), the hexagonal feather and linear scale patterns in birds are under dermal control (Reyss-Brion, 1974). Feather formation predominates over scale formation in the chick, but less so in the duck (Sengel & Pautou, 1969). Retinoic acid favours feather development in scale-forming skin (Dhouailly & Hardy,

1978). Heterotopic and heterospecific dermis/epidermis recombinates demonstrate that the dermis not only induces the formation of cutaneous appendages in conformity with its regional character, but also triggers off the corresponding biosynthesis of keratins (Dhouailly, Rogers & Sengel, 1978). While both 6 day chick beak epidermis and 6 day cephalic skin epidermis combined with 6 day upper beak mesenchyme show normal beak histogenesis *in vitro*, 6 day beak epidermis combined with 7 day back skin dermis forms feathers. The inductive capacity of the beak mesenchyme persists until after hatching (Tonégawa, 1973).

Dermal/epidermal interaction in skin formation is not interfered with by the insertion of a millipore filter (Wessels, 1962), almost certainly because the filter allows the passage of cellular processes (compare tooth formation, p. 256). Dhouailly & Sengel (1975) demonstrated that cellular contact is necessary in dermis/epidermis interaction. Close contacts had actually been observed between dermal cells and the epidermal basal lamina in $6-7\frac{1}{2}$ day chick embryos by Sengel & Rusaouën (1969). We may therefore conclude that cell–matrix (and perhaps cell–cell) interactions play an important role in dermis/epidermis inductive interaction.

Feather pattern formation
Sengel (1964*, 1975*) describes feather pattern formation in the dorsal spinal feather tract of the chick as follows: in the $6\frac{1}{2}$ day embryo the first feather anlage appears mid-dorsally in the lumbar region as soon as the longitudinal band of dense dermis has acquired sufficient width. The next anlagen are formed in front and behind the first one. As progressively more dense dermis becomes available, anlagen are laid down laterally to the mid-dorsal row. This continues till all the dense dermis is used up. The individual anlagen are formed as closely as possible to the previous one(s), thus establishing a hexagonal pattern. In the outgrowing feather germ the epidermis forms the epidermal sheath and the dermis the inner pulpa.

Using quail/chick chimaeras, Mauger (1972*a*) demonstrated that the dermis of the dorsal skin originates exclusively from the somitic dermatomes. The dermal mesenchyme spreads over an area extending from the dorsal midline to halfway down the flank. The lower flank dermis derives from the somatopleural mesoderm. Although the latter normally forms part of the feather-forming skin, it is unable to replace the somitic mesoderm in the formation of the dorsal spinal feather tract. While inversion of the anteroposterior axis of the somitic mesoderm has little effect on the dorsal plumage pattern, medio-lateral inversion leads to the appearance of a featherless notch. The medio-lateral polarity of the feather-forming somitic mesoderm is irreversibly determined at the time of somite segmentation. The pattern specificity along the cephalo-caudal axis, in the form of differences in width of the row pattern, is determined by the somitic mesoderm prior to segmentation; the latter in its turn is dependent on the neural tube (Sengel, 1972; Mauger, 1972*b*, *c*; Linsenmayer, 1972*a*).

Sengel (1958*) demonstrated the essential role of the dermis in the formation of the cutaneous appendages by using heterochronic and heterotopic recombinates of dermis and epidermis. Dhouailly (1967, 1970) confirmed these conclusions using heterospecific recombinates. In reversed dermis/epidermis recombinates Linsenmayer (1972*a*) observed that the temporal pattern of feather primordium formation always conforms to the dermal orientation. Linsenmayer (1972*b*) suggests that the position of newly forming feather primordia is established by some influence of the adjacent 'older' ones. Sengel (1964*) noted that there is a close relationship between the transverse pattern of the feather germs and the metamerism of the axial organs.

Explanted dorsal skin shows a stable morphogenetic pattern when the mid-dorsal row of anlagen is preserved. After dis- and reassociation, dorsal skin forms a new 'primary' mid-dorsal row (Novel, 1973). Novel concluded that a wave of morphogenetic activity moves from the dorsal midline to the lateral edges of the spinal feather tract. D. Davidson (1983*a*, *b*) describes a new method of skin culture on a substratum of hydrated collagen. Stretching of the skin just lateral to the most recently formed row of feather anlagen leads to an increased number of primordia. Since the wave of feather primordia formation passes undisturbed across a cut made before the pattern is formed, Davidson comes to the conclusion that no passage of signals is involved but that the sequential formation of feather anlagen is due to the temporal pattern of skin differentiation, which is established before stage 29. He draws an interesting parallel with the temporal control of somitogenesis in the amphibian embryo, as proposed by Pearson & Elsdale (1979; see p. 232).

Skin glands

The skin glands can be subdivided into purely epidermal glands and epitheliomesenchymal glands. The former are represented by the hatching and cement glands in fish and larval amphibia and the latter by the avian preen gland and the mammalian mammary gland.

The anuran *cement gland* develops from cells of the superficial layer of the ectoderm in the region ventral to the mouth anlage. When late blastula or early gastrula ectoderm of *Xenopus laevis* is subjected to Holtfreter solution containing ammonium chloride, sodiumthiocyanide or urea and subsequently cultured in Barth's solution, it chiefly differentiates into cement gland tissue. Under optimal conditions 80–90% of the superficial cells of the explanted ectoderm form cement gland cells, which are typical mucus-producing cells (Picard, 1975*a*; Grossi & Fascio, 1979). The cells of the animal pole region have the highest competence at the early gastrula stage (stage 10, N. & F.; see ectodermal polarity, p. 249). There is no difference in response between the dorsal and the ventral blastocoelic wall. Endo- and mesoderm cells cannot be transformed into cement gland cells, which therefore represent a typical epidermal differentiation (Picard, 1975*b*).

Hatching gland cells begin to differentiate in the dorsal frontal region of *Rana chensinensis* embryos at stage 17 and acquire their histotypic characteristics at stage 18 (Normal Table of *Rana pipiens*, Taylor & Kollros, 1946), while enzyme synthesis and secretion continue during stages 18–21, after which a gradual regression sets in (Yoshizaki & Katagiri, 1975). The hatching gland cells derive from the neural crest (Yoshizaki, 1979 in *Rana japonica*). Actinomycin D treatment of embryos of stage 13b (N.T. of *Rana japonica*, Tahara, 1959, 1975) inhibits the differentiation of both hatching gland cells and cilia-bearing cells, both cell types being transformed into common epidermal cells. DNA-dependent RNA synthesis is apparently required for the formation of secretory granules and cilia (Yoshizaki, 1976). While superficial cells of the presumptive ectoderm of *Rana japonica* form cement gland cells, cilia-bearing cells and common epidermal cells in standard salt solution, ectoderm treated with lithium chloride forms hatching gland cells and pigment cells, the former with jelly-digesting activity (Yoshizaki, 1979).

In *Salmo gairdneri* hatching gland cells are distributed over the entire head region and are not innervated; the secretion mechanism seems to be regulated by an external factor, possibly oxygen concentration (Hagenmaier, 1974). Denucé (1976) characterised the hatching enzyme in *Gobius jozo*, while Schoots *et al.* (1982) made an electron microscopic analysis of the enzymatic breakdown of the egg envelope in various teleosts.

The formation of the *preen gland* in birds depends on an interaction between dermis and epidermis. The caudal dermis first exerts an inductive action on the overlying epidermis, which thickens and partially invaginates (Gomot, 1961*). This phase is comparable with the initiation of feather formation. Undifferentiated and early differentiating foreign epidermis can also be induced by caudal dermis. The subsequent phase of differentiation, in which the epidermis forms ectodermal cords, requires an interaction between the dermis and the induced epidermis. According to Gomot (1961*) the latter phase resembles submandibular gland formation in the mouse.

In the $12\frac{2}{3}$ day rabbit embryo, a zone of thickened epidermis is present bilaterally between the lateral edges of the somites and the flank. At the beginning of the 13th day it differentiates into the *mammary ridge* extending between the fore and hind limbs. At the end of the 13th and the beginning of the 14th day its anterior part forms the pectoral *mammary gland*, while the posterior part forms three or four more primary buds. The thickened epithelium of the primary buds rapidly sinks into the underlying mesenchyme. The buds grow and lengthen and give rise to the primary galactophoric canals (Propper, 1970). The primary ridge, which has a lower mitotic rate than ordinary epidermis (Balinsky, 1950), is probably formed by epidermal cell migration (Propper, 1970). A mammary ridge can be induced in neutral epidermis by mammary mesenchyme. A mammary ridge once formed is capable of segmentation in the absence of mesenchyme. Primary buds can also sink into morphogenetically neutral mesenchyme. After invagination the epithelial bud acts upon the mesenchyme, inducing it to form a mesenchymal envelope. The mesenchyme finally determines the morphology of the epidermal bud, as in tooth formation (Propper, 1970). Mammary gland epithelium shows a persistent responsiveness to mammary as well as salivary gland mesenchyme (Sakakura, Sakagami & Nishizuka, 1979). Mammary mesenchyme can also induce mammary bud formation in chick or duck epidermis, resulting in typical epithelial ramification and glycogen synthesis (Propper & Gomot, 1973; Propper, 1975).

The endo-mesenchymal interactions

The development of the alimentary canal

In the urodeles the alimentary canal is formed by dorsal closure of the endodermal portion of the archenteron, while in the anurans the entire invaginating archenteron is endodermal from the very beginning (see p. 126). The definitive regional pattern of the endodermal organs is due to the interaction between the yolk mass and the polarised induced endoderm on the one hand, and the cranio-caudally and dorso-ventrally strongly regionalised mesoderm on the other (see p. 249).

Nakamura (1961) established a fate map of the endodermal organ anlagen at the early gastrula stage in *Rana nigromaculata*, and described their localisations in the archenteron at the early neurula, tail bud and early larval stages (see Fig. 35).

T. M. Harris (1964) followed the fate of the so-called polar endoderm cells lying at the base of the vegetal yolk column in *Ambystoma maculatum*. They are translocated to the middle of the archenteric floor during gastrulation. During neurulation they invaginate into the yolk column and migrate posteriorly to the region of the presumptive cloaca. As a consequence a distinct fault is formed in the surface of the archenteric floor, leading to the subsequent complete occlusion of the archenteron. The definitive intestinal canal is formed *de novo* through the centre of the yolk column along the path which the polar endoderm cells took previously. Anteriorly the new cavity communicates with the persisting foregut at the original 'fault' (see Fig. 35).

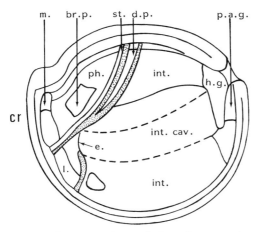

Fig. 35. Fate map of presumptive endodermal areas of *Rana nigromaculata* projected onto a sagittal section of the neurula. br.p., branchial pouches; cr., cranial side; d.p., dorsal pancreas; duodenum (stippled); e., entrance to intestinal cavity (int.cav.); h.g., hind gut; int., small intestine; l., liver; m., mouth; p.a.g., post-anal gut; ph., pharynx; st., stomach (after O. Nakamura, 1961).

Albert (1978*a*, *b*) distinguished five successive phases in the morphogenesis of the alimentary canal in the anuran *Rana dalmatina*: the formation of the endodermal archenteron; a 5 day period of organogenetic nonactivity; the sinking-in of two intraendodermal grooves; the appearance of three morphogenetically active foci; and finally the appearance of small undifferentiated basal cells in the alimentary mucosa. In the absence of the chordomesoderm the digestive system shows no morphogenesis when isolated at stage 17 (late neurula) but normal morphogenesis when isolated after stage 21. Continuation and completion of organogenesis of the alimentary canal requires contact of the narrow, dorsal endodermal cell strip with the chordomesoderm between stages 17 and 21 (N.T. of *Rana dalmatina*, Cambar & Marrot, 1954). Removal of anterior, middle or posterior thirds or of the anterior or posterior two-thirds of the endodermal mass – which chiefly forms nutritive yolk – leaving the narrow, dorsal cell strip intact, leads to complete regulation when performed until stage 26, to partial regulation at stage 27, and to absence of regulation at stage 28 (C. & M.). The anterior and posterior intraendodermal grooves mentioned above appear at stage 29 due to direct contact between the strictly determined outermost endoderm and the mesoderm that provides an unspecific stimulus (Albert, 1978*a*, *b*).

Balinsky (1948) studied the regulatory capacity of the archenteron in the neurula of *Triturus taeniatus* and *Ambystoma mexicanum* by exchanging different regions, and found that stomach formation can also occur heterotopically, while the liver only develops orthotopically. While stomach formation does not seem to require any special induction, the liver develops only where special inductive actions prevail. These observations were confirmed by T. S. Okada (1953 *a*, *b*). T. S. Okada (1957) found that the pharyngeal rudiment in the early neurula of *Triturus pyrrhogaster* and *Hynobius nebulosus* is still pluripotent, its fate being determined by the kind of mesenchyme with which it is confronted. Cranial mesectoderm is most effective in inducing pharyngeal differentiation in presumptive gastric and even in presumptive intestinal endoderm, while lateral plate mesoderm favours intestinal differentiation. The regionalisation of the endoderm is partially due to self-differentiation, since it already has a cranio-caudal polarity. This autonomous regionalisation is locally strengthened by corresponding mesenchymes but suppressed or transformed by others (T. S. Okada, 1957, 1960 *a*, *b*).

C. Takata (1960 *a*) found that in *Triturus pyrrhogaster* a certain antero-posterior regional determination of the endoderm already exists in the gastrula. This may have originated during primary meso-endoderm induction (see Chapter 10, p. 95). It is strengthened during gastrulation and neurulation. Notochord enhances pharynx and suppresses liver differentiation, antero-lateral mesoderm promotes liver, latero-ventral mesoderm promotes intestinal differentiation, and presumptive somite mesoderm promotes liver and intestinal differentiation. While anterior neural fold promotes pharynx formation, posterior neural fold enhances intestinal differentiation (Takata, 1960 *b*). This was confirmed by Mikami & Nishimura (1965) in *Hynobius lichenatus*. Chibon (1969) found that in *Pleurodeles* the different endodermal regions are no longer interchangeable at stage 24 (N.T. of *Pleurodeles waltlii* by Gallien & Durocher, 1957) but local deficiencies can still be regulated up to stage 25.

Boterenbrood & Nieuwkoop (1973), studying meso-endoderm induction in the blastula of *Ambystoma mexicanum* by making recombinates of dorsal, lateral and ventral endodermal yolk mass with animal 'ectoderm', noted a striking correspondence between the regional differentiation of the mesoderm and that of the endodermal structures; the latter often differed markedly from the presumptive significance of the endodermal material.

Using small ^{3}H-thymidine-labelled grafts, Rosenquist (1972) followed endodermal ingression through the anterior one-quarter to one-third of the primitive streak of the chick embryo. Anterior grafts move anteriorly, more lateral ones laterally, and posterior ones posteriorly towards the edge of the area pellucida. Rosenquist (1971) projected the locations of the various regions of the future gut back onto the invaginated endoderm of the definitive primitive streak stage, demonstrating the purse-like closure of the intestinal canal. Jacob (1971) described the first step in liver and pancreas formation as a local thickening of the endodermal epithelium, which he thinks may be caused by local interference with tangential growth.

Le Douarin, Bussonnet & Chaumont (1968) found that the various regions of the endoderm of the 7-somite chick embryo (stage 9, H. & H.) when combined with lateral plate mesoderm differentiate according to their prospective significance into stomodaeal, oesophageal, thyroidic, thymic, hepatic, pancreatic and intestinal endoderm. Sumiya (1976*a*) observed that isolated endodermal epithelium of the $2\frac{1}{2}$ day chick embryo (*c.* stage 17) can differentiate into various endodermal structures in the absence of mesenchyme when enveloped in vitelline membrane (cf. Wolff, 1961); under these culture conditions a PAS-positive basement membrane is formed. Pharynx, oesophagus, stomach, small intestine and yolk sac in the chick self-differentiate from stage 4 onwards, gizzard from stage 5, thyroid and proventriculus from stage 6, and large intestine, liver and dorsal pancreas from stages 8, 9 and 10, respectively (Sumiya & Mizuno, 1976). The anlagen of oesophagus, stomach, pancreas and small intestine are arranged in antero-posterior sequence in the dorsal endoderm (Sumiya, 1976*b*; Ishizuya-Oka, 1983).

While presumptive oesophagus, proventriculus, gizzard and small intestine of the 5 day chick embryo (*c.* stage 27) show homotypic differentiation when associated with homologous mesenchyme, oesophagus and small intestine also show homotypic differentiation with heterologous mesenchyme, but gizzard and proventricular endoderm in that situation mainly differentiate heterotypically. Although 5 day chick and quail endoderm is rather firmly determined, it can still react to varying extents to heterologous mesenchymal stimuli (Yasugi & Mizuno, 1978).

Wolff (1973*) observed that chick *allantoic* endodermal epithelium is still pluripotent and will differentiate into various endodermal structures under the influence of different regions of digestive tract mesenchyme. This was confirmed for oesophageal, proventricular, gizzard and small intestine mesenchyme by Yasugi & Mizuno (1974). They found that the competence of the allantoic endoderm is highest in the 3 day embryo and is gradually lost in older embryos. When allantoic endoderm differentiates in any of these structures it always also forms goblet cells typical for the allantois (Yasugi, 1979).

Andrew (1963) concluded from extirpation experiments that the entero-chromaffin cells of the digestive tract do not derive from the neural crest. However, Fontaine & Le Douarin (1977), using quail/chick chimaeras, could demonstrate that the enteric ganglia, the enterochromaffin cells and the endocrine cells of the gut epithelium are all of neural crest origin.

Mouth formation

(See under ecto-mesenchymal interactions, p. 256.)

The development of the salivary glands

In the 8 day chick embryo the submandibular gland anlagen are small spherical outgrowths from the buccal epithelium into the lower jaw mesenchyme on either side of the tongue. In the 10 day embryo the anlage has developed a tubular extension. Cavitation begins at about the 11th day, branching by the 14th day and secretory activity by the 16th day. The 12 day intact or reassociated epitheliomesenchymal anlage shows auto-differentiation. The 12 day chick salivary gland mesenchyme exerts a similar effect on mouse salivary gland epithelium as mouse mesenchyme. The salivary gland epithelium evidently plays the active role in gland morphogenesis (J. E. Sherman, 1960).

Early submandibular rudiments of the mouse continue morphogenesis *in vitro*, but morphogenesis is slower and more limited than that of older rudiments. Acinus formation is favoured by close association with condensed capsular mesenchyme. The epithelial rudiment separated from its mesenchyme fails to continue morphogenesis, but does so when reassociated with its own mesenchyme (Grobstein, 1953a). The submandibular anlage of the 13 day mouse embryo, which consists of a main duct with a single bud, forms a variable number of acini with intermediate ducts *in vitro*, but much fewer than normal. Rudiments with 4–10 buds develop best *in vitro*. Sublingual and submandibular glands develop more normally when explanted inside their common capsule than when cultured separately. The mesenchyme cells become oriented concentrically around the epithelial acini and groups of acini (Borghese, 1950).

Thirteen day submandibular epithelium combined with 11 day lung or metanephric mesenchyme showed no morphogenesis (Grobstein, 1953b). Grobstein (1962*) concluded that salivary epithelium only responds to its own mesenchyme. This conclusion was contested by Lawson (1974), who observed restricted branching of salivary gland epithelium in lung mesenchyme. Moreover, mesenchyme of accessory sexual glands supports growth, branching and differentiation of salivary gland epithelium. Mesenchyme of the urogenital sinus or preputial gland exhibits an age-dependent effect: mesenchyme taken from 15–16 day embryos supports morphogenesis of salivary gland epithelium but 13–14 day mesenchyme only allows atypical epithelial differentiation (Cumba, 1972). Salivary gland mesenchyme evokes salivary-gland-like branching of mammary gland epithelium when implanted into 14 day mammary gland anlagen (Sakakura, Sakagami & Nishizuka, 1979).

The development of the thyroid

Pharyngeal endoderm of stage 26 *Hynobius* embryos grafted to the belly of a host embryo together with ventral heart mesoderm forms thyroid tissue, demonstrating the local induction of the thyroid anlage in the pharyngeal endoderm by adjacent heart mesoderm (Murakawa, 1960).

In the chick the differentiation capacity of the endodermal thyroid epithelium is attained progressively; it becomes more and more independent of the mesenchymal environment. The determined endodermal anlage can form thyroid tissue when associated with heterologous mesenchyme (Le Lièvre & Le Douarin, 1970).

Reaggregates of 8 day chick thyroid epithelial cells reveal the necessity of association with thyroid capsule cells for normal differentiation. Both young and mature epithelial cells show abnormal histogenesis in response to fibroblasts from heterologous sources. Capsule cells of 16 day thyroid support normal thyroid formation when combined with 8 or 16 day epithelium (Hilfer & Stern, 1971).

The development of the thymus

The thymus gland originates from the endodermal epithelium of the fourth pharyngeal pouch in association with the mesenchyme of the third and fourth branchial arches.

The thymus gland anlage of the 12 day mouse embryo undergoes normal lobulation but remains predominantly epithelial when cultured *in vitro*. It grows profusely and becomes lymphoidal when grafted into the anterior eye chamber of an adult mouse. The epithelial and mesenchymal components only differentiate when reassociated. Lung, submandibular and metanephrogenic mesenchyme can fully replace thymus mesenchyme; limb bud and newborn mouse mesenchyme serve less well, and spinal cord tissue not at all. The epitheliomesenchymal interaction also occurs across a 20 μm thick millipore filter (Auerbach, 1960). Neural crest cells, which colonise the 3rd and 4th branchial arches, form the perivascular mesenchyme of the thymus gland.

Whereas Auerbach (1961) claimed that the lymphoid tissue of the developing thymus originates from the epithelial component of the early rudiment, Le Douarin & Jotereau (1975), using quail/chick chimaeras, could demonstrate that the lymphoid cell population is entirely derived from immigrant blood-borne stem cells, which are chemotactically attracted by 3 and 4 day pharyngeal pouch endoderm. The latter is determined to form the epithelial reticulum of the thymus gland at the 15-somite stage and is then able to attract stem cells, even in heterotopic positions. The attraction the epithelium exerts on lymphoid stem cells lasts for 24 h in the quail and for 36 h in the chick embryo. Shortly before hatching a second wave of lymphocytoblasts invades the thymus. The primitive embryonic lymphoid population is completely renewed around hatching. Competent thymic stem cells are present in the blood also before and after the periods of thymic attraction.

The development of the lungs

Neck mesoderm induces oesophagus and lung anlagen in upper pharyngeal endoderm of stage 26 embryos of *Hynobius* when they are grafted together to the belly of a host embryo of the same age (Murakawa, 1960).

Mouse bronchial mesenchyme associated with tracheal epithelium induces tracheal lung buds with characteristic glycogen deposition in their epithelial cells. Primary bronchial epithelium associated with metanephrogenic mesenchyme ceases its budding activity while glycogen is totally absent from the resting epithelium. Reassociation with bronchial mesenchyme restores budding activity and leads to the reappearance of epithelial glycogen. The mesenchyme thus affects the specific metabolic pattern of the pulmonary epithelium (Alescio & Dani, 1971). Wessels (1970b) observed that a variety of different mesenchymes can evoke supernumerary lung bud formation in tracheal epithelium, but typical branching occurs only in bronchial mesenchyme. The subsequent cytological differentiation of the pulmonary epithelium is dependent on the presence of lung mesenchyme (Dameron, 1972). Lawson (1983) noticed a continuation of the branching process of *c.* 13 day mouse lung epithelium in submandibular mesenchyme as soon as the epithelium had a high proportion of dividing cells

The development of the stomach

In granivorous birds, the stomach consists of the glandular proventriculus and the muscular gizzard. Both are formed from a single endodermal anlage surrounded by mesenchyme. Gland formation starts in the proventriculus at 6 days of incubation. The proventricular epithelium is devoid of glycogen, in contrast to the gizzard epithelium, which is very rich in glycogen. This is probably related to its higher respiration rate (Sigot, 1971). Proventricular epithelium alone or associated with gizzard mesenchyme differentiates into gizzard, and only with proventricular mesenchyme does it form glandular proventricular tissue. Proventricular mesenchyme can also induce gland formation in undetermined epithelium. Direct contact between the two components seems necessary for the interaction, since insertion of a piece of vitelline membrane prevents gland induction (Sigot, 1971).

Proventricular or gizzard mesenchyme cultured on plexiglass leaves residues, probably of extracellular matrix material, that can act as inducer or inhibitor. Proventricular mesenchyme residues induce gland formation in uncommitted proventricular epithelium, but gizzard mesenchyme residues arrest the differentiation of proventricular epithelium. Subsequent reassociation of the latter with proventricular mesenchyme does not restore its competence for gland formation (Sigot & Marin, 1970).

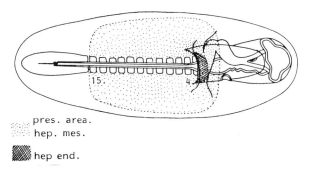

pres. area.
hep. mes.
hep end.

Fig. 36. Ventral view of chick embryo of 15 somites showing the localisation of the presumptive hepatic mesenchyme (pres. area hep.mes.; stippled) and of the hepatic endoderm (hep.end.; hatched). Somites 4 and 15 indicated (after N. le Douarin, 1964).

Different fractions of a chick liver homogenate either inhibit or stimulate rabbit stomach epithelium proliferation (David, 1970).

The development of the liver

In the chick embryo the presumptive hepatic endoderm acquires competence for hepatic cord formation at the 4–5-somite stage under the influence of the pre-cardiac mesoderm. Its subsequent differentiation into cords of hepatocytes is dependent on its association with hepatic mesenchyme derived from the pre-cardiac mesoderm (Le Douarin, 1974; see Fig. 36). The action of the latter is non-specific as it can be replaced by various other mesenchymes, such as ventral lateral plate and metanephrogenic mesenchyme, even from a mouse embryo, but not by head, limb bud or somitic mesenchyme (Le Douarin, 1970*).

In quail embryos Fukuda (1979) detected hepatogenic potencies in anterior endoderm from the 2-somite stage onwards. They gradually become stabilised with increasing age. He could confirm that cardiac mesenchyme is specifically required by uncommitted anterior endoderm to acquire hepatogenic potencies.

The hepatic mesenchyme forms reticulo-endothelial tissue independent of the hepatic parenchyma (Le Douarin, 1970*). Acetylcholinesterases appear precociously in the liver mesenchyme of 3 day chick embryos (c. stage 20). Although this enzymic activity does not depend on the presence of hepatic parenchyma, the latter can induce enzyme activity in somatopleural mesenchyme, which normally does not contain these enzymes, demonstrating the reciprocal character of the epitheliomesen-chymal interaction (Houssaint & Le Douarin, 1971).

Hepatic endoderm separated from hepatic mesenchyme by a Nuclepore filter of 1 μm pore size shows parenchymal differentiation due to the traverse of the filter by cellular processes; 0.6 μm pores, through which no penetration occurs, preclude differentiation. These experiments demonstrate that either cell–cell contact or contact between cell and extracellular matrix (free or membrane-bound) is required for this epitheliomesenchymal interaction (Le Douarin, 1975).

The development of the pancreas

Grobstein (1967*b**) states that the regional determination of the pancreatic anlage in the epithelium of the alimentary canal, which occurs long before the appearance of the pancreatic bud, is not well understood. Kratochwil (1972*) suggests that the neural plate may play a role in the early determination of the pancreatic anlage in the dorsal endoderm. The pancreatic anlage is determined at least 12–16 h before its association with the pancreatic mesenchyme.

Dieterlen-Lièvre (1970) studied the differentiation of the 15 day dorsal and ventral pancreatic buds of the chick *in vitro*. The former differentiates into the typical splenic branch of the pancreas, with large endocrine islets, some exocrine tissue and large areas of granulopoiesis. The latter predominantly forms acinar tissue with only sparse and small endocrine islets and small areas of granulopoiesis.

Golosow & Grobstein (1962) observed organ-specific cellular differentiation of mouse pancreatic epithelium in association with homologous as well as heterologous mesenchyme. However, the pattern of lobulation was mesenchyme dependent. In transfilter experiments, using millipore filters, epithelial growth and morphogenesis, though aberrant, is promoted by salivary mesenchyme, somewhat less by pancreatic mesenchyme, and little or not at all by metanephrogenic mesenchyme. Dieterlen-Lièvre (1970) found that dorsal as well as ventral pancreatic epithelium combined with lung, gizzard or somatopleural mesenchyme differentiates into its respective dorsal or ventral histotypical pattern. Dorsal pancreatic bud epithelium combined with metanephrogenic mesenchyme forms only small pancreatic structures showing intensive granulopoiesis. These experiments demonstrate the early determination of the pancreatic epithelium. Grobstein (1962*) mentions that acinar differentiation can be arrested by enforced spreading of the cultured epithelium.

The regional segregation of the pancreatic anlage into exocrine and endocrine tissue, as well as granulopoiesis in the associated mesenchyme, are also determined early. Dieterlen-Lièvre & Hadorn (1972) analysed the exocrine enzymatic activity of the pancreatic anlage up to hatching and could show that the pancreatic epithelium has the same enzymic activity in association with pulmonary as with pancreatic mesenchyme. This demonstrates the fully determined state of the epithelium of the early pancreatic anlage.

Andrew & Kramer (1979) already suggested that the pancreatic endocrine cells of the $3\frac{3}{4}$ day chick embryo do not derive from the neural crest. Kramer & Andrew (1981), using quail/chick chimaeras, could show definitively that the endocrine A, B and D cells are not of trunk neural crest origin, which is in agreement with Dieterlen-Lièvre & Hadorn's (1972) conclusion.

Evaluation

All ecto-mesenchymal interactions resulting in the formation of the body appendages and the various skin derivatives, as well as all endo-mesenchymal interactions leading to the formation of the various alimentary organs and endodermal derivatives, are characterised by sequential and reciprocal actions of epithelium and mesenchyme. While in some organ anlagen the epithelial component plays the leading role, as e.g. in salivary gland and pancreas development, in other anlagen the mesenchyme (which is of either neural crest or mesodermal origin) exerts the primary and decisive inductive action, as e.g. in limb bud and liver development.

It is likely that extracellular matrix components are directly or indirectly involved in all epitheliomesenchymal interactions. Although direct cell–cell interaction may play a role, as e.g. in stomach and liver development, cell–extracellular-matrix interaction seems to be the more common mechanism (the matrix components being free or membrane-bound). In our opinion matrix–matrix interaction may also be taken into consideration.

All epitheliomesenchymal interactions represent transient or permanent confrontations of cells or tissues engaged in different pathways of cellular differentiation. It is very likely that the extracellular matrix components that seem to play a role in the interactions are ordinary, transient products of epithelial and mesenchymal cellular differentiation.

According to Grobstein (1955*, 1967a*) all epitheliomesenchymal interaction systems are characterised by the presence of extracellular matrix material at the interface between epithelium and mesenchyme. The extracellular matrix secreted by the epithelium differs from that formed by the mesenchyme; the former is sensitive to hyaluronidase and the latter to collagenase, but not vice versa. Grobstein considered the extracellular matrix material, free or membrane-bound, as a likely candidate for the role of mediator in the inductive interaction. He emphasised that the entire process of synthesis of the complex matrix components may be involved in the induction process. Wessels (1970a*) emphasised that in several epitheliomesenchymal interactions the mesenchyme seems to evoke only a reversible differentiation of the target cells: a regulatory alteration in gene usage rather than a change in the determination of cell type.

19

General Evaluation

We have restricted our discussion to early development in order not unduly to expand the length of this volume. It is however evident that the whole of embryonic development, as well as the maintenance of the adult form, is based on a long series of interactions at all levels of organisation: subcellular, cellular and supracellular.

Epigenetic development and inductive interaction represent as it were the two faces of the same coin. 'Epigenesis' implies an increase in complexity of the developing egg and embryo through interaction with its environment and among its constituent parts. However, 'inductive interaction' in addition presupposes a minimal heterogeneity in the egg in order that interaction may occur between its different moieties. This initial heterogeneity represents the indispensable 'preformistic' element in development. Without inductive interaction the structural complexity of the egg cannot develop beyond that already present at the beginning. In this volume we have tried to show that chordate development (including that of the lower chordates) is almost entirely epigenetic, representing a gradual increase in heterogeneity of the developing system due to inductive interactions among its different parts (see Fig. 37).

The epigenetic nature of chordate development

Embryonic development is generally considered to begin with the fertilised egg, at the moment that it undergoes activation of all essential processes. In our opinion it seems more logical to let development start with the formation of the primordial germ cells, which are specially designed as the predecessors of the next generation. The sexual differentiation of the potentially bisexual germ cells usually depends on the male or female nature of the cellular environment of the gonad, and under experimental conditions can be expressed irrespective of the germ cells' own genetic constitution (Blackler, 1970*).

Unlike the spermatozoon, cellular differentiation of the oocyte is not accompanied by loss of potencies but is characterised by the preservation of totipotency. Oocyte formation therefore constitutes an exception to the

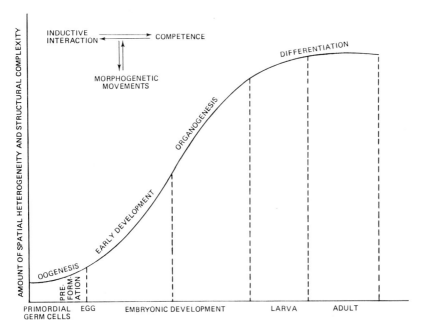

Fig. 37. Attempt to visualise (top left) the relationships between inductive interaction and competence, and between them and morphogenetic movements in epigenetic development, as well as (graph) the relative significance preformation and epigenesis have in the increase in structural complexity during development.

rule that cellular differentiation leads to a restriction of developmental potencies. In the embryo this restriction rule applies to the cytoplasmic machinery and the plasma membrane, but not to the same extent to the nucleus, the 'potencies' of which can be reactivated upon transplantation into egg cytoplasm (Gurdon & Woodland, 1968*, 1969; Gurdon, 1974*). The sperm nucleus must also preserve its totipotency, which comes to full expression when it enters the egg cytoplasm.

The totipotency of the female gamete holds for the egg as a whole but not necessarily for all its separate parts. We have seen that the fertilised amphibian egg consists of two different cytoplasmic moieties; the already more or less firmly determined 'endodermal' vegetal moiety and the still totipotent 'ectodermal' animal moiety (see p. 22). However, such an early segregation is not a common feature of all chordate eggs. In the meroblastic fish, reptilian and avian eggs the segregation of the blastodisc into a more or less firmly determined endodermal cell layer and a still totipotent 'ectodermal' layer occurs later in development, when it already consists of many blastomeres (see p. 64). The nearly alecithal mammalian embryo shows a similar late segregation of the inner cell mass. It may in any case be concluded that in all the vertebrate groups the development of the embryonic anlage starts with the very minimum of spatial heterogeneity, i.e. the presence of only two different moieties. After symmetrisation of the egg or embryonic anlage under the influence of the external factors of sperm entry or gravity, or both, the complexity of the vertebrate embryo is gradually built up as a result of mutual interactions of its constituent parts.

The first step in the increase of multiplicity of the vertebrate egg or embryo is the induction of the 'embryonic meso-endoderm',† which is due to the first large-scale cellular interaction between its two primary moieties. In the amphibian embryo the newly formed meso-endoderm has a tube-like configuration due to the presence of the blastocoelic cavity, which restricts interaction to a peripheral annular region (see p. 100). In the avian embryo embryonic meso-endoderm formation occurs similarly in the ectodermal epiblast under the influence of the underlying secondary hypoblast, but shows a concentric configuration with the presumptive embryonic endoderm surrounded by presumptive embryonic and extraembryonic mesoderm (see p. 116). A similar situation probably exists in the mammalian embryo, although the experimental evidence is still scanty. In all groups meso-endoderm induction in the totipotent 'ectoderm' is accompanied by the suppression of the ectodermal differentiation tendencies that were already present in it.

Holtfreter (1968*) called attention to the fact that inductive interactions and morphogenetic movements alternate with each other; for instance, meso-endoderm induction is followed by gastrulation movements, neural and neural crest induction is succeeded by neurulation movements and neural crest segregation, etc. It seems that inductive interactions often are a prerequisite for morphogenetic movements, but the reverse is also true: morphogenetic movements lead to a confrontation of previously separate parts of the embryo and thus to inductive interaction.

Davenport (1979*) introduced the notion of the emergence of 'instability' in the developing system, which inevitably will give rise to a relocation or readjustment of the different parts or processes into a more stable configuration. Examples are: the initiation of cleavage after the elimination of a cytostatic factor which previously prevented nuclear division and cytokinesis in the strongly enlarging ovarian oocyte; gastrulation after the formation of the embryonic meso-endoderm, which possesses new cellular properties; neurulation after the appearance of the neural plate, which again has new cellular properties, etc. We think the notion may be extended

---

† The vegetal yolk mass of the amphibian egg, representing its initial 'endodermal' moiety, chiefly forms nutritive yolk, while the primary and secondary hypoblast of the chick embryo do not contribute to the embryonic endoderm. Consequently in all forms the true embryonic endoderm is chiefly or exclusively induced in the totipotent 'ectodermal' moiety together with the mesoderm.

to intracellular processes as well. Moreover, the emergence of instabilities may lead to inductive interaction, as e.g. between the 'ageing' animal and vegetal moieties of the amphibian blastula, and between the segregating avian epi- and hypoblast cell layers. We therefore feel that this idea is very fruitful for a better understanding of embryonic development. Chandebois & Faber (1983*) use a similar notion in their concept of developmental readjustments, but Davenport particularly emphasises its inevitability.

The morphogenetic process of gastrulation gives rise to the formation of a triple-layered embryo, thus creating new and extensive possibilities for interaction among the three germ layers. The invagination of the marginal zone of the amphibian blastula is attended with a reversal of its antero-posterior axis. The resulting mesodermal mantle shows strong cranio-caudal and dorso-ventral polarities, partly expressed in its early differentiation; it is the leading germ layer in subsequent development. The overlying ectodermal layer seems to be only weakly polarised cranio-caudally, the underlying endoderm more strongly so. It is the spatial interaction among these three germ layers which is ultimately responsible for the three-dimensional organisation of the vertebrate embryo.

In the amphibians, and probably in all other vertebrates as well, the development of the neural anlage (the fourth large-size moiety of the developing embryo) is the result of two successive inductive actions exerted by the underlying archenteron roof: an activating or neuralising influence, which determines the spatial extension of the neural anlage, and a superimposed transforming or caudalising influence, which changes the emerging prosencephalic differentiation tendencies of the neuralised ecto-derm into tendencies for more caudal segments of the central nervous system (see p. 158).

The question arises whether the epigenetic nature of early vertebrate development also holds for the lower chordates, the uro- and cephalo-chordates. The unfertilised egg of both urochordates and cephalochordates is apparently still isopotential, since any fragment containing the nucleus, if subsequently fertilised, can give rise to a complete miniature embryo (see p. 63). The rapid segregation of different cytoplasmic areas that occurs after fertilisation must be guided by external influences such as sperm entry or gravity, or both, apparently affecting a highly reactive cytoplasmic machinery and plasma membrane system. In the uro- and cephalochordates the spatial segregation of ooplasms not only leads to the formation of animal and vegetal plasms, as in the amphibian egg, but also to that of a dorsal neuro- and chordoplasm and a ventral mesoplasm, so that both the animal–vegetal and the dorso-ventral polarities manifest themselves between fertilisation and the onset of cleavage. We believe that a deeper insight into the process of ooplasmic segregation is an essential prerequisite for a better understanding of embryonic development. The lower chordates seem to be well suited for this analysis.

Although the ascidian embryo shows a rather pronounced 'determinative' or 'mosaic' type of development after the segregation of its different ooplasms, inductive interactions are apparently involved in the development of the larval nervous system (see p. 179). The cephalochordates have an ooplasmic segregation that is very similar to that of the ascidians but are far less 'determinative' in their embryonic development. Neural induction by the notochordal anlage at the advanced gastrula stage has been clearly demonstrated in *Branchiostoma* (T. C. Tung *et al.*, 1962*b*). In addition, we feel that the development of recombinates of animal and vegetal octets of blastomeres of the 32–64-cell stage (T. C. Tung *et al.*, 1959, 1960*a*) can be satisfactorily interpreted as due to mesoderm induction, notwithstanding the early segregation of the chordo- and mesoplasms. Therefore, although we do not yet understand the mechanism of ooplasmic segregation we may safely state that all chordate eggs and embryos have an early development that is to a large extent epigenetic.

Vertebrate organogenesis is the direct consequence of the spatial interaction of the three germ layers, an interaction that is reciprocal in nature. Although reciprocal interaction may be simultaneous, more often it is an alternating process, since the required competences usually develop first in one and then in the other component. The most striking feature of organ development is the sequential character of the interactions with surrounding anlagen. Such interactions may be either transient or more permanent. Examples of transient interactions are those between the just-invaginated and still-uninvaginated portions of the presumptive archenteron roof, or between the liver anlage in the wall of the alimentary canal and the pre-cardiac mesoderm. More permanent interactions are e.g. those between the mesoderm and the ectoderm of the limb anlage, and between the ecto- or endodermal epithelium and the corresponding mesenchyme in feather formation and in lung or thyroid development, respectively. The sequential and reciprocal nature of inductive interactions in organ development has been stressed by many authors. Saxén *et al.* (1976*) characterise inductive tissue interaction as providing a central, vital guiding principle throughout embryogenesis.

Another aspect of organogenesis, which is related to the differences between transient and permanent interaction, should be emphasised: the self-organising capacity of certain organ anlagen once they are determined. This has been very clearly demonstrated in the development of the central nervous system, particularly the prosencephalon (see p. 193). It holds just as well for the somites and for the epithelial components of the pancreas and the mammary gland (see pp. 234 and 261, 270). However, it is not a universal phenomenon; other organ systems such as the skin and various eye structures require a more or less permanent interaction between their different components (see pp. 195, 255).

It is evident from many xenoplastic recombination experiments that the great majority of intra- and intercellular interactions upon which embryonic development is based are neither species- nor class-specific but depend on features of cell function and cellular differentiation common to all chordates. Certain structures which are foreign to a particular species or class can nevertheless be induced by special inductors, e.g. a more or less typical mammary gland can be induced in avian epidermis by mammary mesenchym (see p. 262), and teeth can be induced in avian pharyngeal arch epidermis by mammalian mandibular molar mesenchyme (see p. 257). Since each cell of an organism contains the full genetic information of the species and this cannot be transferred from one cell or tissue to another, cellular differentiation will always be species-specific. Therefore, the genes on which the competence of the avian epidermis for mammary gland or tooth development is based must be present in the avian genome.

Our present insight into embryonic development leads to the conclusion that there is little place for preformation in chordate development, with the exception of the spatial organisation of the egg, in so far as it is based on features derived directly from the primordial germ cells which form the link from one generation to the next. Fig. 37 on p. 273 shows, among other things, the relative role of preformation and epigenesis in the increase of structural complexity during development. Chandebois & Faber (1983*) consider the animal–vegetal polarity of the egg as the only preformistic element of development. However, neither the unfertilised eggs of the uro- and cephalochordates nor the fertilised mammalian egg manifest an internal animal–vegetal polarity. Nevertheless, the egg must have an intrinsic organisation which allows it to react to external stimuli, leading to the animal–vegetal polarisation of the egg or embryo.

It must be emphasised that there are still many gaps in our understanding of the successive steps in the complex chain reactions typical of epigenetic chordate development. Let us only mention here the intracellular interactions occurring during gametogenesis, the mechanism of ooplasmic segregation before or after cleavage, the transfer of dorso-ventral polarity from the animal to the vegetal moiety of the amphibian embryo before or during cleavage, and particularly the specification of the basic pattern of body appendages in fish and tetrapods.

Inductive interaction as the guiding principle in development

The great significance of intracellular processes in inductive interaction in our opinion fully justifies the discussion of the processes of oogenesis, egg and sperm maturation and fertilisation in this volume; indeed, we think these chapters are important for an understanding of inductive interactions in general, the more so since so much more is known of the biochemical events occurring during these preparatory phases than during embryonic development.

The ubiquitous nature of inductive interactions

In the whole of vertebrate development, from the primordial germ cell to the highly structured embryo with its many different organ systems, interaction seems to occur wherever different components (ooplasms, cells, tissues or organ anlagen) of the egg or embryo are juxtaposed during particular phases of development. Lang (1969*) and Horder (1976*) have stated that in any developing system, be it unicellular or multicellular, there must be continuous communication between the different parts, leading to two-way interaction. In our opinion the data presented in the preceding chapters convincingly demonstrate that interaction takes place as soon as adjoining subcellular, cellular or multicellular parts of the egg or embryo differ sufficiently for 'messages' released by one part to be recognised as meaningful 'signals' by the other part. The theoretical assumptions developed in Chapter 2 seem to be valid, including the implicit reciprocity of interactions.

There are two principal mechanisms at work in embryonic diversification. The first is based on 'cytoplasmic differentiation *in situ*', as a result of which adjoining parts of the egg or embryo that proceed along different developmental pathways gradually become so different that interaction must occur; an example is the divergent development of the animal and vegetal moieties of the amphibian egg and embryo leading to meso-endoderm induction. A variant of this first mechanism is 'cytoplasmic segregation' whereby cytoplasmic components become differentially distributed; examples are ooplasmic segregation in uro- and cephalochordate eggs, and the segregation of epi- and hypoblast in the avian blastoderm. The second mechanism is based on the transient or permanent apposition of initially separate parts of the embryo as a consequence of morphogenetic movements. This is the most common mechanism, of which we have met many examples in the preceding chapters.

Interaction does not seem to require the presence of interjacent cell membranes. The striking similarity in morphological and temporal development between the unfertilised, activated anuran egg, which manifests pseudogastrulation phenomena even after removal of the nucleus (see p. 138), and the normally developing embryo strongly suggests that the same or at least similar interactions can take place between the various moieties of the egg or embryo both in the presence and absence of interblastomeric cell membranes. The latter apparently allow the passage of necessary information in an almost unhampered manner. However, this does not mean that the cell membrane does not play an important role in intercellular communication. On the other hand, the cytoplasm of the uncleaved egg must be so highly structured that different cytoplasmic moieties can persist and interaction between them can occur.

The biochemical and biophysical nature of inductive interactions

As already stated repeatedly, inductive interactions result from the appearance of regional differences in the developing system. They must be intimately linked up with both the overall and the regional cellular changes occurring in the system and must therefore be considered against the background of the biochemical and physiological diversification that takes place in it.

Early embryonic development shows a rapidly changing pattern of gene activity. Gene products synthesised at a given stage are different from those formed at the previous or next stage. Those formed in the initial stages of development are for the greater part utilised immediately, whereas those formed at later stages are partially retained in the form of 'luxury' molecules. Transcription begins well in advance of the morphogenetic events for which it is required (see Zalokar, 1964*; Wilde & Crawford, 1968*; E. H. Davidson *et al.*, 1968; E. H. Davidson, 1976*). Organogenesis is characterised by marked changes in enzyme patterns (see Duspiva, 1969*) and by the synthesis of organ-specific isozymes (see Eppenberger, 1975*; Whitt, 1975*). Phenotypic changes are usually if not always accompanied by cell division, but as we have seen there is no convincing evidence for the assumption, made by Holtzer & Mayne (1973*), that a specific cell division would be required for each differentiative event (see Lash & Vasan, 1977*).

Control of synthesis may take place at the level of DNA synthesis, at the transcriptional level or at the translational level. Markert (1960*) still assumed that chromosomes 'differentiate' during development and that their progressive changes represent the essential element of embryonic differentiation. However, Gurdon & Brown (1965) and many others found that chromosomal changes are usually reversible. Gurdon (1968*b*, 1974*) and others offered convincing evidence that differential gene expression during development and differentiation is regulated at the transcriptional level. Translational control only seems to be responsible for quantitative changes in protein synthesis (Monroy, 1965*; Gross, 1967*b*; see also Ebert, 1968*; Asao & Nakamura, 1978*).

Markert (1965*) emphasised that not all information required for development is encoded in the DNA, since there is cytoplasmic continuity between the generations involving the transfer of a complex cellular organisation, of which the DNA is only a part. The interaction between the DNA and the rest of the cell in a sense provides the motive force for development and differentiation. Chandebois & Faber (1983*) compare the cellular DNA to the 'hardware' of the computer and the developmental programme, represented by the cellular organisation of the egg and the subsequent intra- and intercellular interactions, to the 'software'. The complex cellular organisation in our opinion represents the preformistic element in development, as discussed on p. 272.

The hypothesis of Morgan (1934*), already half a century old, which implies that nuclear activity diverges in different parts of the developing organism due to differences in cytoplasmic composition, still seems to account for the known facts. It has been further substantiated by Davidson and coworkers (see E. H. Davidson, 1976*), Gurdon and coworkers (see Gurdon, 1967b*, 1968b*), Brachet (1967a*), Gross (1967a*, b*) and others.

Unfortunately we know little about the regulatory mechanisms involved in nucleocytoplasmic interactions. There seems not to be sufficient specificity in basic proteins such as histones, nor in polymerases, to account for the complex differential activation and repression of genes during development and differentiation (Brachet, 1965*; Rutter *et al.*, 1972*). Moreover, two-thirds of the nuclear proteins are acidic proteins (Paul, 1968*).

How do inductive interactions affect the complex synthetic machinery of the responding cells? It must be said that it is not even known at what level inductive agents act on that machinery. Various suggestions have been made: action through direct unmasking of the DNA, through transcription of specific mRNAs, through unmasking of stored mRNAs, through regulation of protein synthesis and (in certain cases) through control of enzyme activity.

It should be emphasised that the primary target of the inducing factor does not necessarily coincide with the ultimate regulatory site in the synthetic cell machinery; its action may be transmitted to that site via receptors in the cell membrane and intracellular messengers. Lang (1969*) already suggested that the primary candidates for the sites of first interaction in interacting cell systems are receptor proteins in the cell membrane, not the genes themselves. He believes that the common feature of the first interaction between effector and reacting cell is physical rather than chemical in nature. Moreover, it seems rather unlikely that the site of primary interaction would always be the same in different inductive interactions. Structural and permeability changes of the cell membrane may constitute the first steps through which inducing agents initiate a new course of cytodifferentiation (Holtfreter, 1968*). Such membrane changes may secondarily give rise to changes in cell adhesiveness (Holtfreter, 1968*; Grunz, 1976*; Subtelny & Wessels, 1980*). However, it should be stressed that the elucidation of the site of action of the inductive stimulus in both intra- and intercellular interactions still constitutes one of the main problems in developmental biology.

Saxén *et al.* (1965*) and Saxén & Kohonen (1968*) have called attention to the fact that after an inductive action a 'silent period' without any discernible structural changes occurs, during which the expression of the new cell type is apparently prepared. Any interference with transcription during this period, e.g. by actinomycin D treatment, prevents subsequent morphogenesis.

Inductive interactions seem to range in nature from direct cell–cell interaction via tight or gap junctions, as suggested for the interaction between the ureteric bud and the metanephric mesenchyme (Saxén *et al.*, 1976*, 1978*), all the way to interaction at a distance mediated by 'diffusible' factors, as in the case of the differentiation of the mesodermal mantle under the influence of the notochord (T. Yamada, 1940) or the formation of lens fibres under the influence of the retina (A. J. Coulombre, 1965). Although cell–cell interaction involving the transfer of a molecular surface pattern (P. Weiss, 1962*) cannot be excluded, it seems more likely that interaction occurs via the extracellular matrix, which is apparently present from very early stages onwards (see p. 113). In this type of interaction the exchange of information may be either between the cell membranes of one and the matrix of the other component, or between the two matrices as such. Although the first type of interaction is supported by various authors, there are some recent studies on lens and kidney development which plead in favour of a matrix–matrix type of interaction (see pp. 208 and 247). The extracellular matrix consists of many components such as hyaluronate, glycosaminoglycans and glycoproteins as well as collagen. Some of their precursors may initially be present in diffusible form, so that they can cover intercellular distances and pass through the pores of millipore or Nuclepore filters (W. W. Minuth, 1978; Kawakami *et al.*, 1978). We think that the fact that the presence of such extracellular matrix material in filter pores can never be excluded raises doubts about the validity of transfilter experiments.

Although Holtfreter (1968*) still denied a mediating role to extracellular matrix material in inductive interactions, Grobstein (1955*, 1967*b**) had already called attention to its possible role in embryonic induction. This role has since been advanced by many authors and strongly emphasised by Saxén (1980*b**) and Slavkin (1982). Though the majority of these authors studied inductive interactions in later development, in Chapters 10 and 12 (pp. 113, 121 and 168) we have presented evidence that extracellular matrix material may play a role in inductive interactions during early development. Moreover, it has become rather likely that certain heterogenous inductors are identical to extracellular matrix components of differentiated cells (Hoperskaya *et al.*, 1984).

Inductive interactions during early and later embryonic development are not necessarily of the same nature; they may indeed differ significantly since both inducing and reacting systems undergo cellular differentiation (see p. 160). Saxén *et al.* (1976*) suggest that transmission of morphogenetic signals at a distance is restricted to so-called primary induction, representing meso-endodermal as well as neural induction, whereas matrix-mediated inductive actions are restricted to 'secondary' inductive events during organogenesis. We are not convinced that this distinction is so strict, since it has been shown that induction at a distance occurs also in the regional differentiation of the mesodermal mantle and in lens development. Moreover, meso-endoderm induction, the first large-scale cellular interaction in early embryonic development, seems to coincide with the appearance of extracellular matrix material, while matrix material can also act as a heterogenous mesoderm inductor (see p. 114).

The possible nature of inducing factors

Holtfreter (1948*), L. G. Barth & Barth (1969, 1972), Wolpert (1973) and others have suggested that inducing factors may be very simple molecules. McMahon (1974) believes that they may represent normal physiological regulatory molecules, such as cAMP and cGMP, since the fertilised egg already contains all the necessary information for its development. According to MacMahon substances with a high information content would be required only if cells would depend for their further development on *new* information (see also Saxén *et al.*, 1976*).

Changes in ionic composition and pH of the medium can influence cellular differentiation by affecting the competence of the reacting cells (L. G. Barth & Barth, 1974). One of the most striking examples is the mesodermalising and neuralising action of the Li^+ ion, an ion that is completely foreign to the embryo. In our opinion ions may affect the composition of the cell membrane or may alter the configuration of the extracellular matrix.

Intracellular messengers like cAMP, H^+ and Ca^{2+}, which are involved in cellular processes such as exocytosis, contractility, etc. also seem to play a role in induction processes. Their role has been demonstrated in the acrosome reaction of the sperm and in cortical granule exocytosis during egg activation. Ca^{2+} and Mg^{2+} as well as cAMP have a function in oocyte maturation and Ca^{2+} is essential for cortical contraction and wound healing. The significance of intracellular messengers in inductive interaction is supported by, among other things, the observation that the AER/limb-bud-mesenchyme interaction is influenced by cAMP (Kosher *et al.*, 1979), while neural induction can be mimicked by cAMP in amphibians (Wahn *et al.*, 1975, 1976) and birds (Bjerre, 1974).

Although ions and intracellular messengers may be involved in some links of the complex chain process of inductive interaction, we do not believe that natural inductors are just single molecules. We consider suggestions that the Na^+ ion, the heparan sulphate molecule or FSH would represent the neural inductor (suggestions made by L. G. Barth & Barth (1974), Løvtrup (1983*) and Sherbet & Lakshmi (1974*), respectively) as too simplistic in view of the very complex nature of the interacting systems which develop along different pathways of cellular differentiation and in which many different processes are going on simultaneously or successively.

In some cases certain metabolites seem to act as natural inducers in embryonic development, e.g. vitamin A as an inducer of mucous epithelium and phenylalanine as an inducer of pigment cells. Hay (1963) suggests that such metabolites produce their effect upon competent cells by inducing or repressing the activity of regulatory genes. Apart from these special cases it is very unlikely that in the great majority of inductive interactions only a single factor would be involved. The released morphogenetic events are likely to comprise complex chains of reciprocal inductive interactions. Nearly all of these can be mimicked by a great variety of artificial agents, which may act at different levels in the chain reaction. We have seen in the preceding chapters that there is little specificity in inductive actions, such specificity as there is being mainly built into the reaction system, which at a given time can react in only a very restricted number of alternative ways.

We think that it is much more plausible that the inducing factors are common, transient products of cellular differentiation of the interacting systems than that they would be highly specific factors specially produced for particular actions. Normal products of cellular differentiation may vary from ions and simple metabolites to high-molecular extracellular matrix components. As already mentioned on p. 278, intra- and intercellular interactions are caused by the confrontation of parts of the egg or embryo which are engaged in different pathways of differentiation. The products that are formed at the beginning of either developmental pathway may differ from those formed in later phases during the progression of differentiation. This implies that the products involved in early interactions may differ significantly from those formed later. Although at a given time a certain single product may act as the main inducing factor, it is likely that in the majority of inductive interactions not only qualitative but also quantitative differences in the composition of cellular products may affect the metabolism of the adjoining cells.

Recently, the concept of 'determinant' has been reintroduced, indicating a specific role for a given structure or agent in morphogenetic events. We want to caution our readers against the possibly misleading nature of this concept. In the preceding pages we have emphasised that it is much more likely that unspecific, sometimes perhaps purely quantitative differences in the metabolism of the action and reaction system are responsible for their mutual interaction, rather than that the formation of highly specific factors by the action system would cause a specific reaction in the responding cells. Though determinants were originally supposed to be intracellular (genetic) components (Weismann, 1892*), the term is recently used increasingly for intercellular interactions as well. (See the announcement for the UCLA Symposium on 'Molecular Determinants of Animal Form' (Park City, Utah, USA, April 1985).)

We cannot escape the conclusion that a search for specific inductive factors which would be active in particular phases of embryonic development is without much perspective and will remain a search for a 'needle in a haystack', as long as we have no better insight into the physiology and biochemistry of the developing embryo. In conclusion, we think that one of the primary tasks for developmental biology in the coming decades ought to be *the study of the rapidly changing structure and function of interacting embryonic systems engaged in different pathways of cellular differentiation, on the one hand passing through separate phases of responsiveness (competence) and on the other hand producing various transient cellular products that may act as inducing stimuli.*

The significance of competence and its possible nature

The term 'competence' was coined by Waddington in 1932 and defined as the temporary reactivity of a group of cells to a particular inductive stimulus. Holtfreter & Hamburger (1955*) defined competence as the physiological state of a tissue which permits it to react in a morphogenetically specific way to determinative stimuli. In the second definition more emphasis is already placed on the reaction system than in the first.

Attempts at isolation and identification of inductive factors dominated the biochemical approach during the 1940s and 1950s, but in the 1960s and 1970s more and more authors began to recognise the importance of the reaction system (see, among others, Saxén & Toivonen, 1962*; Holtfreter, 1968*; Holtzer, 1968*; Saxén & Kohonen, 1968*; Kratochwil, 1972*; Hay, 1977b*; Gurdon, 1981*). Gurdon stated that the result of an inductive action is determined by the properties of the responding cells. In a recent interview (Mehr, 1982) Holtfreter emphasised again that it is not the inducing substances but the reacting cells that ought to constitute the focus for developmental biochemical research.

In a theoretical article Raven (1938) analysed the relationship between 'reactive power' and 'differentiation tendency', the former representing the reactivity to or competence for an inductive action of a certain tissue, and the latter the emergence of the capacity for autodifferentiation in the reacting cells. Raven stressed the point that increase in 'reactive power' due to the action of an inducing stimulus leads at a certain point to the appearance of a 'differentiation tendency'. In the preceding chapters we have encountered some striking examples. While the cephalic neural crest material of the amphibian neurula only manifests *competence* for cartilage induction by pharyngeal endoderm, avian cephalic neural crest material isolated before migration is already capable of *autodifferentiation* of small cartilage nodules (see pp. 223, 238). Meso- and metanephric mesenchyme autonomously initiate tubulogenesis but cannot develop further without an inductive stimulus from the Wolffian or the ureteric duct: the initial tubules regress (see pp. 244, 246). In the first case the inductive action has obviously elevated the corresponding competence to the level of differentiation tendency, while in the second case it further increases the already existing differentiation tendencies.

A. G. Jacobson (1966*) calls attention to the fact that the appearance or disappearance of competence is often the consequence of a previous inductive or inhibitory action. For instance, lens competence in the cephalic ectoderm would be due to an inductive influence of the pharyngeal endoderm, and liver competence in the alimentary endoderm to an inductive action of the pre-cardiac mesoderm. The suppression of ectodermal differentiation tendencies in the equatorial region of the amphibian blastula is due to an inhibitory influence of first the mesodermal and then the neural inductor (Nieuwkoop, 1973*; Grunz, 1975*). Jacobson speaks of 'predetermination' to indicate the appearance of a given competence or 'bias'. Although we feel this is too static a term, we agree that in the development of many organ systems the appearance of competence can be related to previous inductive interactions. Saxén & Kohonen (1968*) emphasise the role of the entire developmental history of the responding tissue, its responsiveness being the result of a chain of preparatory events. Lopashov & Hoperskaya (1977*) express the same point of view.

In our opinion, however, this cannot be the whole story. The normal 'ageing' or 'ripening' of a tissue can also lead to the emergence of competence; for instance, the emergence of meso-endodermal competence in the isolated animal 'ectodermal' moiety of the amphibian blastula and in the isolated epiblast of the avian blastoderm does not seem to be based on previous inductive interaction. The natural ageing of the ectoderm not only gives rise to the emergence and subsequent decline of meso-endodermal competence but also, slightly later, to neural competence; in its turn this declines to be replaced by placodal competence (see, among others, Nieuwkoop, 1963, 1973*; Davenport, 1979*). It must therefore be concluded that *in the emergence of competence both the natural ageing process and inductive interactions may play a role*, competence for a given inductive influence representing a *transient state* in the physiology of the tissue in question.

Kratochwil (1972*), Wessells (1973*) and Keating & Kennard (1976*) state that 'determination' as it results from inductive interactions represents the ultimate step in the restriction of developmental potentialities, in other words, the confinement to a single commitment of a cell or cell group to generate a particular class of differentiated cells. According to Zwilling (1968*), determination is marked by a dramatic increase in gene activity and subsequently of enzyme activity, but not by its initiation, which would mean that gene activity must already have started during the preceding period of competence. Based on actinomycin D studies Flickinger (1970 *b*) ascribes determination to the accumulation of DNA-like RNA molecules in the cytoplasm of the responding cells. Wilde (1959*) already drew attention to the adaptability of the enzyme complement in competent embryonic cells. Tiedemann *et al.* (1967) emphasised the need for RNA synthesis and Grunz (1968, 1970) that for protein synthesis in the development of competence. J. Cohen (1979*) states that competence has to do with the masking or unmasking of genes and with the initiation of RNA synthesis. Only cells which are competent would have genes in a state that is receptive to the action of the inducer. Histones may play a role in keeping genes inactive, and removal of histones may make the genes susceptible to inducer action. All these ideas are rather theoretical, and Tarin (1978*) rightly states that the molecular and genetic basis of competence as well as of its emergence and decline is still largely untouched. *We believe that the understanding of the nature of competence and of its emergence and decline actually constitutes one of the key problems in developmental biology.* Insight into the nature of competence will much more rapidly lead to an understanding of inductive interaction than the isolation and characterisation of putative inducing factors (see also Gurdon, 1981* and Kratochwil, 1983*).

Zalokar (1964*) already suggested that competence may have to do with a masking or unmasking of genes and with the initiation of RNA synthesis. Hay & Meier (1974), Wessells (1977*) and Hay (1977 *b**) emphasise the fact that in the majority of inductive interactions the inducer does not initiate the protein synthesis specific for the reacting cells; the genetic programme is already present but remains dormant or insignificant in its expression until induction occurs. Recent investigations have brought to light that organ-specific antigens are also present, though at a low level, in tissues outside the organ anlage in question; for instance, lens antigens can be detected in embryonic retina and pigment epithelium as well as in adult retina, pigment epithelium, iris, cornea and adenohypophysis; all these tissues have lens-forming or lens-regenerating capacity (R. M. Clayton, 1982*). These observations led Clayton to suggest that competence implies the presence of a low level of specific mRNAs, inductive interaction then leading to a marked increase in the amount of these specific gene products (see also Chandebois & Faber, 1983*).

A few words ought to be said about the notions of 'instructive' or 'directive' versus 'permissive' or 'supportive' inductive interaction (Kratochwil, 1972*; Wessells, 1973*; Saxén, 1975, 1977*; Saxén *et al.*, 1976*). In our opinion these concepts are closely related to the reactivity of the responding cells or tissues. The higher the competence, the weaker and less specific the inductive action needs to be to raise the competence to the level of the corresponding differentiation tendency; conversely, the lower the competence, the stronger and more specific the inductive action must be.

It must be emphasised that competence, just as induction in the strict sense, always implies a choice from among different developmental pathways, since in the absence of inductive action the cells in question proceed along the pathway already in progress. For instance, when no inductive action occurs the animal cap of the amphibian blastula or the epiblast of the avian embryo will proceed in the direction of epidermal differentiation. In most instances the choice is between two alternative pathways only, the one already in progress and one other pathway. The totipotent cap of the amphibian blastula and the epiblast of the avian embryo possess not only epidermal competence but also meso-endodermal as well as neural competence. The periods of the meso-endodermal and neural competences overlap (Leikola, 1963); meso-endodermal competence prevails in early stages, masking neural competence, while the latter begins to predominate at later stages when meso-endodermal competence declines. Also here we may therefore have to do with a choice between only two pathways, first between epidermal and meso-endodermal and later between epidermal and neural.

It is important to realise that a period of competence represents a 'labile' phase of development, during which different developmental pathways strive for dominance; as long as the choice has not been made the balance between them is still reversible.

Homoiogenetic induction

Homoiogenetic induction implies the propagation of an inductive stimulus through the isopotent cell population of the reaction system. Propagation of the inductive action from cell to cell is not a universal mechanism, as Saxén and coworkers demonstrated for the induction of tubulogenesis in metanephric mesenchyme. In this case only cells that have been in direct contact with the inducer (ureteric bud or spinal cord) can proceed to form tubules. Spreading of the inductive stimulus from cell to cell is nonetheless relatively common, particularly in large-scale inductions such as meso-endodermal and neural induction.

In homoiogenetic induction we have to deal with two problems: that of the propagation of the stimulus itself, and that of how the inductive range of action becomes spatially limited. Propagation of an all-or-none response, as in nerve impulse conduction, cannot account for homoiogenetic induction, since a weak inductive stimulus leads to a small and a strong stimulus to a large induced structure. There must be some form of decrement, but its causes are largely unknown. B. Albers (unpublished results) recently found that the range of homoiogenetic neural induction is restricted by the loss of competence of the reacting ectoderm during the time it takes for the inductive action to propagate from the dorsal midline up to the lateral (and anterior) boundary of the neural plate. The neural inductive action seems to be able to pass through non-competent ectoderm. This latter observation is very interesting but requires careful verification.

Homoiogenetic induction may also account for the spreading of meso-endodermal induction. In the avian embryo the cranio-caudal extension of the primitive streak corresponds to the spatial limits of the inductive capacity of the hypoblast as well as to the regional loss of competence of the epiblast (Azar & Eyal-Giladi, 1981). In the amphibian embryo the propagation of meso-endodermal induction starts at the very early blastula stage, reaches about halfway along the width of the presumptive marginal zone at the early gastrula stage, and declines in its initial form by the middle gastrula stage (see p. 101).

General conclusion

We have seen that the concepts of epigenesis and of inductive interaction are intimately related. Epigenesis is impossible without inductive interaction, and the presence of inductive interactions characterises embryonic development as epigenetic. However, the notion of inductive interaction alone is not fully sufficient to characterise epigenetic development. It is only meaningful when it is realised that the reaction system makes its own contribution, sometimes quite independent of the inducer. Therefore competence must be included in the concept of epigenesis. Inductive interaction and competence are actually complementary principles. Both are in their turn connected with morphogenetic movements (see Fig. 37 on p. 273). Inductive interaction and competence are organ-specific and time-dependent. They, moreover, require spatial contiguity of the interacting tissues (largely provided by morphogenetic movements). Interestingly, the two principles show no species- or class-specificity but have common features for at least the entire subphylum of the vertebrates.

From a theoretical point of view it seems meaningless to separate the inductive action exerted by the action system from the complementary competence of the reaction system, since interaction is essentially based on the differences between the two systems. It therefore does not make much sense to search selectively for inducing factors. This only becomes meaningful when we have better insight into the physiological basis of the corresponding competence. Indeed, the one-sided search for inducing factors has yielded only meagre results in half a century of analysis. *It is only the combined analysis of the biochemistry and physiology of the inducing and the reacting system and, in particular, of the differences between them which are responsible for their interaction that seems promising for the coming decades.* We hope that this volume may stimulate such an integrated approach.

References

Abercrombie, M. (1970). *In Vitro*, **6**, 128–42.

Adamson, E. D. & Woodland, H. R. (1977). *Developmental Biology*, **57**, 136–49.

Adelmann, H. B. (1932). *Journal of Morphology*, **54**, 1–67.

Adelmann, H. B. (1934). *Journal of Experimental Zoology*, **67**, 217–81.

Adelmann, H. B. (1936*). *Quarterly Review of Biology*, **11**, 161–82.

Adelmann, H. B. (1937). *Journal of Experimental Zoology*, **75**, 199–227.

Aimar, C., Delarue, M. & Vilain, C. (1981). *Journal of Embryology and Experimental Morphology*, **64**, 259–74.

Ajiro, K. (1971). *Science Reports of the Tokyo Kyoiku Daigaku*, Sect. B., **14**, 145–76.

Aketa, K. (1973). *Experimental Cell Research*, **80**, 439–41.

Albert, J. (1978*a*). *Bulletin Biologique de la France et de la Belgique*, **112**, 167–205.

Albert, J. (1978*b*). *Archives de Biologie*, **89**, 139–53.

Alescio, T. & Dani, A. M. (1971). *Journal of Embryology and Experimental Morphology*, **25**, 131–40.

Al-Mukhtar, K. A. K. & Webb, A. C. (1971). *Journal of Embryology and Experimental Morphology*, **26**, 195–217.

Amprino, R. (1964). *Archives de Biologie*, **75**, 1047–80.

Amprino, R. (1965*). In *Organogenesis*, eds. R. L. de Haan & H. Ursprung, pp. 255–81. New York: Holt, Rinehart & Winston.

Amprino, R. (1975*a*). *Nova Acta Leopoldina*, **41**, 235–70.

Amprino, R. (1975*b*). *Archives d'Anatomie, d'Histologie et d'Embryologie*, **58**, 29–40.

Amprino, R. (1976). *Archives de Biologie*, **87**, 1–14.

Amprino, R. (1977). *Acta Embryologiae Experimentalis*, **1**, 51–70.

Amprino, R. (1978). *Anatomy and Embryology*, **153**, 305–20.

Amprino, R. & Camosso, M. E. (1966). *Acta Anatomica*, **63**, 363–87.

Ancel, P. & Vintemberger, P. (1948*). *Bulletin Biologique de la France et de la Belgique*, Suppl. 31, 1–182.

Anderson, C. B. & Meier, S. (1981). *Developmental Biology*, **85**, 385–402.

Andrew, A. (1963). *Journal of Embryology and Experimental Morphology*, **11**, 307–24.

Andrew, A. (1964). *Journal of Anatomy*, **98**, 421–8.

Andrew, A. (1969). *Journal of Anatomy*, **105**, 89–101.

Andrew, A. (1970). *Journal of Anatomy*, **107**, 327–36.

Andrew, A. & Kramer, B. (1979). *Journal of Embryology and Experimental Morphology*, **52**, 23–38.

Ansevin, K. D. (1966). *Journal of Morphology*, **118**, 1–10.

Ansevin, K. D. (1969). *Journal of Embryology and Experimental Morphology*, **21**, 383–90.

Araki, M., Yanageda, M. & Okada, T. S. (1979). *Developmental Biology*, **69**, 170–81.

Asahi, K., Born, J., Tiedemann, H. & Tiedemann, H. (1979). *W. Roux' Archives of Developmental Biology*, **187**, 231–44.

Asao, T. & Nakamura, O. (1978*). In *Organiser – a Milestone of a Halfcentury from Spemann*, eds O. Nakamura & S. Toivonen, pp. 283–308. Amsterdam: Elsevier Biomedical Press.

Asashima, M. (1975). *W. Roux' Archives of Developmental Biology*, **177**, 301–8.

Asashima, M. (1980). *W. Roux' Archives of Developmental Biology*, **188**, 123–6.

Asashima, M. & Grunz, H. (1983). *Differentiation*, **23**, 206–12.

Atsumi, T. & Takeuchi, M. (1980). *Development, Growth and Differentiation*, **22**, 133–42.

Auerbach, R. (1960). *Developmental Biology*, **2**, 271–84.

Auerbach, R. (1961). *Developmental Biology*, **3**, 336–54.

Austin, C. R. (1968*). *Ultrastructure of Fertilisation*. New York: Holt, Rinehart & Winston.

290   *Reference list*

Austin, C. R. (1969*). In *Fertilization, Comparative Morphology, Biochemistry and Immunology*, vol. II, eds C. B. Metz & A. Monroy, pp. 437–66. New York: Academic Press.

Austin, C. R. (1978*). *Current Topics in Developmental Biology*, **12**, 1–9.

Ave, K., Kawakami, I. & Sameshima, M. (1968). *Developmental Biology*, **17**, 617–26.

Ave, K., Sasaki, N. & Kawakami, I. (1968a). *Memoirs of the Faculty of Science, Kyushu University*, Ser. E, **5**, 25–38.

Ave, K., Sasaki, N. & Kawakami, I. (1968b). *Memoirs of the Faculty of Science, Kyushu University*, Ser. E, **5**, 39–47.

Avery, G., Chow, M. & Holtzer, H. (1956). *Journal of Experimental Zoology*, **132**, 409–23.

Azar, Y. & Eyal-Giladi, H. (1979). *Journal of Embryology and Experimental Morphology*, **52**, 78–88.

Azar, Y. & Eyal-Giladi, H. (1981). *Journal of Embryology and Experimental Morphology*, **61**, 133–44.

Azar, Y. & Eyal-Giladi, H. (1983). *Journal of Embryology and Experimental Morphology*, **77**, 143–51.

Bachvarova, R. & Davidson, E. H. (1966). *Journal of Experimental Zoology*, **163**, 285–96.

Bachvarova, R., Davidson, E. H., Allfrey, V. G. & Mirsky, A. E. (1966). *Proceedings of the National Academy of Sciences of the United States of America*, **55**, 358–65.

Bagnara, J. T. (1961*). *Bios*, Genova, **32**, 127–31.

Baker, P. C. (1964). *Zeitschrift für Zellforschung*, **64**, 636–54.

Baker, P. C. (1965). *The Journal of Cell Biology*, **24**, 95–116.

Balinsky, B. I. (1933). *W. Roux' Archiv für Entwicklungsmechanik der Organismen*, **130**, 704–46.

Balinsky, B. I. (1948). *W. Roux' Archiv für Entwicklungsmechanik der Organismen*, **143**, 365–95.

Balinsky, B. I. (1950). *Journal of Anatomy*, **84**, 227–35.

Balinsky, B. I. (1957). *Journal of Experimental Zoology*, **135**, 255–300.

Balinsky, B. I. (1961). I.I.E. and Fond. Baselli (1960) Symposium *Germ Cells and Development*, ed. S. Ranzi, pp. 550–63. Pavia: Tipogr. Succ. Tusi.

Balinsky, B. I. (1966). *Acta Embryologiae et Morphologiae Experimentalis*, **9**, 132–54.

Balinsky, B. I. & Devis, R. J. (1963). *Acta Embryologiae et Morphologiae Experimentalis*, **6**, 55–108.

Ballard, W. W. (1966a). *Journal of Experimental Zoology*, **161**, 193–200.

Ballard, W. W. (1966b). *Journal of Experimental Zoology*, **161**, 201–10.

Ballard, W. W. (1966c). *Journal of Experimental Zoology*, **161**, 211–20.

Ballard, W. W. (1973a). *Journal of Experimental Zoology*, **184**, 27–48.

Ballard, W. W. (1973b). *Journal of Experimental Zoology*, **184**, 49–74.

Ballard, W. W. (1973c*). *Revue Roumaine de Biologie*, Sér. Zoologie, **18**, 119–35.

Ballard, W. W. (1976*). *BioScience*, **26**, 36–8.

Ballard, W. W. (1981). *The American Zoologist*, **21**, 391–9.

Ballard, W. W. & Ginzburg, A. S. (1980). *Journal of Experimental Zoology*, **213**, 69–103.

Ballester, C. D. (1966). *Anales del Desarrollo*, **13**, 313–25.

Baltus, E. & Brachet, J. (1962) *Biochimica et Biophysica Acta*, **61**, 157–63.

Baltus, E., Brachet, J., Hanocq-Quertier, J. & Hubert, E. (1973). *Differentiation*, **1**, 127–43.

Bancroft, M. & Bellairs, R. (1976). *Journal of Embryology and Experimental Morphology*, **35**, 383–401.

Ban-Holtfreter, H. (1965). PhD thesis Ann Arbor, Mich., 251 p.

Barbieri, F. D., Sanchez, S. S. & Del Pino, E. J. (1980). *Journal of Embryology and Experimental Morphology*, **57**, 95–106.

Barondes, S. H. (1980*). In *The Cell surface: Mediator of Developmental Processes*, eds
S. Subtelny & N. K. Wessels, pp. 349–63. New York: Academic Press.

Bartelmez, G. W. (1954). *Contributions to Embryology*, **35**, 56–71.

Barth, L. G. (1964). *The American Zoologist*, **4**, Abstract no. 199.

Barth, L. G. (1965). *Biological Bulletin*, **129**, 471–81.

Barth, L. G. (1966). *Biological Bulletin*, **131**, 415–26.

Barth, L. G. & Barth, L. J. (1963). *Biological Bulletin*, **124**, 125–40.

Barth, L. G. & Barth, L. J. (1964). *Biological Bulletin*, **127**, 413–27.

Barth, L. G. & Barth, L. J. (1967). *Biological Bulletin*, **133**, 495–501.

Barth, L. G. & Barth, L. J. (1968). *Journal of Embryology and Experimental Morphology*,
19, 387–96.

Barth, L. G. & Barth, L. J. (1969). *Developmental Biology*, **20**, 236–62.

Barth, L. G. & Barth, L. J. (1972). *Developmental Biology*, **28**, 18–34.

Barth, L. G. & Barth, L. J. (1974). *Developmental Biology*, **39**, 1–22.

Barth, L. G. & Sze, L. C. (1951). *Experimental Cell Research*, **2**, 608–14.

Barth, L. G. & Sze, L. C. (1953). *Physiological Zoology*, **26**, 205–11.

Barth, L. J. & Barth, L. G. (1974). *Biological Bulletin*, **146**, 313–25.

Battikh, H. K. (1971). *Archives d'Anatomie, d'Histologie et d'Embryologie Normale et
Expérimentale*, **54**, 113–22.

Beal, C. M. & Dixon, K. E. (1975). *Journal of Experimental Zoology*, **192**, 277–83.

Becker, U. (1960). *W. Roux' Archiv für Entwicklungsmechanik der Organismen*, **152**,
339–72.

Bedford, J. M. & Cooper, G. W. (1978*). In *Cell Surface Reviews*, vol. 5, *Membrane
Fusion*, eds G. Poste & G. L. Nicolson, pp. 65–125. Amsterdam: North Holland
Publishing Company.

Bedford, J. M., Cooper, G. W. & Calvin, H. I. (1972*). In Symposium 1971 *The Genetics
of the Spermatozoon*, eds R. A. Beatty & S. Glueckson-Waelsch, pp. 69–89. Edinburgh:
Department of Genetics, University of Edinburgh.

Beetschen, J. C. (1970). *Compte Rendu de l'Académie des Sciences*, Paris, Sér. D, **270**,
855–8.

Beetschen, J. C. (1979a). *Compte Rendu de l'Académie des Sciences*, Paris, Sér. D, **288**,
643–6.

Beetschen, J. C. (1979b). *Archives d'Anatomie Microscopique et de Morphologie
Expérimentale*, **68**, 211–12.

Beetschen, J. C. & Buisan, J. J. (1977). *Mémoires de la Société Zoologique de France*,
Symposion L. Gallien, **41**, 139–52.

Belanger, A. M. & Schuetz, A. W. (1975). *Developmental Biology*, **45**, 371–81.

Bell, E. (1960). *Experimental Cell Research*, **20**, 373–83.

Bellairs, R. (1963). *Journal of Embryology and Experimental Morphology*, **11**, 697–714.

Bellairs, R. (1974a*). In *The Cell in Medical Science*; vol. 2: *Cellular Genetics,
Development and Cellular Specialisation*, eds F. Beck & J. B. Lloyd, pp. 249–82.
London: Academic Press.

Bellairs, R. (1974b*). In *The Cell in Medical Science*; vol. 2: *Cellular Genetics,
Development and Cellular Specialisation*, eds F. Beck & J. B. Lloyd, pp. 283–312.
London: Academic Press.

Bellairs, R. (1979). *Journal of Embryology and Experimental Morphology*, **51**, 227–43.

Bellairs, R., Curtis, A. S. G. & Sanders, E. J. (1978). *Journal of Embryology and
Experimental Morphology*, **46**, 207–13.

Bellairs, R., Ireland, G. W., Sanders, E. J. & Stern, C. D. (1981). *Journal of Embryology
and Experimental Morphology*, **61**, 15–33.

Bellairs, R. & Portch, P. A. (1977*). In *Vertebrate Limb and Somite Morphogenesis*, eds
D. A. Ede, J. R. Hinchliffe & M. Balls, pp. 449–63. Cambridge University Press.

Bellairs, R., Sanders, E. J. & Portch, P. A. (1980). *Journal of Embryology and Experimental Morphology,* **56**, 41–58.

Bellairs, R. & Veini, M. (1980). *Journal of Embryology and Experimental Morphology* **55**, 93–108.

Bellé, R., Boyer, J. & Ozon, R. (1982). *Developmental Biology,* **90**, 315–19.

Beloussov, L. V., Dorfman, J. G. & Cherdantzev, V. G. (1975). *Journal of Embryology and Experimental Morphology,* **34**, 559–74.

Beloussov, L. V. & Petrov, K. V. (1983). *Ontogenez,* **14**, 21–9.

Benbow, R. M. & Ford, C. C. (1975). *Proceedings of the National Academy of Sciences of the United States of America,* **72**, 2437–41.

Benbow, R. M., Jaenje, H., White, S. H., Breaux, C. B., Krause, M. R., Ford, M. R. & Laskey, R. A. (1977*). In *International Cell Biology 1976–1977,* eds B. P. Brinkley & K. R. Porter, pp. 453–63. New York: Rockefeller University Press.

Benford, H. H. & Namenwirth, M. (1974). *Developmental Biology,* **39**, 172–6.

Bennett, M. V. L., Spray, D. C. & Harris, A. L. (1981). *The American Zoologist,* **21**, 413–27.

Benoit, J. A. A. (1964). *Archives d'Anatomie Microscopique et de Morphologie Expérimentale,* **53**, 357–66.

Benos, D. J. & Biggers, J. D. (1981*). In *Fertilisation and Embryonic Development in Vitro,* eds L. Mastroianni & J. D. Biggers, pp. 285–99. New York: Plenum Press.

Bergink, E. W. & Wallace, R. A. (1974). *Journal of Biological Chemistry,* **249**, 2897–903.

Bergquist, H. (1963*). *Academia Republicii Populare Romine, Studii si Cercetări Stinte Medicale,* **10**, 1–17.

Bijtel, J. H. (1931). *W. Roux' Archiv für Entwicklungsmechanik der Organismen,* **125**, 448–86.

Bijtel, J. H. (1936). *W. Roux' Archiv für Entwicklungsmechanik der Organismen,* **134**, 262–83.

Bijtel, J. H. (1958). *Nederlands Tijdschrift voor Geneeskunde,* **102**, Abstract no. 33.

Billett, F. S. & Adam, E. (1976). *Journal of Embryology and Experimental Morphology,* **33**, 697–710.

Bird, A. P. & Birnstiel, M. (1971). *Biochimica et Biophysica Acta,* **247**, 157–63.

Bishop-Calame, S. (1966*). *Archives d'Anatomie Microscopique et de Morphologie Expérimentale,* **55**, 215–309.

Bjerre, B. (1974). *Experientia,* **30**, 534–5.

Black, J. B. & Patterson, P. H. (1980*). *Current Topics in Developmental Biology,* **15**, 27–40.

Blackler, A. W. (1970*). *Current Topics in Developmental Biology,* **5**, 71–87.

Bluemink, J. G. (1970). *Journal of Ultrastructure Research,* **32**, 142–66.

Bluemink, J. G. (1971*a*). *Cytobiologie,* **3**, 176–87.

Bluemink, J. G. (1971*b*). *Zeitschrift für Zellforschung,* **121**, 102–26.

Bluemink, J. G. (1972). *Journal of Ultrastructure Research,* **41**, 95–114.

Bluemink, J. G. (1978*). In *Cytochalasins – Biochemical and Cell-biological Aspects, Frontiers of Biology,* **46**, 114–42.

Bluemink, J. G. & De Laat, S. W. (1973). *The Journal of Cell Biology,* **59**, 89–108.

Bluemink, J. G. & De Laat, S. W. (1977*). In *The Synthesis, Assembly and Turnover of Cell Surface Components,* eds G. Poste & G. L. Nicolson. *Cell Surface Reviews,* **4**, 403–61. Amsterdam: Elsevier/North Holland Publishing Company.

Bluemink, J. G. & Hoperskaya, O. A. (1975). *W. Roux's Archives of Developmental Biology,* **177**, 75–9.

Bluemink, J. G. & Tertoolen, L. G. J. (1978). *Developmental Biology,* **62**, 334–43.

Bluemink, J. G., Tertoolen, L. G. J., Ververgaert, P. H. J. T. & Verkley, A. J. (1976). *Biochimica et Biophysica Acta,* **443**, 143–55.

Bolender, D. L., Seliger, W. G. & Markwald, R. R. (1980). *The Anatomical Record*, **196**, 401–12.

Bolzern, A., Leonardi, M. C., de Bernardi, F., Maci, R. & Ranzi, S. (1979). *Accademia Nazionale dei Lincei, Rendiconte della Classe di Scienze fisiche, Mathematiche e Naturali*, Ser. VIII, **64**, 621–5.

Borghese, E. (1950). *Journal of Anatomy*, **84**, 287–302.

Born, G. (1885). *Archiv für mikroskopische Anatomie und Entwicklungsmechanik*, **24**, 475–545.

Born, J., Geithe, H. P., Tiedemann, H., Tiedemann, H. & Kocher-Becker, U. (1972). *Zeitschrift für Physiologische Chemie*, **353**, 1075–84.

Born, J., Grunz, H., Tiedemann, H. & Tiedemann, H. (1980). *W. Roux's Archives of Developmental Biology*, **189**, 47–56.

Bosco, L., Filoni, S. & Connata, S. (1979). *Journal of Experimental Zoology*, **209**, 261–82.

Bose, A. & Chatterjee, A. (1964). *Die Naturwissenschaften*, **10**, 249–50.

Bose, A. & Medda, J. (1965). *Folia Biologica*, **13**, 289–95.

Boterenbrood, E. C. (1962). PhD thesis, University of Utrecht, 96 p.

Boterenbrood, E. C. (1970). *Journal of Embryology and Experimental Morphology*, **23**, 751–9.

Boterenbrood, E. C., Narraway, J. M. & Hara, K. (1983). *W. Roux's Archives of Developmental Biology*, **192**, 216–21.

Boterenbrood, E. C. & Nieuwkoop, P. D. (1973). *W. Roux' Archiv für Entwicklungsmechanik der Organismen*, **173**, 319–32.

Boucaut, J. C. (1973). *Differentiation*, **1**, 413–18.

Boucaut, J. C. & Darribere, T. (1983*a*). *Cell Differentiation*, **12**, 77–83.

Boucaut, J. C. & Darribere, T. (1983*b*). *Cell and Tissue Research*, **234**, 135–45.

Bounoure, L. (1939*). *L'Origine des Cellules Reproductrices et le problème de la lignée germinale*, 271 pp. Paris: Gauthier-Villars.

Brachet, J. (1942). *Acta Biologica Belgica*, **2**, 16–19.

Brachet, J. (1942). *Bulletin de l'Academie Royale de Belgique, Classe des Sciences*, Sér. V, **29**, 707–18.

Brachet, J. (1962). *Nature* (London), **193**, 87–8.

Brachet, J. (1965*). *Progress in Biophysical and Molecular Biology*, **15**, 97–127. Oxford: Pergamon Press.

Brachet, J. (1967*a**). In *De l'Embryologie Experimentale à la Biologie Moléculaire*, ed. E. Wolff, pp. 5–22. Paris: Dunod.

Brachet, J. (1967*b**). In *De l'Embryologie Experimentale à la Biologie Moléculaire*, ed. E. Wolff, pp. 23–42. Paris: Dunod.

Brachet, J.(1967*c*). *Experimental Cell Research*, **48**, 233–6.

Brachet, J. (1972*). In *Cell Differentiation: the Proceedings of the first International Conference on Cell Differentiation*, eds R. Harris, P. Allin & D. Viza, pp. 3–9. Copenhagen: Munksgaard.

Brachet, J. (1974*). *Annales de Biologie*, **13**, 403–34.

Brachet, J. (1977*). *Current Topics in Developmental Biology*, **11**, 133–86.

Brachet, J. (1978*). In *Control of Differentiation*, eds L. Saxén, S. Nordling & J. Wartiovaara, *Medical Biology*, **56**, 304–9.

Brachet, J. (1980*). *Comparative Biochemistry and Physiology*, **67** B, 367–72.

Brachet, J., Baltus, E., Pays-de Schutter, A., Hanocq-Quertier, J. & Hubert, E. (1974). *Molecular and Cellular Biochemistry*, **3**, 189–205.

Brachet, J., Baltus, E., Pays-de Schutter, A., Hanocq-Quertier, J., Hubert, E. & Steinert, G. (1975). *Proceedings of the National Academy of Sciences of the United States of America*, **72**, 1574–8.

Brachet, J., Hanocq, F. & van Gansen, P. (1970). *Developmental Biology*, **21**, 157–95.

Brachet, J. & Hubert, E. (1972). *Journal of Embryology and Experimental Morphology*, **27**, 121–45.

Brachet, J., Pays-de Schutter, A. & Hubert, E. (1975). *Differentiation*, **3**, 3–14.

Brannigan, M. & Fabian, B. C. (1980). *Lecture notes on Biomathematics*, no. 33, 131–48.

Branton, D. (1980*). In *The Cell Surface: Mediator of Developmental Processes*, eds S. Subtelny & N. K. Wessels, pp. 3–7. New York: Academic Press.

Brick, I., Schaeffer, B. E., Schaeffer, H. E. & Gennaro, J. F. (1974). *Annals of the New York Academy of Sciences*, **238**, 390–407.

Briggs, R., Green, E. U. & King, T. J. (1951). *Journal of Experimental Zoology*, **116**, 455–500.

Briggs, R. & King, T. J. (1953). *Journal of Experimental Zoology*, **122**, 485–506.

Bronner-Fraser, M. E. & Cohen, A. M. (1980*). *Current Topics in Developmental Biology*, **15**, 1–26.

Bronner-Fraser, M., Sieber-Blum, M. & Cohen, A. M. (1980). *Journal of Comparative Neurology*, **193**, 423–34.

Brown, D. D. (1966*). *National Cancer Institute Monograph*, no. 23, 297–309.

Brown, D. D. & David, I. B. (1968). *Science*, **160**, 272–80.

Brown, D. D. & Littna, E. (1964). *Journal of Molecular Biology*, **8**, 669–87.

Brown, D. D. & Littna, E. (1966*a*). *Journal of Molecular Biology*, **20**, 81–94.

Brown, D. D. & Littna, E. (1966*b*). *Journal of Molecular Biology*, **20**, 95–112.

Brun, R. B. & Garson, J. A. (1983). *Journal of Embryology and Experimental Morphology*, **74**, 275–95.

Brustis, J. J. (1976). *Compte Rendu de l'Académie des Sciences*, Paris, Sér. D, **283**, 379–82.

Brustis, J. J. (1978). *Compte Rendu de l'Académie des Sciences*, Paris, Sér. D, **287**, 1153–5.

Brustis, J. J. (1979). *Archives de Biologie*, **90**, 261–72.

Burgess, D. R. & Schroeder, T. E. (1979*). *Methods and Achievements in Experimental Pathology*, **8**, 171–89.

Burnside, M. B. (1973). *The American Zoologist*, **13**, 989–1006.

Burnside, M. B. & Jacobson, A. G. (1968). *Developmental Biology*, **18**, 537–52.

Bustuoabad, O. & Pisanó, A. (1971). *Acta Embryologiae Experimentalis*, **3**, 225–37.

Butros, J. M. (1963*a*). *Journal of Experimental Zoology*, **152**, 57–66.

Butros, J. M. (1963*b*). *Journal of Experimental Zoology*, **154**, 125–34.

Butros, J. M. (1965). *Journal of Embryology and Experimental Morphology*, **13**, 119–28.

Butros, J. M. (1967). *Journal of Embryology and Experimental Morphology*, **17**, 119–30.

Byers, B. & Porter, K. R. (1964). *Proceedings of the National Academy of Sciences of the United States of America*, **52**, 1091–9.

Byskov, A. G. (1982*). In *Reproduction in Mammals*, vol. I, *Germ Cells and Fertilization*, eds C. R. Austin & R. V. Short, pp. 1–16. Cambridge University Press.

Bytinski-Salz, H. (1937). *Archivio Italiano di Anatomia e di Embriologia*, **39**, 177–228.

Cambar, R. & Gipouloux, J. D. (1970*a*). *Compte Rendu de l'Académie des Sciences*, Paris, Sér. D, **270**, 1359–61.

Cambar, R. & Gipouloux, J. D. (1970*b*). *Compte Rendu de l'Academie des Sciences*, Paris, Sér D, **270**, 1607–9.

Cambar, R. & Marrot, B. (1954). *Bulletin Biologique de la France et de la Belgique*, **88**, 168–77.

Camosso, M. E. & Roncali, L. (1968). *Acta Embryologiae et Morphologiae Experimentalis*, **10**, 243–63.

Campanella, C. & Andreuccetti, P. (1977). *Gamete Research*, **3**, 99–114.

Campbell, J. C. (1965). *Journal of Embryology and Experimental Morphology*, **13**, 171–9.

Campbell, R. D. (1967). *Proceedings of the National Academy of Sciences of the United States of America*, **58**, 1422–9.

Capco, D. G. (1982). *Journal of Experimental Zoology*, **219**, 147–54.

Capco, D. G. & Jeffery, W. R. (1981). *Nature* (London), **294**, 255–6.

Capco, D. G. & Jeffery, W. R. (1982). *Developmental Biology*, **89**, 1–12.

Capuron, A. & Maufroid, J. P. (1981). *Archives d'Anatomie Microscopique et de Morphologie Expérimentale*, **70**, 219–26.

Cardellini, P., Milan, F. & Sala, M. (1982). *Acta Embryologiae et Morphologiae Experimentalis*, **3**, p. VIII/IX.

Carroll, E. J. & Epel, D. (1975). *Developmental Biology*, **44**, 22–32.

Cassin, C. & Capuron, A. (1979). *Journal de Biologie Buccale*, **7**, 61–76.

Chambers, E. L. (1975). *The Journal of Cell Biology*, **67**, 60a, Abstr. no. 120.

Chambers, E. L. (1976). *Journal of Experimental Zoology*, **197**, 149–54.

Chan, S. T. H. (1977*). In *Handbook of Sexology*, eds H. Musaph & J. Money, pp. 91–105. Amsterdam: Exerpta Medica, Biological and Medical Press.

Chan, S. T. H. & Wai-Sum, O. (1981*). In *Mechanisms of Sex Differentiation in Animals and Man*, eds C. R. Austin & R. G. Edwards, pp. 55–111. London: Academic Press.

Chanconic, M. & Clairambault, P. (1975). *Compte Rendu de l'Académie des Sciences*, Paris, Sér. D, **280**, 475–8.

Chandebois, R. (1981). *Acta Biotheoretica*, **30**, 143–69.

Chandebois, R. & Faber, J. (1983*). *Automation in Animal Development*, 204 p. Basel: S. Karger Press.

Charbonneau, M. & Picheral, B. (1983). *Development, Growth and Differentiation*, **25**, 23–37.

Chevallier, A. (1972). *Journal of Embryology and Experimental Morphology*, **27**, 603–14.

Chevallier, A., Kieny, M. & Mauger, A. (1976). *Compte Rendu de l'Académie des Sciences*, Paris, Sér. D, **282**, 309–11.

Chiang, W. (1964). *Acta Biologiae Experimentalis Sinica*, **9**, 18–25.

Chibon, P. (1966). *Mémoires de la Société Zoologique de France*, **36**, 4–107.

Chibon, P. (1967). *Journal of Embryology and Experimental Morphology*, **18**, 343–58.

Chibon, P. (1969). *Annales d'Embryologie et de Morphogenèse*, **2**, 307–15.

Chibon, P. (1970). *Journal of Embryology and Experimental Morphology*, **24**, 479–96.

Chibon, P. (1974*). *Annales de Biologie*, **13**, 459–80.

Choudhury, S. & Khare, M. K. (1978). *Indian Journal of Experimental Biology*, **16**, 555–7.

Christ, B. (1970). *Verhandlungen der anatomischem Gesellschaft*, **64**, 555–64.

Christ B. (1971). *Verhandlungen der anatomischen Gesellschaft*, **65**, 255–61.

Christ, B., Jacob, H. J. & Jacob, M. (1972). *Zeitschrift für Anatomie und Entwicklungsgeschichte*, **138**, 82–97.

Christ, B., Jacob, H. J. & Jacob, M. (1974). *Verhandlungen der anatomischen Gesellschaft*, **68**, 573–9.

Christ, B., Jacob, H. J. & Jacob, M. (1979). *Experientia*, **35**, 1376–8.

Chuang, H. H. (1938). *Biologisches Zentralblatt*, **58**, 472–80.

Chuang, H. H. (1939). *W. Roux' Archiv für Entwicklungsmechanik der Organismen*, **139**, 556–638.

Chuang, H. H. (1940). *W. Roux' Archiv für Entwicklungsmechanik der Organismen*, **140**, 25–38.

Chuang, H. H. (1955). *Chinese Journal of Experimental Biology*, **4**, 151–86.

Chuang, H. H. (1963). *Acta Biologiae Experimentalis Sinica*, **8**, 370–87.

Chuang, H. H. & Tseng Mi-Pai (1956*a*). *Scientia Sinica*, **6**, 668–708.

Chuang, H. H. & Tseng Mi-Pai (1956*b*). *Acta Biologiae Experimentalis Sinica*, **5**, 289–371.

Chulitskaia, E. V. (1962). *Doklady Akademii Nauk SSSR, Embryol.*, **144**, 245–7.

Chulitskaia, E. V. (1967). *Doklady Akademii Nauk SSSR, Embryol.*, **173**, 1473–6.

Chulitskaia, E. V. (1970). *Journal of Embryology and Experimental Morphology*, **23**, 359–74.

Chulitskaia, E. V. & Felgengauer, P. E. (1977). *Ontogenez*, **8**, 305–8.

Chung, H.-M. & Malacinski, G. M. (1975). *Proceedings of the National Academy of Sciences of the United States of America*, **72**, 1235–9.

Chung, H.-M. & Malacinski, G. M. (1980). *Developmental Biology*, **80**, 120–33.

Chung, H.-M. & Malacinski, G. M. (1981). *Differentiation*, **18**, 185–9.

Chung, H.-M. & Malacinski, G. M. (1983). *Journal of Embryology and Experimental Morphology*, **73**, 207–20.

Chung, S. H. & Cooke, J. (1978). *Proceedings of the Royal Society*, London, Ser. B, **201**, 335–73.

Cigada, M. L., Maci, R. & de Bernardi, F. (1968). *Rendiconti della Istituto Lombardo – Accademia di Scienze e Lettere, Classe di Scienze*, **102**, 213–26.

Cigada-Leonardi, M., de Bernardi, F., Fascio, U., Maci, R., Ranzi, S. & Sotgia, C. (1980). *Abstracts of papers read at the XIVth International Embryological Conference*, in Patras, Greece, September 11–17, 1980, p. 49, Abstr. no. 22.

Clairambault, P. (1968). *Archives de Biologie*, **79**, 537–78.

Clairambault, P. (1971). *Acta Embryologiae Experimentalis*, 61–92.

Clarke, W. M. & Fowler, I. (1960). *Developmental Biology*, **2**, 155–72.

Clavert, A. (1974). *Compte Rendu des Séances de la Société de Biologie*, **168**, 807–11.

Clavert, J. (1961). *Bulletin de la Société Zoologique de France*, **86**, 381–401.

Clavert, J. (1962*). *Advances in Morphogenesis*, **2**, 27–60.

Clavert, J. & Filogamo, G. (1957). *Compte Rendu des Séances de la Société de Biologie*, **151**, 1740–2.

Clavert, J. & Filogamo, G. (1959). *Compte Rendu de l'Association des Anatomistes*, **46**, 165–9.

Clavert, J. & Zahnd, J. P. (1955). *Compte Rendu des Séances de la Société de Biologie*, **149**, 1650.

Clayton, M. B. & Dixon, K. E. (1975). *Journal of Experimental Zoology*, **192**, 277–83.

Clayton, R. M. (1982*). In *Stability and Switching in Cellular Differentiation*, eds R. M. Clayton & D. E. S. Truman, pp. 23–48. New York: Plenum Press.

Clayton, R. M., Thomson, I. & de Pomerai, D. I. (1979). *Nature*, **282**, 628–9.

Clayton, R. M., Truman, D. E. S. & Bird, A. P. (1982*). In *Stability and Switching in Cellular Differentiation*, eds R. M. Clayton & D. E. S. Truman, pp. 327–30. New York: Plenum Press.

Cleine, J. H. (1983). PhD thesis, Flinders University, S. Australia, 154 p.

Cleine, J. H., Boorman, A. P. & Dixon, K. E. (1982). *Abstracts of papers read at the XVth International Embryological Conference*, in Straszbourg, June 1982, p. 40, Abstr. no. 49.

Cochard, P. & le Douarin, N. M. (1982*). *Scandinavian Journal of Gastroenterology*, **17**, suppl. 71, 1–14.

Coggins, L. W. (1973). *Journal of Cell Science*, **12**, 71–93.

Cohen, A. M. (1972). *Journal of Experimental Zoology*, **179**, 167–82.

Cohen, A. M. & Konigsberg, I. C. (1975). *Developmental Biology*, **46**, 262–80.

Cohen, J. (1979*). In *Maternal effects in Development*, 4th Symposium of the British Society of Developmental Biology, eds D. R. Newth & M. Balls, pp. 1–28. Cambridge University Press.

Cohen, S. (1965*). In *Developmental and Metabolic Control Mechanisms and Metaplasia*, 19th Annual Symposium of Fundamental Cancer Research, **19**, 251–72. Baltimore: Williams & Wilkins.

Collins, F. (1976). *Developmental Biology*, **49**, 281–94.

Collins, F. & Epel, D. (1977). *Experimental Cell Research*, **106**, 211–22.

Conklin, E. G. (1905*a*). *Journal of the Academy of Natural Sciences of Philadelphia*, **13**, 1–119.

Conklin, E. G. (1905*b*). *Biological Bulletin*, **8**, 205–30.

Conklin, E. G. (1932). *Journal of Morphology*, **54**, 69–151.

Conklin, E. G. (1933). *Journal of Experimental Zoology*, **64**, 303–51.

Cooke, J. (1972*a*). *Journal of Embryology and Experimental Morphology*, **28**, 13–26.

Cooke, J. (1972*b*). *Journal of Embryology and Experimental Morphology*, **28**, 27–46.

Cooke, J. (1972*c*). *Journal of Embryology and Experimental Morphology*, **28**, 47–56.

Cooke, J. (1973*a*). *Journal of Embryology and Experimental Morphology*, **30**, 49–62.

Cooke, J. (1973*b*). *Nature* (London), **242**, 55–7.

Cooke, J. (1973*c*). *Journal of Embryology and Experimental Morphology*, **30**, 283–300.

Cooke, J. (1975*a*). *Journal of Embryology and Experimental Morphology*, **33**, 147–57.

Cooke, J. (1975*b*). *Nature* (London), **254**, 196–9.

Cooke, J. (1979*a*). *Journal of Embryology and Experimental Morphology*, **51**, 165–82.

Cooke, J. (1979*b*). *Journal of Embryology and Experimental Morphology*, **53**, 269–89.

Cooke, J. (1980*). *Current Topics in Developmental Biology, Neural Development: Part I. Emergence of Specificity in Neural Histogenesis*, **15**, 373–407.

Cooke, J. (1981). *Nature* (London), **290**, 775–9.

Cooper, G. W. (1965). *Developmental Biology*, **12**, 185–212.

Cooper, G. W. & Bedford, J. M. (1971). *Journal of Reproduction and Fertility*, **25**, 431–6.

Cooper, M. S. & Keller, R. E. (1984). *Proceedings of the National Academy of Sciences of the United States of America*, **81**, 160–7.

Corner, M. A. (1966). *Experientia*, **22**, 188–90.

Corsin, J. (1972). *Archives d'Anatomie Microscopique et de Morphologie Expérimentale*, **61**, 47–60.

Corsin, J. (1975). *Journal of Embryology and Experimental Morphology*, **33**, 335–42.

Coulombre, A. J. (1965). *Investigative Ophthalmology*, **4**, 411–19.

Coulombre, J. L. & Coulombre, A. J. (1963). *Science*, **142**, 1489–90.

Criley, B. B. (1969). *Journal of Morphology*, **128**, 465–502.

Crippa, M. (1970). *Nature* (London), **227**, 1138–40.

Critchley, D. R., England, M. A. & Wakely, J. (1979). *Nature* (London), **280**, 498–500.

Croisille, Y. (1963*). *Annales de Biologie*, **2**, 155–77.

Croisille, Y., Gumpel-Pinot, M. & Martin, J. (1976*). In *Organ Culture in Biomedical Research*, Symposium of the British Society for Cell Biology, eds M. Balls & M. A. Monnickendam, pp. 95–109. Cambridge University Press.

Cross, N. L. & Elinson, R. P. (1978). *The American Zoologist*, **18**, 642 (Abstract).

Cross, N. L. & Elinson, R. P. (1980). *Developmental Biology*, **75**, 187–98.

Cuevas, P. (1977). *Experientia*, **33**, 660–1.

Cuevas, P. & Orts Llorca, F. (1974). *Acta Anatomica*, **89**, 423–30.

Cumba, G. R. (1972). *The Anatomical Record*, **173**, 205–12.

Curtis, A. S. G. (1960). *Journal of Embryology and Experimental Morphology*, **8**, 163–73.

Curtis, A. S. G. (1961). *Experimental Cell Research*, suppl. **8**, 107–22.

Curtis, A. S. G. (1962*a*). *Journal of Embryology and Experimental Morphology*, **10**, 410–22.

Curtis, A. S. G. (1962*b*). *Journal of Embryology and Experimental Morphology*, **10**, 451–63.

Cusimano-Carollo, T. (1972). *Acta Embryologiae Experimentalis*, 289–322.

Czołowska, R. (1969). *Journal of Embryology and Experimental Morphology*, **22**, 229–51.

Dalcq, A. (1947). *Acta Anatomica*, **4**, 100–7.

Dalcq, A. & Dollander, A. (1948). *Compte Rendu des Séances de la Société de Biologie*, **142**, 1307–12.

Dalcq, A. & Pasteels, J. (1937). *Archives de Biologie*, **48**, 670–712.

Dalcq, A. & Pasteels, J. (1938). *Bulletin de l'Académie Royale de Belgique, Classe des Sciences*, Sér. VI, **3**, 261–308.

Dameron, F. (1972). *Bulletin de la Société Zoologique de France*, **97**, 497–503.

Dan, J. C. (1967*). In *Fertilization: Comparative Morphology, Biochemistry and Immunology*, eds C. B. Metz & A. Monroy, pp. 237–93. New York: Academic Press.

Dan, K. & Kojima, M. K. (1963). *Journal of Experimental Biology*, **40**, 7–14.
Dasgupta, J. D. & Singh, U. N. (1981). *W. Roux's Archives of Developmental Biology*, **190**, 358–60.
Dasgupta, S. & Kung-Ho, C. (1971). *Experimental Cell Research*, **65**, 463–6.
Davenport, R. (1979*). *An Outline of Animal Development*, 412 p. Reading, Mass.: Addison Wesley.
David, D. (1970). *Annales d'Embryologie et de Morphogenèse*, **3**, 337–53.
Davidson, D. (1983a). *Journal of Embryology and Experimental Morphology*, **74**, 245–59.
Davidson, D. (1983b). *Journal of Embryology and Experimental Morphology*, **74**, 261–73.
Davidson, E. H. (1976*). *Gene Activity in Early Development*, 452 p. New York: Academic Press.
Davidson, E. H., Crippa, M., Kramer, E. R. & Mirsky, A. E. (1966). *Proceedings of the National Academy of Sciences of the United States of America*, **56**, 856–63.
Davidson, E. H., Crippa, M. & Mirsky, A. E. (1968). *Proceedings of the National Academy of Sciences of the United States of America*, **60**, 152–9.
Davidson, E. H., Haslett, G. W., Finney, R. J., Allfrey, V. G. & Mirsky, A. E. (1965). *Proceedings of the National Academy of Sciences of the United States of America*, **54**, 696–704.
Davis, E. M. & Trinkaus, J. P. (1981). *Journal of Embryology and Experimental Morphology*, **63**, 29–51.
Dawid, I. B. (1966). *Proceedings of the National Academy of Sciences of the United States of America*, **56**, 269–76.
De Bernardi Laria, F., Maci, R. L., Cigada, M. L. & Ranzi, S. (1977). *Rendiconti dell' Accademia di Scienze e Lettere, Classe de Scienze B*, **111**, 69–79.
Decker, R. S. (1981). *Developmental Biology*, **81**, 12–22.
Decker, R. S. & Friend, D. S. (1974). *The Journal of Cell Biology*, **62**, 32–47.
De Graaff, A. B. (1960). *Acta Embryologiae et Morphologiae Experimentalis*, **3**, 40–52.
De Haan, R. L. (1964). *Journal of Experimental Zoology*, **157**, 127–38.
De Laat, S. W. (1972). *Abstracts of papers read at the IVth International Biophysical Congress*, Moscow, Sections IX–XV, pp. 413.
De Laat, S. W. (1975). PhD thesis, University of Utrecht, 57 p.
De Laat, S. W. & Barts, P. W. J. A. (1976). *Journal of Membrane Biology*, **27**, 131–51.
De Laat, S. W., Barts, P. W. J. A. & Bakker, M. I. (1976). *Journal of Membrane Biology*, **27**, 109–29.
De Laat, S. W. & Bluemink, J. G. (1974). *The Journal of Cell Biology*, **60**, 529–40.
De Laat, S. W., Luchtel, D. & Bluemink, J. G. (1973). *Developmental Biology*, **31**, 163–77.
De Laat, S. W., Wouters, W., Marques Da Silva Pimenta Guarda, M. M. & Da Silva Guarda, M. A. (1975). *Experimental Cell Research*, **91**, 15–30.
Denis, H. (1964). *Developmental Biology*, **9**, 435–57.
Denis, H. (1968*). *Advances in Morphogenesis*, **7**, 115–50.
Denis, H. (1974*). In *Chemical Zoology: Amphibia and Reptilia*, eds M. Florkin & B. T. Scheer, vol. 9, 3–22. San Francisco: Academic Press.
Denis-Domini, S., Baccetti, B. & Monroy, A. (1976). *Journal of Ultrastructure Research*, **57**, 104–12.
Denucé, J. M. (1976). *Archives Internationales de Physiologie et de Biochémie*, **84**, 1067–8.
Deol, M. S. (1966). *Nature* (London), **209**, 219–20.
De Pomerai, D. I., Pritchard, D. J. & Clayton, R. M. (1977). *Developmental Biology*, **60**, 416–27.
Desnitsky, A. G. (1974). *Vestnik Leningradskogo Gosudarstvennogo Universiteta*, **21**, 18–21.
Desnitsky, A. G. (1978). *Ontogenez*, **9**, 197–200.
Desnitsky, A. G. (1979). *Ontogenez*, **10**, 3–12.
Destrée, O. H. J. (1975). PhD thesis, University of Amsterdam, 114 p.

Dettlaff, T. A. (1964*). *Advances in Morphogenesis*, **3**, 323–62.
Dettlaff, T. A. (1966). *Journal of Embryology and Experimental Morphology*, **16**, 183–95.
Dettlaff, T. A. (1983). *Journal of Embryology and Experimental Morphology*, **75**, 67–86.
Dettlaff, T. A., Felgengauer, P. E. & Chulitskaia, E. V. (1977). *Ontogenez*, **8**, 478–86.
Dettlaff, T. A. & Ginzburg, A. S. (1954*). *The Embryonic Development of the Sturgeons (Acipenser stellatus, A. gülderstädti colchicus and Huso huso) in Connection with Problems of Breeding*, ed. S. G. Kryzhanovskii, 216 p. Moscow: USSR Academic Science Press (translated from Russian).
Dettlaff, T. A., Ginzburg, A. S. & Shmalgauzen, O. I. (1981*). *Development of True Sturgeons; Egg Maturation, Fertilization, Embryonic and Larval Development*, 224 p. Moscow: Nauka.
Dettlaff, T. A. & Skoblina, M. N. (1969). *Annales d'Embryologie et de Morphogenèse*, suppl. 1, 133–51.
Deuchar, E. M. (1966*). *Biochemical Aspects of Amphibian Development*, 206 p. London: Methuen & Co.
Deuchar, E. M. (1970). *Developmental Biology*, **22**, 185–99.
Deuchar, E. M. (1971). *Acta Embryologiae Experimentalis*, 93–101.
Deuchar, E. M. (1972*). *Biological Reviews*, **47**, 37–112.
Deuchar, E. M. (1973*). *Advances in Morphogenesis*, **10**, 173–225.
Deuchar, E. M. & Burgess, A. M. C. (1967). *Journal of Embryology and Experimental Morphology*, **17**, 239–58.
Devillers, Ch. (1951). *Compte Rendu de l'Association des Anatomistes*, Nancy, 38ᵉ Réunion, **67**, 418–25.
Devillers, Ch. (1956*). *Annales de Biologie*, **32**, 437–56.
Devillers, Ch. (1961*). *Advances in Morphogenesis*, **1**, 379–428.
Dhouailly, D. (1967). *Journal of Embryology and Experimental Morphology*, **18**, 389–400.
Dhouailly, D. (1970). *Journal of Embryology and Experimental Morphology*, **24**, 73–94.
Dhouailly, D. (1977*). *Frontiers in Matrix Biology*, **4**, 86–121.
Dhouailly, D. & Hardy, M. H. (1978). *W. Roux's Archives of Developmental Biology*, **185**, 195–200.
Dhouailly, D. & Kieny, M. (1972). *Developmental Biology*, **28**, 162–75.
Dhouailly, D., Rogers, G. E. & Sengel, P. (1978). *Developmental Biology*, **65**, 58–68.
Dhouailly, D. & Sengel, P. (1975). *Compte Rendu de l'Academie des Sciences*, Paris, Sér. D, **281**, 1007–10.
Dicaprio, R. A., French, A. S. & Sanders, E. J. (1976). *Journal of Membrane Biology*, **27**, 393–408.
Dictus, W. J. A. G., van Zoelen, E. J. J., Tetteroo, P. A. T., Tertoolen, L. G. J., de Laat, S. W. & Bluemink, J. G. (1984). *Developmental Biology*, **101**, 201–11.
Dieterlen-Lièvre, F. (1970). *Developmental Biology*, **22**, 138–56.
Dieterlen-Lièvre, F. & Hadorn, H. B. (1972). *W. Roux' Archiv für Entwicklungsmechanik der Organismen*, **170**, 175–84.
Dollander, A. (1950). *Archives de Biologie*, **61**, 1–111.
Dollander, A. (1957). *Compte Rendu des Séances de la Société de Biologie*, Paris, **151**, 977–9.
Dollander, A. (1961*). *Proceedings 1st European Congress of Anatomy*, 1960, 274–306.
Dollander, A. (1962). *Archives d'Anatomie, d'Histologie et d'Embryologie*, **44**, suppl. 93–103.
Dollander, A. & Melnotte, J. P. (1952). *Compte Rendu des Séances de la Société de Biologie*, **146**, 1614–16.
Dollander, A. & Vivier, F. (1954). *Compte Rendu des Séances de la Société de Biologie*, **148**, 135–6.
Doucet-de Bruïne, M. H. M. (1973). *W. Roux' Archiv für Entwicklungsmechanik der Organismen*, **173**, 136–63.

Drews, U., Kocher-Becker, U. & Drews, U. (1972). *W. Roux' Archiv für Entwicklungsmechanik der Organismen*, **171**, 17–37.

Dreyer, C., Wang, Y. H., Wedlich, D. & Hausen, P. (1983). In *Current Problems in Germ Cell Differentiation*, eds A. McLaren & C. C. Wylie, pp. 329–51. Cambridge University Press.

Duband, J. D. & Thiery, J. P. (1982). *Developmental Biology*, **94**, 337–50.

Dumont, J. N. (1972). *Journal of Morphology*, **136**, 153–80.

Duncan, C. J. (1979). *Experientia*, **35**, 817.

Duprat, A. M., Gualandris, L. & Rougé, P. (1982). *Journal of Embryology and Experimental Morphology*, **70**, 171–87.

Duprat, A. M., Mathieu, C. & Buisan, J. J. (1977). *Differentiation*, **9**, 161–7.

Duspiva, F. (1962*a**). *Verhandlungen der Deutschen Zoologischen Gesellschaft Zoologischer Anzeiger*, Suppl. **25**, 210–50.

Duspiva, F. (1962*b**). *13^{es} Colloquium Gesellschaft für Physiologische Chemie*, Mosbach/Baden, 205–45.

Duspiva, F. (1969*). *Naturwissenschaftliche Rundschau*, **22**, 191–202.

Dziadek, M. (1979). *Journal of Embryology and Experimental Morphology*, **53**, 367–79.

Eakin, R. M. & Lehmann, F. E. (1957). *W. Roux' Archiv für Entwicklungsmechanik der Organismen*, **150**, 177–98.

Ebert, J. D. (1968*). *Current Topics in Developmental Biology*, 3, XV–XXV.

Eddy, E. M. & Shapiro, B. M. (1976). *The Journal of Cell Biology*, **71**, 35–48.

Edelman, G. M. (1983). *Science*, **219**, 450–7.

Edelman, G. M., Gallin, W. J., Delouvée, A., Cunningham, B. A. & Thiery, J. P. (1983). *Proceedings of the National Academy of Sciences of the United States of America*, **80**, 4384–8.

Eisenberg, S. & van Alten, P. J. (1964). *Acta Embryologiae et Morphologiae Experimentalis*, **7**, 61–70.

Elinson, R. P. (1975). *Developmental Biology*, **47**, 257–68.

Elinson, R. P. (1980*). In *The Cell Surface: Mediator of Developmental Processes*, eds S. Subtelny & N. K. Wessels, pp. 217–34. New York: Academic Press.

Elinson, R. P. & Manes, M. E. (1978). *Developmental Biology*, **63**, 67–75.

Ellinger, M. S. (1978). *Developmental Biology*, **65**, 81–9.

Ellinger, M. S. & Murphy, J. A. (1979*a*). *Journal of Embryology and Experimental Morphology*, **59**, 249–61.

Ellinger, M. S. & Murphy, J. A. (1979*b*). *Journal of Anatomy*, **129**, 361–76.

Elsdale, T. & Pearson, M. (1979). *Journal of Embryology and Experimental Morphology*, **53**, 245–67.

Emanuelsson, H. (1961). *Acta Physiologica Scandinavica*, **52**, 211–33.

England, M. A. (1973). *Experientia*, **29**, 1267–8.

England, M. A. (1974). *Experientia*, **30**, 808–9.

England, M. A. (1981*a*). *Biblioteca Anatomica*, **19**, 131–5.

England, M. A. (1981*b*). *Journal of Microscopy*, **123**, 133–46.

England, M. A. & Cowper, S. V. (1975). *Experientia*, **31**, 1449–51.

England, M. A. & Cowper, S. V. (1976). *Experientia*, **32**, 1578–80.

Engländer, H. (1962*a*). *W. Roux' Archiv für Entwicklungsmechanik der Organismen*, **154**, 124–42.

Engländer, H. (1962*b*). *W. Roux' Archiv für Entwicklungsmechanik der Organismen*, **154**, 134–59.

Engländer, H. & Johnen, A. G. (1967). *W. Roux' Archiv für Entwicklungsmechanik der Organismen*, **159**, 346–56.

Epel, D. (1978*). *Current Topics in Developmental Biology*, **19**, 185–216.

Epel, D., Cross, N. L. & Epel, N. (1977). *Development, Growth and Differentiation*, **13**, 15–21.

Epel, D., Steinhardt, R., Humphreys, T. & Mazia, D. (1974). *Developmental Biology*, **40**, 245–55.

Epel, D. & Vacquier, V. D. (1978*). In *Cell Surface Reviews*, vol. 5, *Membrane Fusion*, eds G. Poste & G. L. Nicolson, pp. 1–63. Amsterdam: Elsevier/North Holland Biomedical Press.

Eppenberger, H. M. (1975*). In *Biochemistry of Animal Development*, vol. 3, *Molecular Aspects of Animal Development*, ed. R. Weber, pp. 217–55. New York: Academic Press.

Epperlein, H. H. (1974). *Differentiation*, **2**, 151–67.

Epperlein, H. H. (1978*a*). *Zoon, a Journal of Zoology*, **6**, 123–7.

Epperlein, H. H. (1976*b*). *Differentiation*, **11**, 109–23.

Epperlein, H. H. & Lehmann, R. (1975). *Differentiation*, **4**, 159–74.

Errick, J. E. & Saunders, J. W. (1976). *Developmental Biology*, **50**, 26–34.

Estensen, R. D., Rosenberg, M. & Sheridan, J. D. (1971). *Science*, **173**, 356–8.

Etheridge, A. L. (1968). *Journal of Experimental Zoology*, **169**, 357–70.

Etheridge, A. L. (1972). *W. Roux' Archiv für Entwicklungsmechanik der Organismen*, **169**, 268–70.

Evans, D. & Birnstiel, M. L. (1968). *Biochimica et Biophysica Acta*, **166**, 274–6.

Eyal-Giladi, H. (1954). *Archives de Biologie*, **65**, 179–259.

Eyal-Giladi, H. (1969). *Journal of Embryology and Experimental Morphology*, **21**, 177–92.

Eyal-Giladi, H. (1970*a*). *Journal of Embryology and Experimental Morphology*, **23**, 739–49.

Eyal-Giladi, H. (1970*b*). *Annales d'Embryologie et de Morphogenèse*, **3**, 133–43.

Eyal-Giladi, H. & Fabian, B. C. (1980). *Developmental Biology*, **77**, 228–32.

Eyal-Giladi, H., Farbiasz, I., Ostrovsky, D. & Hochmann, J. (1975*b*). *Developmental Biology*, **45**, 358–65.

Eyal-Giladi, H., Ginsburg, M. & Farbarov, A. (1981). *Journal of Embryology and Experimental Morphology*, **65**, 139–47.

Eyal-Giladi, H. & Kochav, S. (1976). *Developmental Biology*, **49**, 321–37.

Eyal-Giladi, H., Kochav, S. & Yerushalmi, S. (1975*a*). *Differentiation*, **4**, 57–60.

Eyal-Giladi, H. & Spratt, N. T. (1964). *The American Zoologist*, **4**, 428.

Eyal-Giladi, H. & Spratt, N. T. (1965). *Journal of Embryology and Experimental Morphology*, **13**, 267–73.

Eyal-Giladi, H. & Wolk, M. (1970). *W. Roux' Archiv für Entwickslungsmechanik der Organismen*, **165**, 226–41.

Fabian, B. & Eyal-Giladi, H. (1981). *Journal of Embryology and Experimental Morphology*, **64**, 11–22.

Fallon, J. F. & Kelley, R. O. (1977). *Journal of Embryology and Experimental Morphology*, **41**, 223–32.

Fankhauser, G. (1925). *W. Roux' Archiv für Entwicklungsmechanik der Organismen*, **105**, 501–80.

Fankhauser, G. (1930). *W. Roux' Archiv für Entwicklungsmechanik der Organismen*, **122**, 671–735.

Fankhauser, G. (1932). *Journal of Experimental Zoology*, **62**, 185–235.

Fankhauser, G. (1948). *Annals of the New York Academy of Sciences*, **49**, 684–708.

Farinella-Ferruzza, N. (1961). *La Ricerca Scientifica*, **31**, II B, 37–41.

Faulhaber, I. (1972). *W. Roux' Archiv für Entwicklungsmechanik der Organismen*, **171**, 87–108.

Faulhaber, I. & Geithe, H. P. (1972). *Revue Suisse de Zoologie*, **79**, 103–17.

Faulhaber, I. & Lyra, L. (1974). *W. Roux' Archiv für Entwicklungsmechanik der Organismen*, **176**, 151–7.

Fautrez-Firlefyn, N. & Fautrez, J. (1967*). *International Review of Cytology*, **22**, 171–204.

Fawcett, D. W. (1972*). *Proceedings of the International Symposium: The Genetics of the*

Spermatozoon, eds R. A. Beatty & S. Glücksohn-Waelsch, pp. 37–68. Edinburgh: Dept. of Genetics, University of Edinburgh.

Fawcett, D. W. (1975*). In *Developmental Biology of Reproduction*, eds C. L. Markert & J. Papaconstantinou, pp. 25–53. New York: Academic Press.

Fell, H. B. & Mellanby, E. (1953). *Journal of Physiology*, **119**, 470–88.

Fernández, M. S. (1979). *Archives d'Anatomie microscopique et de Morphologie Expérimentale*, **68**, 214–15.

Fernández, M. S. & Izquierdo, L. (1980). *Anatomy and Embryology*, **160**, 77–81.

Ferrand, R. (1972). *Archives de Biologie*, **83**, 297–371.

Ferris, W. & Bagnara, J. T. (1960). *The Anatomical Record*, **137**, 355 (Abstract).

Ficq, A. (1972). *Experimental Cell Research*, **73**, 242–8.

Filoni, S., Bosco, L., Cioni, C. & Venturini, G. (1983). *Experientia*, **39**, 315–17.

Finnigan, C. V. & Briggin, W. P. (1966). *Journal of Embryology and Experimental Morphology*, **15**, 1–14.

Flickinger, R. A. (1963*). *The American Zoologist*, **3**, 209–21.

Flickinger, R. A. (1969). *Experimental Cell Research*, **55**, 422–3.

Flickinger, R. A. (1970*a*). *Experientia*, **26**, 778–9.

Flickinger, R. A. (1970*b*). *Developmental Biology*, Suppl. 4, 12–41.

Flickinger, R. A. (1972). *W. Roux' Archiv für Entwicklungsmechanik der Organismen*, **171**, 256–8.

Flickinger, R. A. (1980). *W. Roux's Archives of Developmental Biology*, **188**, 9–12.

Flickinger, R. A. & Daniel, J. C. (1972). *W. Roux' Archiv für Entwicklungsmechanik der Organismen*, **169**, 350–2.

Flickinger, R. A., Daniel, J. C. & Greene, R. F. (1970). *Nature* (London), **228**, 557–9.

Flickinger, R. A., Freedman, M. L. & Stambrook, P. J. (1967*a*). *Developmental Biology*, **16**, 457–73.

Flickinger, R. A., Greene, R. F., Kohl, D. M. & Miyagi, M. (1966). *Proceedings of the National Academy of Sciences of the United States of America*, **56**, 1712–18.

Flickinger, R. A., Miyagi, M., Moser, C. R. & Rollins, E. (1967*b*). *Developmental Biology*, **15**, 414–31.

Flickinger, R. A., Moser, C. R. & Rollins, E. (1967*c*). *Experimental Cell Research*, **46**, 78–88.

Flickinger, R. A. & Stone, G. (1960). *Experimental Cell Research*, **21**, 541–7.

Flower, M. & Grobstein, C. (1967). *Developmental Biology*, **15**, 193–205.

Fontaine, J. & le Douarin, N. M. (1977). *Journal of Embryology and Experimental Morphology*, **41**, 209–22.

Ford, P. J. (1979*). In *Maternal Effects in Development*, eds D. R. Newth & M. Balls, pp. 81–110. Cambridge University Press.

Forman, D. & Slack, J. M. W. (1980). *Nature* (London), **286**, 492–4.

Fraser, R. C. (1960). *Journal of Experimental Zoology*, **145**, 151–67.

Fukuda, S. (1979). *Journal of Embryology and Experimental Morphology*, **52**, 49–62.

Fuldner, D. & von Ubisch, L. (1965). *W. Roux' Archiv für Entwicklungsmechanik der Organismen*, **155**, 693–700.

Fullilove, S. L. (1970). *Journal of Experimental Zoology*, **175**, 323–6.

Gall, J. C. (1968). *Proceedings of the National Academy of Sciences of the United States of America*, **60**, 553–60.

Gall, L., Picheral, B. & Gounon, P. (1983). *Biologie Cellulaire*, **47**, 331–42.

Gallera, J. (1961). *Revue Suisse de Zoologie*, **68**, 311–30.

Gallera, J. (1965). *Experientia*, **21**, 218–20.

Gallera, J. (1966*a*). *Bulletin de l'Association des Anatomistes*, 50ième Réunion, Lausanne, **132**, 433–7.

Gallera, J. (1966*b*). *Acta Anatomica*, **63**, 388–97.

Gallera, J. (1966*c*). *Revue Suisse de Zoologie*, **73**, 492–503.

Gallera, J. (1967). *Experientia,* **23**, 461–4.

Gallera, J. (1968). *Revue Suisse de Zoologie,* **75**, 227–34.

Gallera, J. (1969). *Acta Embryologiae Experimentalis,* 5–16.

Gallera, J. (1970*a*). *Experientia,* **26**, 886–7.

Gallera, J. (1970*b*). *Experientia,* **26**, 1353–4.

Gallera, J. (1970*c*). *Journal of Embryology and Experimental Morphology,* **23**, 473–89.

Gallera, J. (1971). *Archives de Biologie,* **82**, 85–102.

Gallera, J. (1972). *Experientia,* **28**, 1217–18.

Gallera, J. (1974*a*). *Archives de Biologie,* **85**, 399–413.

Gallera, J. (1974*b*). *Archives de Biologie,* **85**, 415–25.

Gallera, J. (1975). *Experientia,* **37**, 584–5.

Gallera, J. & Dicenta, C. (1966). *Revue Suisse de Zoologie,* **73**, 43–54.

Gallera, J. & Ivanov, I. (1964). *Journal of Embryology and Experimental Morphology,* **12**, 693–711.

Gallera, J. & Nicolet, G. (1969). *Journal of Embryology and Experimental Morphology,* **21**, 105–18.

Gallera, J., Nicolet, G. & Baumann, M. (1968). *Journal of Embryology and Experimental Morphology,* **19**, 439–50.

Gallien, L. & Durocher, M. (1957). *Bulletin Biologique de la France et de la Belgique,* **61**, 97–114.

Gardner, R. L. (1970*). *Advances in the Biosciences,* **6**, 279–301.

Gardner, R. L. (1972). *Journal of Embryology and Experimental Morphology,* **28**, 279–312.

Gardner, R. L. (1978*). In *Results and Problems in Cell Differentiation,* ed. W. J. Gehring, vol. 9, 205–41. Berlin: Springer Verlag.

Gardner, R. L. (1982). *Journal of Embryology and Experimental Morphology,* **68**, 175–98.

Gardner, R. L. & Papaioannou, V. E. (1975*). In *Early Development of Mammals,* eds M. Balls & A. E. Wild, pp. 107–32. Cambridge University Press.

Gardner, R. L., Papaioannou, V. E. & Barton, S. C. (1973). *Journal of Embryology and Experimental Morphology,* **30**, 561–72.

Gautier, J. & Beetschen, J.-C. (1983). *W. Roux's Archives of Developmental Biology,* **192**, 196–9.

Gaze, R. M. & Watson, W. E. (1968). In *Growth of the Nervous System,* eds G. E. W. Wolstenholme & M. O'Connor, pp. 53–76. Boston: Little, Brown & Co.

Gebhardt, D. O. E. & Nieuwkoop, P. D. (1964). *Journal of Embryology and Experimental Morphology,* **12**, 317–31.

Genis-Galvez, J. M. (1964). *Bulletin de l'Association des Anatomistes,* 49ième Réunion, Madrid, pp. 624–53.

Genis-Galvez, J. M. (1966). *Nature* (London), **210**, 209–10.

Genis-Galvez, J. M., Santos-Gutierrez, L. & Rios-Gonzalez, A. (1967). *Experimental Eye Research,* **6**, 48–56.

Gerhart, J. C. (1980*). In *Biological Regulation and Development,* vol. 2. *Molecular Organisation and Cell Function,* ed. R. F. Goldberger, pp. 133–316. New York: Plenum Press.

Gerhart, J., Black, S., Gimlich, R. & Scharf, S. (1983*). In *Time, Space and Pattern in Embryonic Development,* MBL Lectures in Biology, eds W. R. Jeffery & R. A. Raff, vol. 2, 261–86. New York: A. R. Liss Inc.

Gerhart, J., Ubbels, G., Black, S., Hara, K. & Kirschner, M. (1981). *Nature* (London), **292**, 511–16.

Geuskens, M. & Tencer, R. (1979*a*). *Journal of Cell Science,* **37**, 48–58.

Geuskens, M. & Tencer, R. (1979*b*). *Journal of Cell Science,* **37**, 59–67.

Gierer, A. & Meinhardt, H. (1972). *Kybernetik,* **12**, 30–9.

Gierer, A. & Meinhardt, H. (1974). *Lectures on Mathematics and Life Sciences,* **7**, 163–83.

Gilkey, J. C. (1981). *The American Zoologist,* **21**, 359–75.

Gilula, N. B. (1980*). In *The Cell surface: Mediator of Developmental Processes*, eds S. Subtelny & N. K. Wessels, pp. 23–41. New York: Academic Press.

Gimlich, R. L. & Cooke, J. (1983). *Nature* (London), **306**, 571–3.

Gingell, D. (1970). *Journal of Embryology and Experimental Morphology*, **23**, 583–609.

Gingell, D. & Palmer, J. F. (1968). *Nature* (London), **217**, 98–102.

Gingle, A. R. & Robertson, A. (1979). *Journal of Embryology and Experimental Morphology*, **53**, 353–65.

Ginzburg, A. S. (1971). *Ontogenez*, **2**, 645–8. (English translation: *Soviet Journal of Developmental Biology*, **2**, 515–18.)

Gipouloux, J. D. & Cambar, R. (1961). *Compte Rendu des Séances de l'Académie des Sciences*, Paris, **252**, 3643–5.

Gipouloux, J. D. & Cambar, R. (1974). *Bulletin Biologique de la France et de la Belgique*, **108**, 31–40.

Gipouloux, J. D. & Delbos, M. (1977). *Journal of Embryology and Experimental Morphology*, **41**, 259–68.

Gipouloux, J. D. & Hakim, J. (1978). *Journal of Embryology and Experimental Morphology*, **43**, 137–46.

Giudice, G. (1973*). In *The Cell Cycle in Development and Differentiation*, eds M. Balls & F. S. Billett, pp. 203–13. Cambridge University Press.

Glabe, C. G. & Vacquier, V. D. (1977). *Journal of Cell Biology*, **75**, 410–21.

Glade, R. W., Burrill, E. M. & Falk, R. J. (1967). *Growth*, **31**, 231–49.

Glimelius, B. & Weston, J. A. (1981). *Cell Differentiation*, **10**, 57–67.

Glücksmann, A. (1965). *Archives de Biologie*, **76**, 419–37.

Goettert, L. (1966). *W. Roux' Archiv für Entwicklungsmechanik der Organismen*, **157**, 75–100.

Goldberg, S. (1976). *Journal of Comparative Neurology*, **168**, 379–92.

Golosow, N. & Grobstein, C. (1962). *Developmental Biology*, **4**, 242–55.

Gomot, L. (1961*). In *La Culture Organotypique: Associations et Dissociations d'Organes en Culture in Vitro*, eds E. Wolff, J. A. A. Benoit, K. Haffen & F. Bermann. *Colloques Internationaux du Centre National de la Recherche Scientifique*, Paris, **101**, 117–32.

Gonzalo-Sanz, L. M. (1972). *Acta Anatomica*, **81**, 396–408.

Gordon, M., Dandekar, P. V. & Bartoszewicz, W. (1974). *Journal of Reproduction and Fertility*, **36**, 211–14.

Gordon, R. & Jacobson, A. G. (1978). *Scientific American*, **238**, 80–7.

Graham, C. F. (1971*). In *Control Mechanisms of Growth and Differentiation*, eds D. D. Davies & M. Balls, pp. 371–8. 25th Symposium of the Society of Experimental Biology. Cambridge University Press.

Graham, C. F. (1974*). *Biological Reviews*, **49**, 399–422.

Graham, C. F. & Morgan, R. W. (1966). *Developmental Biology*, **14**, 439–60.

Granholm, N. H. (1970). *The American Zoologist*, **10** (Abstract, no. 3).

Grant, Ph. (1969*). In *Biology of Amphibian Tumors*, ed. M. Mizell, pp. 43–51. New York: Springer Verlag.

Grant, Ph. & Wacaster, J. F. (1972). *Developmental Biology*, **28**, 454–71.

Grant, Ph. & Youngdale, P. (1974). *Journal of Experimental Zoology*, **190**, 289–96.

Green, H., Goldberg, B., Schwartz, M. & Brown, D. D. (1968). *Developmental Biology*, **18**, 391–400.

Grey, R. D., Wolf, D. P. & Hedrick, J. L. (1974). *Developmental Biology*, **36**, 44–61.

Grey, R. D., Bastiani, M. J., Webb, D. J. & Schertel, E. R. (1982). *Developmental Biology*, **89**, 475–84.

Grinfeld, S. & Beetschen, J. C. (1982). *W. Roux's Archives of Developmental Biology*, **191**, 215–21.

Grobstein, C. (1953*a*). *Journal of Morphology*, **93**, 19–44.

Grobstein, C. (1953*b*). *Journal of Experimental Zoology*, **124**, 383–414.

Grobstein, C. (1955*). In *Aspects of Synthesis and Order in Growth*, ed. D. Rudnick, **10**, 233–56. Princeton: Princeton University Press.

Grobstein, C. (1961). *Experimental Cell Research*, **8**, 234–45.

Grobstein, C. (1962*). *Journal of Cellular and Comparative Physiology*, **60**, suppl. 1, 35–48.

Grobstein, C. (1967*a**). *Ciba Foundation Symposium on Cell Differentiation*, eds A. V. S. de Reuck & J. Knight, pp. 131–6. London: Churchill Std.

Grobstein, C. (1967*b**). In *Second Decennial Review Conference on Cell, Tissue and Organ Culture, National Cancer Institute Monographs*, **26**, 279–99.

Grobstein, C. & Holtzer, H. (1955). *Journal of Experimental Zoology*, **128**, 333–58.

Gross, P. R. (1967*a**). *New England Journal of Medicine*, **276**, 1230–47.

Gross, P. R. (1967*b**). *Current Topics in Developmental Biology*, **2**, 1–47.

Grossi, C. S. & Fascio, U. (1979). *Rendiconti dell'Academia Nazionale dei Lincei, Classe di Scienze Fisiche, Matemetiche e Naturali*, Ser. 8, **66**, 435–41.

Grunz, H. (1968). *W. Roux' Archiv für Entwicklungsmechanik der Organismen*, **160**, 244–74.

Grunz, H. (1969). *W. Roux' Archiv für Entwicklungsmechanik der Organismen*, **163**, 184–96.

Grunz, H. (1970). *W. Roux' Archiv für Entwicklungsmechanik der Organismen*, **165**, 91–102.

Grunz, H. (1972). *W. Roux' Archiv für Entwicklungsmechanik der Organismen*, **169**, 41–55.

Grunz, H. (1973). *W. Roux' Archiv für Entwicklungsmechanik der Organismen*, **173**, 283–93.

Grunz, H. (1975*). In *New Approaches to the Evaluation of Abnormal Embryonic Development*, eds D. Neubert & H. J. Merker, pp. 792–803. Stuttgart: Georg Thieme Publishers.

Grunz, H. (1976*). *Proceedings of the sixth European Congress on Electronmicroscopy*, ed. Y. Ben-Shaul, pp. 571–3. Jerusalem: Tol International Publishing Company.

Grunz, H. (1977). *W. Roux's Archives of Developmental Biology*, **181**, 267–77.

Grunz, H. (1978*). *Ontogenez*, **9**, 323–32.

Grunz, H. (1983). *W. Roux's Archives of Developmental Biology*, **192**, 130–7.

Grunz, H., Multier-Lajous, A. M., Herbst, R. & Arkenberg, G. (1975). *W. Roux's Archives of Developmental Biology*, **178**, 277–84.

Grunz, H. & Staubach, J. (1979*a*). *Differentiation*, **14**, 59–65.

Grunz, H. & Staubach, J. (1979*b*). *W. Roux's Archives of Developmental Biology*, **186**, 77–80.

Grunz, H. & Tiedemann, H. (1977). *W. Roux's Archives of Developmental Biology*, **181**, 261–5.

Gualandris, L. & Duprat, A. M. (1981). *Differentiation*, **29**, 270–2.

Gualandris, L., Rougé, P. & Duprat, A. M. (1983). *Journal of Embryology and Experimental Morphology*, **77**, 183–200.

Gumpel-Pinot, M. (1980). *Journal of Embryology and Experimental Morphology*, **59**, 157–73.

Gunia, K. K. & Tumanishvili, G. D. (1972). *Ontogenez*, **3**, 264–74.

Gurdon, J. B. (1967*a*). *Proceedings of the National Academy of Sciences of the United States of America*, **58**, 545–52.

Gurdon, J. B. (1967*b**). In *Heritage from Mendel*, Proceedings of the Mendel Centennial Symposium of the Genetical Society of America, 1965, eds R. A. Brink & E. D. Styles, pp. 203–41. Madison: University of Wisconsin Press.

Gurdon, J. B. (1968*a*). *Journal of Embryology and Experimental Morphology*, **20**, 401–14.

Gurdon, J. B. (1968*b**). In *Essays in Biochemistry; DNA and RNA synthesis in Embryos*, eds P. N. Campbell & G. D. Greville, **4**, 25–68. New York: Academic Press (for the Biochemical Society).

Gurdon, J. B. (1969). In *Communication in Development, Symposium of the Society of Developmental Biology*, ed. A. Lang, *Developmental Biology*, suppl. 3, 59–82. New York: Academic Press.

Gurdon, J. B. (1974*). *The Control of Gene Expression in Animal Development*, 160 p. Oxford: Clarendon Press.

Gurdon, J. B. (1977*). *Proceedings of the Royal Society. London, The Croonion Lecture*, 1976, Ser. B, **198**, 211–47.

Gurdon, J. B. (1981*). In *Developmental Biology Using Purified Genes, ICN–UCLA–Symposium on Molecular and Cellular Biology*, ed. D. D. Brown, **23**, 1–10. New York: Academic Press.

Gurdon, J. B. & Brown, D. D. (1965). *Journal of Molecular Biology*, **12**, 27–35.

Gurdon, J. B. & Woodland, H. R. (1968*). *Biological Reviews*, **43**, 233–67.

Gurdon, J. B. & Woodland, H. R. (1969). *Proceedings of Royal Society, London*, Ser. B, **173**, 99–111.

Gwatkin, R. B. L. (1976*). In *The Cell Surface in Animal Embryogenesis and Development*, eds G. Poste & G. L. Nicolson, pp. 1–54. Amsterdam: Elsevier/North Holland Biomedical Press.

Hagenmaier, H. E. (1974). *Zeitschrift für Morphologie und Okologie der Tiere*, **79**, 233–44.

Hakim, J. & Gipouloux, J. D. (1978). *Journal of Embryology and Experimental Morphology*, **44**, 113–19.

Hall, B. K. (1977*). *Advances in Anatomy, Embryology and Cell Biology*, **53**, 3–50.

Hall, B. K. & Tremaine, R. (1979). *The Anatomical Record*, **194**, 469–76.

Hama, T. (1978*). In *Organiser – A Milestone of a Half-Century from Spemann*, eds O. Nakamura & S. Toivonen, pp. 71–90. Amsterdam: Elsevier/North Holland Biomedical Press.

Hamburger, V. (1969). *Experientia*, **25**, 1121–5.

Hamburger, V. & Hamilton, H. L. (1951). *Journal of Morphology*, **88**, 49–92.

Hammond, W. S. (1974). *The American Journal of Anatomy*, **141**, 303–16.

Hanocq, F., Pays-de Schutter, A., Hubert, A. & Brachet, J. (1974). *Differentiation*, **2**, 75–90.

Hara, K. (1961). PhD thesis, University of Utrecht, 44 p.

Hara, K. (1971). *W. Roux' Archiv für Entwicklungsmechanik der Organismen*, **167**, 183–6.

Hara, K. (1977). *W. Roux's Archives of Developmental Biology*, **181**, 73–87.

Hara, K. (1978*). In *Organiser – A Milestone of Half-Century from Spemann*, eds O. Nakamura & S. Toivonen, pp. 221–65. Amsterdam: Elsevier/North Holland Biomedical Press.

Hara, K. & Tydeman, P. (1979). *W. Roux's Archives of Developmental Biology*, **186**, 91–4.

Hara, K., Tydeman, P. & Hengst, R. T. M. (1977). *W. Roux's Archives of Developmental Biology*, **181**, 189–92.

Hara, K., Tydeman, P. & Kirschner, M. (1980). *Proceedings of the National Academy of Sciences of the United States of America*, **77**, 462–6.

Harris, H. L. & Zalik, S. E. (1982). *W. Roux's Archives of Developmental Biology*, **191**, 208–10.

Harris, P. (1978*). In *Cell Cycle Regulation*, ed. J. R. Jeter Jr, pp. 75–104. New York: Academic Press.

Harris, T. M. (1964). *Developmental Biology*, **10**, 247–68.

Harrison, R. G. (1969). In *Organization and Development of the Embryo*, ed. S. Wilens, pp. 166–214. New Haven: Yale University Press.

Hart, N. H. & Yu, S.-F. (1980). *Journal of Experimental Zoology*, **213**, 137–59.

Hassell, J. R., Greenberg, J. H. & Johnston, M. C. (1977). *Journal of Embryology and Experimental Morphology*, **39**, 267–71.

Hata, R. I. & Slavkin, H. C. (1978). *Proceedings of the National Academy of Sciences of the United States of America*, **75**, 2790–4.

Hay, E. D. (1963). *New England Journal of Medicine*, **268**, 1114–22.

Hay, E. D. (1968*). In *Epithelial–Mesenchymal Interactions*, eds R. Fleischmajer & R. E. Billingham, pp. 31–55. Baltimore: Williams & Wilkins.

Hay, E. D. (1977*a**). In *Cell and Tissue Interactions*, eds J. W. Lash & M. M. Burger, pp. 115–37. New York: Raven Press.

Hay, E. D. (1977*b**). In *Procedings of the 5th International Conference on Birth Defects*, Montreal, 1977, eds J. W. Littlefield & J. de Grouchy, pp. 126–40. Amsterdam: Excerpta Medica.

Hay, E. D. & Meier, S. (1974). *Journal of Cell Biology*, **62**, 889–98.

Hay, E. D. & Meier, S. (1976). *Developmental Biology*, **52**, 141–57.

Hayashi, Y. (1959*a*). *Developmental Biology*, **1**, 247–68.

Hayashi, Y. (1959*b*). *Developmental Biology*, **1**, 343–63.

Head, J. F., Mader, S. & Kaminer, B. (1979). *Journal of Cell Biology*, **80**, 211–18.

Heatley, N. G. & Lindahl, P. E. (1937). *Proceedings of Royal Society of London*, Ser. B, **122**, 395–402.

Hebard, C. N. & Herold, R. C. (1967). *Experimental Cell Research*, **46**, 533–70.

Heidemann, S. R. & Kirschner, M. W. (1978). *Journal of Experimental Zoology*, **204**, 431–44.

Hendrix, R. W. & Zwaan, J. (1974). *Differentiation*, **2**, 357–62.

Herkovits, J. (1977). *Revista del Instituto di Cibernética de la Sociedad Científica Argentina*, An. II, 5–14.

Herkovits, J. (1978). *Medicina*, **38**, 60–6.

Herkovits, J., Bustuoabad, O. D. & Pisanó, A. (1977). *Acta Embryologiae Experimentalis*, 195–205.

Herkovits, J. & Ubbels, G. A. (1979). *Journal of Embryology and Experimental Morphology*, **51**, 155–64.

Hermann, J., Bellé, R., Tso, J. & Ozon, R. (1983). *Cell Differentiation*, **13**, 143–8.

Hertwig, O. (1876). *Morphologische Jahrbücher, Zeitschrift für Anatomie und Entwicklungs Geschichte*, **1**, 347–434.

Hertwig, O. (1877). *Morphologische Jahrbücher, Zeitschrift für Anatomie und Entwicklungs Geschichte*, **3**, 1–86.

Hertwig, O. (1878). *Morphologische Jahrbücher Zeitschrift für Anatomie und Entwicklungs Geschichte*, **4**, 156–75.

Hilfer, S. R. & Stern, M. (1971). *Journal of Experimental Zoology*, **178**, 293–306.

Hillman, N. & Hillman, R. (1967). *Science*, **155**, 1563–5.

Hillman, N. & Niu, M. C. (1963). *Proceedings of the National Academy of Sciences of the United States of America*, **50**, 483–93.

Hillman, N., Sherman, M. I. & Graham, C. (1972). *Journal of Embryology and Experimental Morphology*, **28**, 263–78.

Hiramoto, Y. (1974). *Experimental Cell Research*, **89**, 320–6.

Hirano, S. & Shirai, T. (1982). *Development, Growth and Differentiation*, **24**, 395 (abstract B35).

Hirose, G. & Jacobson, M. (1979). *Developmental Biology*, **71**, 191–202.

His, W. (1901). *Archiv für Anatomie und Physiologie*, Leipzig, *Anatomische Abteilung*, 307–37.

Hoessels, E. L. M. J. (1957). PhD Thesis, University of Utrecht, 71 p.

Hoffman, S. & Edelman, G. M. (1983). *Proceedings of the National Academy of Sciences of the United States of America*, **80**, 5762–6.

Hogan, B. L. M. & Tilly, R. (1981). *Journal of Embryology and Experimental Morphology*, **62**, 379–94.

Hogarth, P. J. (1978*). *Biology of Reproduction*, 189 p. Glasgow: Blackie.

Hollinger, T. G. & Schuetz, A. W. (1976). *Journal of Cell Biology*, **71**, 395–401.

Hollyfield, J. G. (1968). *Developmental Biology*, **18**, 163–79.

Hollyfield, J. G. (1971). *Developmental Biology*, **24**, 264–86.

Holtfreter, J. (1933*a*). *W. Roux' Archiv für Entwicklungsmechanik der Organismen*, **127**, 619–775.

Holtfreter, J. (1933*b*). *W. Roux' Archiv für Entwicklungsmechanik der Organismen*, **129**, 669–793.

Holtfreter, J. (1934*a*). *W. Roux' Archiv für Entwicklungsmechanik der Organismen*, **132**, 225–306.

Holtfreter, J. (1934*b*). *W. Roux' Archiv für Entwicklungsmechanik der Organismen*, **132**, 307–83.

Holtfreter, J. (1939). *Archiv für Zellforschung*, **23**, 169–209.

Holtfreter, J. (1943*a*). *Journal of Experimental Zoology*, **93**, 251–323.

Holtfreter, J. (1943*b*). *Journal of Experimental Zoology*, **94**, 261–318.

Holtfreter, J. (1944*a*). *Journal of Experimental Zoology*, **95**, 171–212.

Holtfreter, J. (1944*b*). *Journal of Experimental Zoology*, **95**, 307–40.

Holtfreter, J. (1945). *Journal of Experimental Zoology*, **98**, 161–209.

Holtfreter, J. (1946). *Journal of Morphology*, **79**, 27–62.

Holtfreter, J. (1947*a*). *Journal of Morphology*, **80**, 25–56.

Holtfreter, J. (1947*b*). *Journal of Morphology*, **80**, 57–92.

Holtfreter, J. (1947*c*). *Journal of Experimental Zoology*, **106**, 197–222.

Holtfreter, J. (1948*). *Symposium of the Society of Experimental Biology*, **2**, 17–48.

Holtfreter, J. (1955). *Experimental Cell Research*, suppl. 3, 188–209.

Holtfreter, J. (1966*). In *McGraw-Hill Yearbook of Science and Technology*, pp. 143–6. New York City: McGraw-Hill.

Holtfreter, J. (1968*). In *Epithelial-Mesenchymal Interactions*, eds R. Fleischmajer & R. E. Billingham, pp. 1–30. Baltimore: Williams & Wilkins.

Holtfreter, J. & Hamburger, V. (1955*). In *Analysis of Development*, eds B. H. Willier, P. A. Weiss & V. Hamburger, pp. 230–96. London: W. B. Saunders.

Holtzer, H. (1968*). In *Epithelial Mesenchymal Interactions*, eds R. Fleischmajer & R. E. Billingham, pp. 152–64. Baltimore: Williams & Wilkins.

Holtzer, H. & Abbott, J. (1968*). In *The Stability of the Differentiated State*, ed. H. Ursprung, vol. 1, pp. 1–16. Berlin: Springer Verlag.

Holtzer, H., Dienstman, S., Biehl, J. & Holtzer, S. (1975*). In *Extracellular Matrix Influences on Gene Expression*, eds H. C. Slavkin & R. C. Gruelich, pp. 253–7. New York: Academic Press.

Holtzer, H. & Detwiler, S. R. (1953). *Journal of Experimental Zoology*, **123**, 335–70.

Holtzer, H. & Matheson, D. W. (1970*). In *Chemistry and Molecular Biology of the Intercellular Matrix*, vol. 3, *Structural Organization and Function of the Matrix*, ed. E. A. Balazs, pp. 1753–70. New York: Academic Press.

Holtzer, H. & Mayne, R. (1973*). In *Pathobiology of Development*, eds E. V. D. Perrin & M. J. Finegold, pp. 52–64. Baltimore: Williams & Wilkins.

Hoperskaya, O. A. (1968). *Doklady Akademii Nauk SSSR*, **180**, 1012–15.

Hoperskaya, O. A. (1972). *W. Roux' Archiv für Entwicklungsmechanik der Organismen*, **171**, 1–16.

Hoperskaya, O. A. (1976). *W. Roux's Archives of Developmental Biology*, **180**, 213–27.

Hoperskaya, O. A. (1978). *W. Roux's Archives of Developmental Biology*, **184**, 15–28.

Hoperskaya, O. A. (1981). *Differentiation*, **20**, 104–16.

Hoperskaya, O. A., Bogdanov, H. E., Zaitseva, L. N. & Golubeva, O. N. (1984). *Differentiation* (in press.).

Hoperskaya, O. A. & Golubeva, O. N. (1980). *Journal of Embryology and Experimental Morphology*, **60**, 173–88.

Hoperskaya, O. A. & Golubeva, O. N. (1982). *Development, Growth and Differentiation,* **24**, 245–57.

Hoperskaya, O. A., Zaitzev, I. & Golubeva, O. N. (1982). *Development, Growth and Differentiation,* **24**, 259–63.

Horder, T. (1976*). In *The Developmental Biology of Plants and Animals,* eds C. F. Graham & P. F. Wareing, pp. 169–98. Oxford: Blackwell.

Hornbruch, A., Summerbell, D. & Wolpert, L. (1979). *Journal of Embryology and Experimental Morphology,* **51**, 51–62.

Hörstadius, S. (1939*). *Biological Reviews,* **14**, 132–79.

Hörstadius, S. & Selman, S. (1946). *Nova Acta Regiae Societatis Scientiarum Upsaliensis,* Ser. IV, **13**, 1–170.

Hoskins, D. D. & Casillas, E. R. (1975*). In *Molecular Mechanisms of Gonadal Hormone Action,* eds J. A. Thomas & R. L. Sinhal, vol. 1, 293–324. Baltimore: University Park Press.

Houssaint, E. & le Douarin, N. (1971). *Journal of Embryology and Experimental Morphology,* **26**, 481–95.

Hubbert, W. T. & Miller, W. J. (1974*). *Journal of Cellular and Comparative Physiology,* **84**, 429–44.

Hubert, J. (1962). *Archives d'Anatomie Microscopique et de Morphologie Expérimentale,* **51**, 11–26.

Huff, R. & Preston, J. T. (1965). *Texas Journal of Science,* **17**, 206–12.

Humphrey, R. R. (1966). *Developmental Biology,* **13**, 57–76.

Hunt, H. H. (1961). *Developmental Biology,* **3**, 175–209.

Hunter, R. H. F. (1976). *Journal of Anatomy,* **122**, 43–59.

Hurmerinta, K. & Thesleff, I. (1982). *Cell Differentiation,* **11**, 107–13.

Ignatieva, G. M. (1960*a*). *Doklady Akademii Nauk SSSR* (Biological Sciences, Section Embryology), **134**, 233–6.

Ignatieva, G. M. (1960*b*). *Doklady Akademii Nauk SSSR* (Biological Sciences, Section Embryology), **134**, 706–10.

Ignatieva, G. M. (1961). *Doklady Akademii Nauk SSSR* (Biological Sciences, Section Embryology), **139**, 503–6.

Ignatieva, G. M. (1962). *Doklady Akademii Nauk SSSR* (Biological Sciences, Section Embryology), **139**, 588–91.

Ignatieva, G. M. (1963). *Doklady Akademii Nauk SSSR* (Biological Sciences, Section Embryology), **151**, 973–6.

Ignatieva, G. M. (1968). *Doklady Akademii Nauk SSSR* (Biological Sciences, Section Embryology), **179**, 1005–8.

Ignatieva, G. M. & Rott, N. N. (1970). *W. Roux' Archiv für Entwicklungsmechanik der Organismen,* **165**, 103–9.

Ikeda, A. & Zwaan, J. (1966). *Investigative Ophthalmology,* **5**, 402–12.

Ikenishi, K. & Nieuwkoop, P. D. (1978). *Development, Growth and Differentiation,* **20**, 1–9.

Ikushima, N. (1961). *Japanese Journal of Zoology,* **13**, 117–39.

Ikushima, N. & Maruyama, S. (1971). *Journal of Embryology and Experimental Morphology,* **25**, 163–76.

Inoue, K. (1962). *Developmental Biology,* **4**, 321–38.

Ishizuya-Oka, A. (1983). *W. Roux's Archives of Developmental Biology,* **192**, 171–8.

Ito, K. & Takeuchi, T. (1982). *Development, Growth and Differentiation,* **24**, 395 (abstract B34).

Ito, S. (1972). *Development, Growth and Differentiation,* **14**, 217–27.

Ito, S. & Hori, N. (1966). *Journal of General Physiology,* **49**, 1019–27.

Ito, S. & Ikematsu, Y. (1980). *Development, Growth and Differentiation,* **22**, 247–56.

Ito, S. & Loewenstein, W. R. (1969). *Developmental Biology*, **19**, 228–43.

Ito, S., Sato, E. & Loewenstein, W. R. (1974*a*). *Journal of Membrane Biology*, **19**, 305–37.

Ito, S., Sato, E. & Loewenstein, W. R. (1974*b*). *Journal of Membrane Biology*, **19**, 339–55.

Ito, S., Yamashita, Y. & Ahsako, F. (1977*a*). *Annotationes Zoologicae Japonenses*, **50**, 131–8.

Ito, S., Yamashita, Y., & Ahsako, F. (1977*b*). *The Zoological Magazine*, **86**, 254–8.

Itoh, Y., Ide, H. & Hama, T. (1980). *Cell and Tissue Research*, **209**, 353–64.

Iyeiri, S. & Kawakami, I. (1962). *Memoirs of the Faculty of Science, Kyushu University*, Ser. E, **3**, 117–35.

Jackson, J. F., Clayton, R. M., Williamson, R., Thomson, I., Truman, D. E. S. & de Pomerai, D. I. (1978). *Developmental Biology*, **65**, 383–95.

Jacob. H. J. (1971). *Verhandlungen der anatomischen Gesellschaft*, **65**, 271–8.

Jacob, H. J., Christ, B. & Jacob, M. (1974). *Verhandlungen der anatomischen Gesellschaft*, **68**, 581–9.

Jacob, H. J., Christ, B. & Jacob, M. (1975). *Verhandlungen der anatomischen Gesellschaft*, **69**, 271–4.

Jacob, H. J., Christ, B., Jacob, M. & Ahlström, P. (1976). *Acta Anatomica*, **94**, 204–20.

Jacob, H. J., Jacob, M. & Christ, B. (1981). *European Journal of Cell Biology*, **24**, suppl. p. 11 (abstract).

Jacobelli, S., Hanocq, J., Baltus, E. & Brachet, J. (1974). *Differentiation*, **2**, 129–35.

Jacobson, A. G. (1963*a*). *Journal of Experimental Zoology*, **154**, 273–83.

Jacobson, A. G. (1963*b*). *Journal of Experimental Zoology*, **154**, 285–91.

Jacobson, A. G. (1963*c*). *Journal of Experimental Zoology*, **154**, 293–304.

Jacobson, A. G. (1966*). *Science*, **152**, 25–32.

Jacobson, A. G. (1978). *Zoon, a Journal of Zoology*, **6**, 13–21.

Jacobson, A. G. (1980). *The American Zoologist*, **20**, 669–77.

Jacobson, A. G. & Duncan, J. T. (1968). *Journal of Experimental Zoology*, **167**, 79–103.

Jacobson, A. G. & Gordon, R. (1976*a*). *Journal of Experimental Zoology*, **197**, 191–246.

Jacobson, A. G. & Gordon, R. (1976*b*). *Journal of Supramolecular Structure*, **5**, 371–80.

Jacobson, C. O. (1959). *Journal of Embryology and Experimental Morphology*, **7**, 1–21.

Jacobson, C. O. (1962). *Zoologiska Bidrag från Uppsala*, **35**, 433–49.

Jacobson, C. O. (1964*a*). *Zoologiska Bidrag från Uppsala*, **36**, 73–160.

Jacobson, C. O. (1964*b**). *Acta Universitatis Upsaliensis*, PhD thesis, 15 p. (Summary).

Jacobson, C. O. (1968). *Journal of Experimental Zoology*, **168**, 125–36.

Jacobson, C. O. (1970). *Journal of Embryology and Experimental Morphology*, **23**, 463–71.

Jacobson, C. O. & Jacobson, A. G. (1973). *Zoon, a Journal of Zoology*, **1**, 17–21.

Jacobson, C. O. & Löfberg, J. (1969). *Zooligiska Bidrag från Uppsala*, **38**, 133–9.

Jacobson, M. (1981*a*). *Journal of Neuroscience*, **1**, 918–22.

Jacobson, M. (1981*b*). *Journal of Neuroscience*, **1**, 923–7.

Jacobson, M. (1982*). In *Neural Development (Current Topics in Neurobiology)*, ed. N. C. Spitzer, pp. 45–99. New York: Plenum Press.

Jacobson, M. (1984). *Developmental Biology*, **102**, 122–9.

Jacobson, M. & Hirose, G. (1981). *Journal of Neuroscience*, **1**, 271–84.

Jaffe, L. A. (1976). *Nature* (London), **261**, 68–71.

Jaffe, L. A. & Robinson, K. R. (1978). *Developmental Biology*, **62**, 215–28.

Jaffe, L. F. (1969*). In *Communication in Development, 28th Symposium of the Society of Developmental Biology*, ed. A. Lang, *Developmental Biology*, suppl. 3, 83–111. New York: Academic Press.

Jaffe, L. F. & Nucitelli, R. (1977*). *Annual Review of Biophysics and Bioengineering*, **6**, 445–76.

Jaffe, L. F. & Stern, C. D. (1979). *Science*, **206**, 569–70.

Janeczek, J., John, M., Born, J., Hoppe, P., Tiedemann, H. & Tiedemann, H. (1983). *W. Roux's Archives of Developmental Biology*, **192**, 45–7.

Janeczek, J., John, M., Born, J., Tiedemann, H. & Tiedemann, H. (1984). *W. Roux's Archives of Developmental Biology*, **193**, 1–12.

Jeffery, W. R. & Capco, D. G. (1978). *Developmental Biology*, **67**, 152–66.

Jeffery, W. R. & Meier, S. (1983). *Developmental Biology*, **96**, 125–43.

Jelínek, R. & Friebová, Z. (1966). *Nature* (London), **209**, 822–3.

Jelínek, R., Siechert, V. & Klika, E. (1969). *Folia Morphologica*, **17**, 355–67.

John, M., Born, J., Tiedemann, H. & Tiedemann, H. (1984). *W. Roux's Archives of Developmental Biology*, **193**, 13–18.

Johnen, A. G. (1961). *W. Roux' Archiv für Entwicklungsmechanik der Organismen*, **153**, 1–13.

Johnen, A. G. (1964a). *W. Roux' Archiv für Entwicklungsmechanik der Organismen*, **155**, 302–13.

Johnen, A. G. (1964b). *W. Roux' Archiv für Entwicklungsmechanik der Organismen*, **155**, 314–41.

Johnen, A. G. (1970). *W. Roux' Archiv für Entwicklungsmechanik der Organismen*, **165**, 150–62.

Johnen, A. G. & Albers, B. (1978). *Medical Biology*, **56**, 317–20.

Johnen, A. G. & Engländer, H. (1967). *W. Roux' Archiv für Entwicklungsmechanik der Organismen*, **159**, 357–64.

Johnson, J. D., Epel, D. & Paul, M. (1976). *Nature* (London), **262**, 661–4.

Johnson, K. E. (1970). *Journal of Experimental Zoology*, **175**, 391–428.

Johnson, K. E. (1972). *Journal of Experimental Zoology*, **179**, 227–38.

Johnson, K. E. (1976). *Experimental Cell Research*, **101**, 71–7.

Johnson, K. E. (1977a). *Journal of Cell Science*, **25**, 313–22.

Johnson, K. E. (1977b). *Journal of Cell Science*, **25**, 323–34.

Johnson, K. E. (1977c). *Journal of Cell Science*, **25**, 335–54.

Johnson, K. E. (1977d). *Journal of Experimental Zoology*, **199**, 137–42.

Johnson, K. E. (1978). *Journal of Cell Science*, **32**, 109–36.

Johnson, K. E. & Adelman, M. R. (1981). *Journal of Cell Science*, **49**, 205–16.

Johnson, K. E. & Smith, E. P. (1976). *Experimental Cell Research*, **101**, 63–70.

Johnson, K. E. & Smith, E. P. (1977). *Cell Differentiation*, **5**, 301–9.

Johnston, M. C. (1966). *The Anatomical Record*, **156**, 143–55.

Johnston, M. C., Noden, D. M., Hazelton, R. D., Coulombre, J. L. & Coulombre, A. J. (1979). *Experimental Cell Research*, **29**, 27–43.

Johnston, R. N. & Paul, M. (1977). *Developmental Biology*, **57**, 364–74.

Jumah, H. & Stanisstreet, M. (1982). *Acta Embryologiae et Morphologiae Experimentalis*, 155–71.

Kageyama, T. (1980). *Development, Growth and Differentiation*, **22**, 659–68.

Kageyama, T. (1982). *Journal of Experimental Zoology*, **219**, 241–56.

Källén, B. (1965a). *Progress in Brain Research*, **14**, 77–96.

Källén, B. (1965b*). In *Organogenesis*, eds R. H. de Haan & H. Ursprung, pp. 107–28. New York: Holt, Rinehart & Winston.

Källén, B. (1968*). In Casey Holker Memorial Lecture, *Hydrocephalus and Spina Bifida*, Suppl. 16, 44–53.

Kallenbach, E. & Piesco, P. (1979). In *Tooth Morphogenesis and Differentiation: a Workshop*, 1978, ed. J. V. Ruch, pp. 163–74. Paris: SNPND and *Journal de Biologie Buccale*, **6**, 229–40.

Kalt, M. R. (1971a). *Journal of Embryology and Experimental Morphology*, **26**, 37–49.

Kalt, M. R. (1971b). *Journal of Embryology and Experimental Morphology*, **26**, 51–66.

Kalt, M. R. (1973). *Zeitschrift für Zellforschung und Mikroskopische Anatomie*, **138**, 41–62.

Kalt, M. R. (1976). *Journal of Experimental Zoology*, **195**, 393–408.

Kaneda, T. (1980). *Development, Growth and Differentiation*, **22**, 841–9.
Kaneda, T. (1981). *Development, Growth and Differentiation*, **23**, 553–64.
Kaneda, T. & Hama, T. (1979). *W. Roux's Archives of Developmental Biology*, **187**, 25–34.
Kaneda, T. & Suzuki, A. S. (1983). *W. Roux's Archives of Developmental Biology*, **192**, 8–12.
Kaprio, E. A. & Tähkä, S. (1978). *Medical Biology*, **56**, 321–7.
Karfunkel, P. (1971). *Developmental Biology*, **25**, 30–56.
Karfunkel, P. (1977). *W. Roux's Archives of Developmental Biology*, **181**, 31–40.
Karkinen-Jääskeläinen, M. (1978a). *Differentiation*, **12**, 31–7.
Karkinen-Jääskeläinen, M. (1978b). *Journal of Embryology and Experimental Morphology*, **44**, 167–79.
Katagiri, C. & Moriya, M. (1976). *Developmental Biology*, **50**, 235–51.
Kato, K. I. (1957). *Memoirs of the College of Science, Kyoto University*, Ser. B, **24**, 165–70.
Kato, K. I. (1958). *Memoirs of the College of Science, Kyoto University*, Ser. B, **25**, 2–10.
Kato, K. I. (1959). *Memoirs of the College of Science, Kyoto University*, Ser. B, **26**, 1–7.
Kato, K. I. (1963a). *Memoirs of the College of Science, Kyoto University*, Ser. B, **30**, 21–8.
Kato, K. I. (1963b). *Memoirs of the College of Science, Kyoto University*, Ser. B, **30**, 29–39.
Katoh, A. K. (1962). *Experimental Cell Research*, **27**, 427–30.
Katoh, Y. (1975). *Development, Growth and Differentiation*, **17**, 143–52.
Kaufman, M. (1979). *American Journal of Anatomy*, **155**, 425–44.
Kawakami, I. (1958). *Gann: Japanese Journal of Cancer Research*, **49**, 177–91.
Kawakami, I. (1976). *Journal of Embryology and Experimental Morphology*, **36**, 315–20.
Kawakami, I., Ave, K., Sasaki, N. & Sameshima, M. (1966*). In *Nucleic Acid Metabolism, Cell Differentiation and Cancer Growth*, eds E. V. Cowdry, & S. Seno, pp. 143–51. Oxford: Pergamon Press.
Kawakami, I. & Iyeiri, S. (1963). *Japanese Journal of Zoology*, **14**, 49–59.
Kawakami, I. & Iyeiri, S. (1964). *Experimental Cell Research*, **33**, 516–22.
Kawakami, I., Noda, S., Kurihara, K. & Okuma, K. (1977). *W. Roux's Archives of Developmental Biology*, **182**, 1–7.
Kawakami, I. & Sasaki, N. (1978*). In *Organizer – A Milestone of a Half-Century from Spemann*, eds O. Nakamura & S. Toivonen, pp. 157–78. Amsterdam: Elsevier/North Holland Biomedical Press.
Kawakami, I., Sasaki, N., Sato, A. & Osaka, N. (1978). *Development, Growth and Differentiation*, **20**, 353–61.
Kawakami, I., Watanabe, H., Ave, K. & Iyeiri, S. (1969). *Embryologia*, **10**, 231–41.
Keating, M. J. & Kennard, C. (1976*). In *The Amphibian Visual System; a Multidisciplinery Approach*, ed. K. V. Fite, pp. 267–315. New York: Academic Press.
Keefe, R. E. (1973a). *Journal of Experimental Zoology*, **184**, 185–206.
Keefe, J. R. (1973b). *Journal of Experimental Zoology*, **184**, 207–32.
Keefe, J. R. (1973c). *Journal of Experimental Zoology*, **184**, 233–8.
Keefe, J. R. (1973d). *Journal of Experimental Zoology*, **184**, 239–58.
Keller, R. E. (1975). *Developmental Biology*, **42**, 222–41.
Keller, R. E. (1976). *Developmental Biology*, **51**, 118–37.
Keller, R. E. (1978). *Journal of Morphology*, **157**, 233–48.
Keller, R. E. (1980). *Journal of Embryology and Experimental Morphology*, **60**, 201–34.
Keller, R. E. (1981). *Journal of Experimental Zoology*, **216**, 81–101.
Keller, R. E. (1984). *The American Zoologist*, (in press).
Keller, R. E. & Spieth, J. (1984). *Journal of Experimental Zoology*, **229**, 109–26.
Keller, R. E. & Schoenwolf, G. C. (1977). *W. Roux's Archives of Developmental Biology*, **182**, 165–86.
Keller, R. E. & Trinkaus, J. P. (1982). *Journal of Cell Biology*, **95**, 325a (abstract).
Kelley, R. O. (1969). *Journal of Experimental Zoology*, **172**, 153–80.

Kelley, R. O. & Fallon, J. F. (1976). *Developmental Biology*, **51**, 241–56.

Kessel, R. G., Beams, H. W. & Shih, C. Y. (1974). *American Journal of Anatomy*, **141**, 341–60.

Khare, M. K. & Choudhury, S. (1984). *Development, Growth and Differentiation*, in press.

Kieny, M. (1960). *Journal of Embryology and Experimental Morphology*, **8**, 457–67.

Kieny, M. (1967). *Revue d'Anatomie et de Morphologie Expérimentale*, **39**, 5–37.

Kieny, M. (1971). *Annales d'Embryologie et de Morphogenèse*, **4**, 281–98.

Kieny, M. & Brugal, M. (1977). *Archives d'Anatomie Microscopique et de Morphologie Expérimentale*, **66**, 235–52.

Kieny, M., Mauger, A. & Sengel, P. (1972). *Developmental Biology*, **28**, 142–61.

Kirschner, M. W., Butner, K. A., Newport, J. W., Black, S. D., Scharf, S. R. & Gerhart, J. C. (1981). *Netherlands Journal of Zoology*, **31**, 50–77.

Kirschner, M. W., Gerhart, J. C., Hara, K. & Ubbels, G. A. (1980*). In *The Cell Surface: Mediator of Developmental Processes*, eds S. Subtelny & N. K. Wessels, pp. 187–215. *38th Symposium of the Society of Developmental Biology*. New York: Academic Press.

Kirschner, M. W. & Hara, K. (1980). *Mikroskopie*, **36**, 12–15.

Kirzon, S. S., Averkina, R. F. & Vyazov, O. E. (1969). *Byulleten' Eksperimental'noĭ Biologii i Meditsinȳ*, **67**, 46–50.

Klag, J. J. & Ubbels, G. A. (1975). *Differentiation*, **3**, 15–20.

Klika, E., Myslivečková, A. & Rychter, Z. (1980). *Folio Morphologica*, **28**, 192–6.

Kobayakawa, Y. & Kubota, H. Y. (1981). *Journal of Embryology and Experimental Morphology*, **62**, 83–94.

Koch, W. E. (1967). *Journal of Experimental Zoology*, **165**, 155–70.

Koch, W. E. & Grobstein, C. (1963). *Developmental Biology*, **7**, 303–23.

Kochav, S. & Eyal-Giladi, H. (1971). *Science*, **171**, 1027–9.

Kochav, S., Ginsburg, M. & Eyal-Giladi, H. (1980). *Developmental Biology*, **79**, 296–308.

Kocher-Becker, U. & Tiedemann, H. (1971). *Nature* (London), **233**, 65–6.

Kocher-Becker, U., Tiedemann, H. & Tiedemann, H. (1965). *Science*, **147**, 167–9.

Koebke, J. (1976). *Verhandlungen der Anatomischen Gesellschaft*, **70**, 849–53.

Koebke, J. (1977). *Zeitschrift für mikroskopisch-anatomische Forschung*, **91**, 215–28.

Koecke, H. U. (1960). *W. Roux' Archiv für Entwicklungsmechanik der Organismen*, **151**, 612–59.

Koecke, H. U. (1964). *Studium Generale*, **17**, 288–322. Berlin: Springer Verlag.

Kohonen, J. (1963). *Annales Zoologici Societatis Zoologico-Botanicae Fennicae, Vanamo*, **25**, 1–21.

Kojima, M. K. (1972). *Development, Growth and Differentiation*, **14**, 301–10.

Kollar, E. J. (1972*). *The American Zoologist*, **12**, 125–35.

Kollar, E. J. & Fischer, C. (1980). *Science*, **207**, 993–5.

Komazaki, Sh. (1982). *Development, Growth and Differentiation*, **24**, 491–9.

Komazaki, Sh. (1983). *Development, Growth and Differentiation*, **25**, 181–92.

Kosher, R. A. (1976). *Developmental Biology*, **53**, 265–76.

Kosher, R. A. & Lash, J. W. (1975). *Developmental Biology*, **42**, 362–78.

Kosher, R. A., Savage, M. P. & Chan, S. Ch. (1979). *Journal of Experimental Zoology*, **209**, 221–8.

Kosher, R. A. & Searls, R. L. (1973). *Developmental Biology*, **32**, 50–68.

Kramer, B. & Andrew, A. (1981). *General and Comparative Endocrinology*, **44**, 279–87.

Kratochwil, K. (1972*). In *Tissue Interactions in Carcinogenesis*, ed. D. Tarin, pp. 1–47. London: Academic Press.

Kratochwil, K. (1983*). In *Cell Interactions and Development; Molecular mechanisms*, ed. K. M. Yamada, pp. 99–122. New York: J. Wiley & Sons.

Kriegel, H. (1961). *Acta Biologica et Medica Germanica*, **6**, 312–21.

Kubo, K. & Wright, D. A. (1977). *The American Zoologist*, **17** (Abstract no. 590).

Kubota, H. Y. (1981). *Experimental Cell Research*, **133**, 137–48.

Kubota, H. Y. (1983). *Development, Growth and Differentiation,* **25**, 404 (abstract).
Kubota, H. Y. & Durston, A. J. (1978). *Journal of Embryology and Experimental Morphology,* **44**, 71–80.
Kubota, T. (1966). *Journal of Experimental Biology,* **44**, 545–52.
Kubota, T. (1967). *Journal of Embryology and Experimental Morphology,* **17**, 331–40.
Kubota, T. (1969). *Journal of Embryology and Experimental Morphology,* **21**, 119–29.
Kubota, T. (1979*a*). *Journal of Cell Science,* **37**, 39–45.
Kubota, T. (1979*b*). *Development, Growth and Differentiation,* **21**, 155–9.
Kudo, Sh. (1983). *Development, Growth and Differentiation,* **25**, 163–70.
Kumar, A. & Warner, J. R. (1972). *Journal of Molecular Biology,* **63**, 233–46.
Kunz, W. & Schäfer, U. (1978*). *Oogenese und Spermatogenese,* 98 p. Jena: Gustav Fischer Verlag.
Kurihara, K. & Sasaki, N. (1981). *Development, Growth and Differentiation,* **23**, 361–9.
Kurrat, H. J. (1974). PhD thesis, Universität Köln, 91 p.
Kurrat, H. J. (1977). *Biologisches Zentralblatt,* **96**, 79–93.
Kurrat, H. J. (1978). *Biologisches Zentralblatt,* **97**, 153–62.
Kuusi, T. (1959). *Archivum Societatis Zoologicae-Botanicae Fennicae 'Vanamo',* **13**, 97–105.
Kuusi, T. (1960). *Archivum Societatis Zoologicae-Botanicae Fennicae 'Vanamo',* **14**, 4–28.
Kuusi, T. (1961). *Archivum Societatis Zoologicae-Botanicae Fennicae 'Vanamo',* **15**, 1–21.
Kvavilashvili, I. Sh., Bozhkova, U. P., Kafiani, K. A. & Chailkhyan, L. M. (1972). *Soviet Journal of Developmental Biology,* **2**, 170–2.
Kvavilashvili, I. Sh., Chikvashvili, Sh. D., Gelashvili, N. A. & Gogiberidze, L. N. (1977). *Ontogenez,* **8**, 180–2.
Lakshmi, M. S. (1962). *Journal of Embryology and Experimental Morphology,* **10**, 383–8.
Lakshmi, M. S. & Mulherkar, L. (1963). *Experientia,* **19**, 155–7.
Lakshmi, M. S. & Sherbet, G. V. (1962). *Naturwissenschaften,* **49**, 501–2.
Lakshmi, M. S. & Sherbet, G. V. (1964). *Naturwissenschaften,* **51**, 64–5.
Lallier, R. (1960). *Experientia,* **16**, 117–20.
Lamprecht, S. A., Zor, U., Tsafriri, A. & Lindner, H. R. (1973). *Journal of Endocrinology,* **57**, 217–33.
Landerman, R. (1967). *Developmental Biology,* **16**, 341–67.
Landström, U. (1977*a*). PhD thesis, University of Umeå, 116 p.
Landström, U. (1977*b*). *Journal of Embryology and Experimental Morphology,* **41**, 23–32.
Landström, U. & Løvtrup, S. (1975). *Journal of Embryology and Experimental Morphology,* **33**, 879–95.
Landström, U. & Løvtrup, S. (1977). *Acta Embryologiae et Morphologiae Experimentalis,* 171–7.
Landström, U. & Løvtrup, S. (1979). *Journal of Embryology and Experimental Morphology,* **54**, 113–30.
Landström, U., Løvtrup-Rein, H. & Løvtrup, S. (1975). *Cell Differentiation,* **4**, 313–25.
Landström, U., Løvtrup-Rein, H. & Løvtrup, S. (1976). *Journal of Embryology and Experimental Morphology,* **36**, 343–54.
Lang, A. (1969*). *28th Symposium of the Society of Developmental Biology, Developmental Biology,* suppl. 3, 244–50.
Langman, J. (1956). *Acta Morphologica Neerlando-Scandinavica,* **1**, 81–92.
Langman, J. (1959). *Journal of Embryology and Experimental Morphology,* **7**, 193–202.
Langman, J. & Nelson, G. R. (1968). *Journal of Embryology and Experimental Morphology,* **19**, 217–26.
Lanot, R. (1971). *Journal of Embryology and Experimental Morphology,* **26**, 1–20.
Lash, J. W. (1968*a**). In *Epithelial–Mesenchymal Interactions,* eds R. Fleischmajer & R. F. Billingham, pp. 165–72. Baltimore: Williams & Wilkins.
Lash, J. W. (1968*b*). *Journal of Cell Physiology,* **72**, suppl. 35–46.

Lash, J. W., Glick, M. C. & Madden, J. W. (1964). *National Cancer Institute Monographs*, **13**, Symposium on *Metabolic Control Mechanisms in Animal Cells* (1963), pp. 39–49.

Lash, J. W., Holtzer, H. & Whitehouse, M. W. (1960). *Developmental Biology*, **2**, 76–89.

Lash, J. W., Holtzer, S. & Holtzer, H. (1957). *Experimental Cell Research*, **13**, 292–303.

Lash, J. W., Ovadia, M. & Vasan, N. S. (1978). *Medical Biology*, **56**, 333–8.

Lash, J. W. & Vasan, N. S. (1977*). In *Cell and Tissue Interactions*, eds J. W. Lash & M. M. Burger, pp. 101–13. New York: Raven Press.

Laskey, R. A. (1983). *Nature* (London), **302**, 290–1.

Laskey, R. A., Gurdon, J. B. & Trendelenburg, M. (1979*). In *Maternal Effects in Development*, eds D. R. Newth & M. Balls, p. 65–80. Cambridge University Press.

Latzis, R. V. & Saraeva, N. Y. (1978). *Ontogenez*, **9**, 524–7.

Lau, E. C. & Ruch, J. V. (1983). *Differentiation*, **23**, 234–42.

Lawson, K. A. (1974). *Journal of Embryology and Experimental Morphology*, **32**, 469–93.

Lawson, K. A. (1983). *Journal of Embryology and Experimental Morphology*, **74**, 183–206.

Leblanc, J. & Brick, I. (1981). *Journal of Embryology and Experimental Morphology*, **61**, 145–63.

Le Douarin, N. (1964). *Journal of Embryology and Experimental Morphology*, **12**, 651–64.

Le Douarin, N. (1970*). *Année de Biologie*, **9**, 335–48.

Le Douarin, N. (1973). *Developmental Biology*, **30**, 217–22.

Le Douarin, N. (1974). *Année de Biologie*, **13**, 101–9.

Le Douarin, N. (1975). *Medical Biology*, **53**, 427–55.

Le Douarin, N. (1979*). *La Recherche*, **10**, 137–46.

Le Douarin, N. (1980a*). *Nature* (London), **286**, 663–9.

Le Douarin, N. (1980b*). *Current Topics in Developmental Biology, Neural Development*, part II, *Neural Development in Model Systems*, **16**, 32–85.

Le Douarin, N., Bussonnet, C. & Chaumont, F. (1968). *Annales d'Embryologie et de Morphogenèse*, **1**, 29–39.

Le Douarin, N. & Jotereau, F. V. (1975). *Journal of Experimental Medicine*, **142**, 17–40.

Le Douarin, N., Le Lièvre, C. S., Schweizer, G. & Ziller, C. M. (1979*). In *Cell Lineage, Stem Cells and Cell Determination*, ed. N. le Douarin, pp. 353–65. Amsterdam: Elsevier/North Holland Biomedical Press.

Le Douarin, N., Smith, J., Teillet, M. A., Le Lièvre, C. S. & Ziller, C. M. (1980*). In *Trends in Neurosciences*, **3**, 39–42.

Le Douarin, N. & Teillet, M. A. (1973). *Compte Rendu de l'Academie des Sciences, Paris*, **277**, 1929–32.

Le Douarin, N. & Teillet, M. A. (1974). *Developmental Biology*, **41**, 162–84.

Le Douarin, N., Teillet, M. A. & Le Lièvre, C. S. (1977*). In *Cell and Tissue Interactions*, eds J. W. Lash & M. M. Burger, pp. 11–27. New York: Raven Press.

Le Douarin, N., Teillet, M. A., Ziller, C. M. & Smith, J. (1978). *Proceedings of the National Academy of Sciences of the United States of America*. **75**, 2030–4.

Lee, H. (1973). *Experientia*, **29**, 332–4.

Lee, H. (1976). *Developmental Biology*, **48**, 392–9.

Lee, H. & Niu, M. C. (1973). In *The Role of RNA in Reproduction and Development*, eds M. C. Niu & S. J. Segal, pp. 137–54. Amsterdam: North Holland.

Lee, H. & Redmond, J. J. (1975). *Experientia*, **31**, 353–4.

Lee, H., Sheffield, J. B. & Nagele, R. G. (1978). *Journal of Experimental Zoology*, **204**, 137–54.

Lee, H., Sheffield, J. B., Nagele, R. G. & Kalmus, G. W. (1976). *Journal of Experimental Zoology*, **198**, 261–6.

Legros, F. & Brachet, J. (1965). *Journal of Embryology and Experimental Morphology*, **13**, 195–206.

Lehtonen, E., Jalanko, H., Laitinen, L., Miettinen, A., Ekblom, P. & Saxén, L. (1983). *W. Roux's Archives of Developmental Biology*, **192**, 145–51.

Leibel, W. S. (1976). *Journal of Experimental Zoology*, **196**, 85–104.

Leikola, A. (1963). *Annales Zoologici Societatis Zoologico-Botanicae Fennicae, Vanamo*, **25**, 1–50.

Leikola, A. (1965). *Experientia*, **21**, 458–9.

Leikola, A. (1976*a***). *Experientia*, **32**, 269–77.

Leikola, A. (1976*b***). *Folio Morphologica*, **24**, 155–72.

Leikola, A. (1977). *Proceedings of the International Union of Physiological Sciences*, Paris, **13**, abstract no. 1293.

Leikola, A. (1978). *Medical Biology*, **56**, 339–43.

Leikola, A. & McCallion, D. J. (1967). *Experientia*, **23**, 869–70.

Le Lièvre, C. (1971). *Bulletin de l'Association des Anatomistes*, 56ᵉ Congrès, 575–83.

Le Lièvre, C. (1974). *Journal of Embryology and Experimental Morphology*, **31**, 453–77.

Le Lièvre, C. & Le Douarin, N. (1970). *Année de Biologie*, **9**, 285–91.

Le Lièvre, C. & Le Douarin, N. (1975). *Journal of Embryology and Experimental Morphology*, **34**, 125–54.

Le Lièvre, C., Schweizer, G. G., Ziller, C. M. & Le Douarin, N. (1980). *Developmental Biology*, **77**, 362–78.

Lemanski, L. F. (1978). *Birth Defects*, **14**, 179–203.

Lemanski, L. F., Marx, B. S. & Hill, C. S. (1977). *Science*, **196**, 894–6.

Leonard, R. A., Hoffner, N. J. & Diberardino, M. A. (1982). *Developmental Biology*, **92**, 343–55.

Lepanto, L. (1965). *Ricerca Scientifica e Ricostruzione*, **35**, II B, 398–407.

Letourneau, P. C., Ray, P. N. & Bernfield, M. R. (1980*). In *Biological Regulation and Development*, vol. 2, *Molecular Organization and Cell Function*, ed. F. Goldberger, pp. 339–76. New York: Plenum Press.

Leussink, J. A. (1970). *Netherlands Journal of Zoology*, **20**, 1–79.

Levak-Svajger, B. & Svajger, A. (1974). *Journal of Embryology and Experimental Morphology*, **32**, 445–59.

Leyhausen, Cl. (1982). PhD thesis, University of Cologne, 218 p.

Linsenmayer, T. F. (1972*a*). *Developmental Biology*, **27**, 244–71.

Linsenmayer, T. F. (1972*b*). *Developmental Biology*, **29**, 16–18.

Lipton, B. H. & Jacobson, A. G. (1974). *Developmental Biology*, **38**, 91–103.

Loeffler, C. A. & Johnston, M. C. (1964). *Journal of Embryology and Experimental Morphology*, **12**, 407–24.

Loewenstein, W. R. (1967*a*). *Developmental Biology*, **15**, 503–20.

Loewenstein, W. R. (1967*b*). *Journal of Colloid and Interface Science*, **25**, 34–46.

Löfberg, J. (1974). *Developmental Biology*, **36**, 311–29.

Löfberg, J. & Ahlfors, K. (1978). *Zoon, a Journal of Zoology*, **6**, 87–101.

Löfberg, J. & Jacobson, C. O. (1974). *Zoon, a Journal of Zoology*, **2**, 85–98.

Lohmann, K. (1972). *W. Roux' Archiv für Entwicklungsmechanik der Organismen*, **169**, 1–40.

Lohmann, K. (1979). *Histochemie*, **63**, 47–56.

Lohmann, K. & Vahs, W. (1969). *Experientia*, **25**, 1315–16.

Lohmann, K. & Schubert, L. (1977). *Experientia*, **33**, 1518–19.

Longo, F. J. (1973*). *Biology of Reproduction*, **9**, 149–215.

Lopashov, G. V. (1961). *Annales Zoologici Societatis Zoologico-Botanicae Fennicae, Vanamo*, **22**, 1–17.

Lopashov, G. V. (1965*). In *Mechanisms of Cellular Differentiation and Induction*, pp. 242–70. Moscow: Nauka.

Lopashov, G. V. (1977). *Differentiation*, **9**, 131–7.

Lopashov, G. V. & Hoperskaya, O. A. (1967*a***). In *Morphologicheskie i Klimicheskie Czmenenija v. Processe Razvitija Kleski, Riga, Izdat, Zinatne*, pp. 7–16.

Lopashov, G. V. & Hoperskaya, O. A. (1967*b**). *Doklady Akademii Nauk SSSR.*, **175**, 962–5.

Lopashov, G. V. & Hoperskaya, O. A. (1977*). *Ontogenez*, **8**, 563–81.

Lopashov, G. V. & Stroeva, O. G. (1964*). *Development of the Eye: Experimental Studies*, 177 p. Jerusalem: S. Monson; New York: Davey.

Lopo, A. C. & Vacquier, V. D. (1981*). In *Fertilization and Embryonic Development in Vitro*, eds L. Mastroianni & J. D. Biggers, pp. 201–32. New York: Plenum Press.

Løvtrup, S. (1962). *Journal of Experimental Zoology*, **151**, 79–84.

Løvtrup, S. (1965*a**). *W. Roux' Archiv für Entwicklungsmechanik der Organismen*, **156**, 204–48.

Løvtrup, S. (1965*b**). *Acta Zoologica*, **46**, 119–65.

Løvtrup, S. (1965*c**). *Acta Universitatis Gothoburgensis, Zoologica Gothoburgensia*, **1**, 1–139.

Løvtrup, S. (1975). *Canadian Journal of Zoology*, **53**, 473–9.

Løvtrup, S. (1983*). *Biological Reviews*, **58**, 91–130.

Løvtrup, S., Landström, U. & Løvtrup-Rein, H. (1978*). *Biological Reviews*, **53**, 1–42.

Løvtrup S. & Perris, R. (1983). *Cell Differentiation*, **12**, 171–6.

Løvtrup-Rein, H., Landström, U. & Løvtrup, S. (1978). *Cell Differentiation*, **7**, 131–8.

Luchtel, D., Bluemink, J. G. & de Laat, S. W. (1976). *Journal of Ultrastructure Research*, **54**, 406–19.

Luckenbill, L. M. (1971). *Experimental Cell Research*, **66**, 263–7.

Lundmark, C., Shih, J., Tibbetts, P. & Keller, R. (1984). *Journal of Embryology and Experimental Morphology*, (in press).

Luther, W. (1937). *W. Roux' Archiv für Entwicklungsmechanik der Organismen*, **135**, 359–83.

Luther, W. (1938). *W. Roux' Archiv für Entwicklungsmechanik der Organismen*, **137**, 404–34.

Lutz, H. (1962). *Archives d'Anatomie, d'Histologie et d'Embryologie*, **44**, 167–77.

Lutz, H. (1964). *Archives d'Anatomie, d'Histologie et d'Embryologie*, **47**, 531–8.

Lutz, H. (1965). *Annales de la Faculté des Sciences*, **26**, 71–84.

Lutz, H., Departout, M., Hubert, J. & Pieau, C. (1963). *Developmental Biology*, **6**, 23–44.

MacCabe, J. A., Errick, J. & Saunders, J. W. (1974). *Developmental Biology*, **39**, 69–82.

MacCabe, J. A., MacCabe, A. B., Abbott, U. K. & McCarrey, J. R. (1975). *Journal of Experimental Zoology*, **191**, 383–94.

MacCabe, J. A. & Parker, B. W. (1975). *Developmental Biology*, **45**, 349–57.

MacCabe, J. A. & Parker, B. W. (1976). *Developmental Biology*, **54**, 297–303.

MacCabe, J. A. & Parker, B. W. (1979). *Journal of Embryology and Experimental Morphology*, **53**, 67–73.

McCallion, D. J. & Leikola, A. (1967). *Annales Zoologici Societatis Zoologico-Botanicae Fennicae*, **4**, 588–91.

McCallion, D. J. & Shinde, V. A. (1973). *Experientia*, **29**, 321–2.

McDevitt, D. S. & Clayton, R. M. (1979). *Journal of Embryology and Experimental Morphology*, **50**, 31–45.

McKeehan, M. S. (1951). *Journal of Experimental Zoology*, **117**, 31–64.

McKeehan, M. S. (1956). *American Journal of Anatomy*, **99**, 131–56.

McKeehan, M. S. (1958). *The Anatomical Record*, **132**, 297–306.

McLaren, A. & Wylie, C. C. (1983*). *Current Problems in Germ Cell Differentiation*, 401 p. Cambridge University Press.

MacLoughlin, C. B. (1961*a*). *Journal of Embryology and Experimental Morphology*, **9**, 370–84.

MacLoughlin, C. B. (1961*b*). *Journal of Embryology and Experimental Morphology*, **9**, 385–409.

MacLoughlin, C. B. (1961c). In *La Culture Organotypique; Associations et Dissociations d'Organes en Culture in Vitro*, eds E. Wolff, J. A. A. Benoit, K. Haffen & F. Bermann, pp. 145–54. Paris: Centre Nationale de la Recherche Scientifique.

McMahon, D. (1974). *Science*, **185**, 1012–21.

Mager, W. (1972). PhD thesis, University of Cologne, 62 p.

Mahowald, A. P. & Hennen, S. (1971). *Developmental Biology*, **24**, 37–53.

Maisel, H. & Langman, J. (1961). *Journal of Embryology and Experimental Morphology*, **9**, 191–201.

Mak, L. L. (1978). *Developmental Biology*, **65**, 435–46.

Malacinski, G. M. (1972). *Cell Differentiation*, **1**, 253–64.

Malacinski, G. M. (1974). *Cell Differentiation*, **3**, 31–44.

Malacinski, G. M. (1985*). In *Program in Molecular, Cellular and Developmental Biology*, (in press).

Malacinski, G. M. (1985*). In *Biology of Fertilization*, eds A. Monroy & C. B. Metz, (in press). New York: Academic Press.

Malacinski, G. M., Allis, C. D. & Chung, H. M. (1974). *Journal of Experimental Zoology*, **189**, 249–54.

Malacinski, G. M., Benford, H. & Chung, H. M. (1975). *Journal of Experimental Zoology*, **191**, 97–110.

Malacinski, G. M., Brothers, A. J. & Chung, H. M. (1977). *Developmental Biology*, **56**, 24–39.

Malacinski, G. M. & Chung, H. M. (1981). *Journal of Morphology*, **169**, 149–59.

Malacinski, G. M., Chung, H. M. & Asashima, M. (1980). *Developmental Biology*, **77**, 449–62.

Malacinski, G. M., Chung, H. M. & Woo Youn, B. (1978a). *Experientia*, **34**, 883–4.

Malacinski, G. M., Ryan, B. & Chung, H. M. (1978b). *Differentiation*, **10**, 101–7.

Malacinski, G. M. & Spieth, J. (1978*). In *Maternal Effects in Development*, eds D. R. Newth & M. Balls, pp. 241–67. Cambridge University Press.

Malacinski, G. M. & Woo Youn, B. (1981a). *Developmental Biology*, **88**, 352–7.

Malacinski, G. M. & Woo Youn, B. (1981b). *Netherlands Journal of Zoology*, **31**, 38–49.

Maleyvar, R. P. & Lowery, R. (1973*). In *The Cell Cycle in Development and Differentiation*, eds M. Balls & F. S. Billett, pp. 249–55. Cambridge University Press.

Maleyvar, R. P. & Lowery, R. (1976). *Cytobios*, **17**, 21–30.

Maleyvar, R. P. & Lowery, R. (1981). *Cytobios*, **32**, 97–105.

Maller, J., Poccia, D., Nishioka, D., Kidd, P. & Gerhart, J. (1976). *Experimental Cell Research*, **99**, 285–94.

Malpoix, P., Quertier, J. & Brachet, J. (1963). *Journal of Embryology and Experimental Morphology*, **11**, 155–66.

Manasek, F. J. (1976). *Journal of Molecular and Cellular Cardiology*, **8**, 389–402.

Manes, M. E. & Barbieri, F. D. (1976). *Developmental Biology*, **53**, 138–41.

Manes, M. E. & Barbieri, F. D. (1977). *Journal of Embryology and Experimental Morphology*, **40**, 187–97.

Manes, M. E. & Elinson, R. P. (1980). *W. Roux's Archives of Developmental Biology*, **189**, 73–6.

Manes, M. E., Elinson, R. P. & Barbieri, F. D. (1978). *W. Roux's Archives of Developmental Biology*, **185**, 99–104.

Mansour, A. M. & Niu, M. C. (1965). *Proceedings of the National Academy of Sciences of the United States of America*, **53**, 764–70.

Markert, C. L. (1960*). *National Cancer Institute Monographs*, **2**, 2–17.

Markert, C. L. (1965*). *Proceedings of the 16th International Congress of Zoology* (1963), 229–58.

Markman, B. (1958). *Acta Zoologica*, **39**, 103–15.

Martin, A. H. (1971). *Acta Embryologiae et Morphologiae Experimentalis*, 9–16.

Martin, A. H. (1977). *Acta Embryologiae et Morphologiae Experimentalis*, 305–13.
Martin, C. (1976). *Journal of Embryology and Experimental Morphology*, **35**, 485–98.
Martz, E. & Steinberg, M. S. (1973). *Journal of Cell Physiology*, **81**, 25–38.
Masui, Y. (1960a). *Memoirs of the Konan University, Science Series*, **4**, 65–78.
Masui, Y. (1960b). *Memoirs of the Konan University, Science Series*, **4**, 79–102.
Masui, Y. (1960c). *Memoirs of the Konan University, Science Series*, **4**, 103–14.
Masui, Y. (1961). *Experientia*, **17**, 458–9.
Masui, Y. (1966). *Journal of Embryology and Experimental Morphology*, **15**, 372–86.
Masui, Y. (1967). *Journal of Experimental Zoology*, **166**, 365–76.
Masui, Y. & Clarke, H. J. (1979*). *International Reviews of Cytology*, **57**, 185–282.
Masui, Y., Forer, A. & Zimmerman, A. M. (1978). *Journal of Cell Science*, **31**, 117–35.
Masui, Y. & Markert, C. L. (1971). *Journal of Experimental Zoology*, **177**, 129–46.
Masui, Y., Meyerhof, P. G., Miller, M. A. & Wasserman, W. J. (1977). *Differentiation*, **9**, 49–57.
Matsuda, M. (1980). *Journal of Embryology and Experimental Morphology*, **60**, 163–71.
Matsuda, M. & Kajishima, T. (1978). *Acta Embryologiae et Morphologiae Experimentalis*, 309–17.
Matsuda, M. & Kajishima, T. (1980). *Acta Embryologiae et Morphologiae Experimentalis*, 165–74.
Maufroid, J. P. & Capuron, A. (1977). *Compte Rendu de l'Academie des Sciences*, Paris, Sér. D, **284**, 1713–16.
Mauger, A. (1972a). *Journal of Embryology and Experimental Morphology*, **28**, 313–41.
Mauger, A. (1972b). *Journal of Embryology and Experimental Morphology*, **28**, 343–66.
Mauger, A. (1972c). *W. Roux' Archiv für Entwicklungsmechanik der Organismen*, **170**, 244–66.
Mehr, J. (1982). *Rochester Review*, Autumn 1982, 7 p.
Meier, S. & Hay, E. D. (1974a). *Proceedings of the National Academy of Sciences of the United States of America*, **71**, 2310–13.
Meier, S. & Hay, E. D. (1974b). *Developmental Biology*, **38**, 249–70.
Meier, S. & Hay, E. D. (1975). *Journal of Cell Biology*, **66**, 275–91.
Menkes, B. & Sandor, S. (1969). *Revue Roumaine d'Embryologie et de Cytologie, sér. Embryologie*, **6**, 65–71.
Menkes, B., Sandor, S. & Elias, S. (1968). *Revue Roumaine d'Embryologie et de Cytologie, sér. Embryologie*, **5**, 131–7.
Merriam, R. W. (1971a). *Experimental Cell Research*, **68**, 75–80.
Merriam, R. W. (1971b). *Experimental Cell Research*, **68**, 81–7.
Merriam, R. W. & Christensen, K. (1983). *Journal of Embryology and Experimental Morphology*, **75**, 11–20.
Merriam, R. W. & Sauterer, R. A. (1983). *Journal of Embryology and Experimental Morphology*, **76**, 51–65.
Merriam, R. W., Sauterer, R. A. & Christensen, K. (1983). *Developmental Biology*, **95**, 439–46.
Messier, P. E. (1978). *Experientia*, **34**, 289–96.
Messier, P. E. & Sequin, C. (1978). *Journal of Embryology and Experimental Morphology*, **44**, 281–95.
Metz, Ch. P. (1978*). *Current Topics in Developmental Biology*, **12**, 107–47.
Meyerhof, P. G. & Masui, Y. (1977). *Developmental Biology*, **61**, 214–29.
Meyerhof, P. G. & Masui, Y. (1979a). *Experimental Cell Research*, **123**, 345–53.
Meyerhof, P. G. & Masui, Y. (1979b). *Developmental Biology*, **72**, 182–7.
Mezger-Freed, L. & Oppenheimer, J. M. (1965). *Developmental Biology*, **11**, 385–401.
Michael, M. I. (1968). *Alexandria Medical Journal*, **14**, 53–64.
Michael, M. I. & Nieuwkoop, P. D. (1967). *Verslagen van de Koninklijke Nederlandse Akademie van Wetenschappen*, serie C , **70**, 272–9.

Mikami, Y. & Nishimura, K. (1965). *Journal of the Mie Medical Association*, **7**, 93–104.

Mikawa, T. & Hiroshe, G. (1982). *Development, Growth and Differentiation*, **24**, 395 (abstract no. B 33).

Miller, R. L. (1983*). In *Reproductive Biology of Invertebrates*, eds K. G. Adiyodi & R. G. Adiyodi, vol. 1, pp. 99–109. New York: Wiley & Sons.

Minuth, M. & Grunz, H. (1980). *Cell Differentiation*, **9**, 229–38.

Minuth, W. W. (1977). PhD thesis, University of Cologne, 84 p.

Minuth, W. W. (1978). *Medical Biology*, **56**, 349–54.

Mitolo, V., Ferrannini, E. & Franchini, G. (1968). *Acta Embryologiae et Morphologiae Experimentalis*, **10**, 302–15.

Mitolo, V., Jirillo, E. & Neri, V. (1970). *Zeitschrift für Anatomie und Entwicklungsgeschichte*, **130**, 9–22.

Mitrani, E. & Eyal-Giladi, H. (1981). *Nature* (London), **289**, 800–2.

Mitrani, E. & Eyal-Giladi, H. (1982). *Differentiation*, **21**, 56–61.

Mitrani, E., Shimoni, Y. & Eyal-Giladi, H. (1983). *Journal of Embryology and Experimental Morphology*, **75**, 21–30.

Miura, Y. & Wilt, F. H. (1969). *Developmental Biology*, **19**, 201–11.

Miyagawa, N. & Suzuki, A. (1969). *Kumamoto Journal of Science*, Ser. B, **9**, 109–15.

Miyayama, Y. & Fujimoto, T. (1977). *Okajimas Folia Anatomica Japonica*, **54**, 97–120.

Mizuno, T. (1970). *Compte Rendu de l'Académie des Sciences*, Paris, **271**, 2190–2.

Mizuno, T. (1972). *Journal of Embryology and Experimental Morphology*, **28**, 117–32.

Mizuno, T. & Katoh, Y. (1972). *Proceedings of the Japanese Academy*, **48**, 522–7.

Modak, S. P. (1965). *Experientia*, **21**, 273.

Modak, S. P. (1966). *Revue Suisse de Zoologie*, **73**, 877–908.

Model, P. G. (1978). *Brain Research*, **153**, 135–43.

Model, P. G. & Wurzelmann, S. (1982). *Developmental Brain Research*, **3**, 123–9.

Monroy, A. (1965*). *Chemistry and Physiology of Fertilisation*, 150 p. New York: Holt, Rinehart & Winston.

Monroy, A. (1976*). in *Tests of Teratogenicity in Vitro*, eds J. D. Ebert & M. Marois, pp. 25–36. Amsterdam: North Holland Publishing Company.

Monroy, A. (1979). *Differentiation*, **13**, 23–4.

Monroy, A. & Baccetti, B. (1975). *Journal of Ultrastructure Research*, **50**, 131–42.

Monroy, A., Baccetti, B. & Denis-Domini, S. (1976). *Developmental Biology*, **49**, 250–9.

Monroy, A., Ortolani, G. O'Dell, D. & Millonig, G. (1973). *Nature* (London), **42**, 409–10.

Moran, D. J. (1974). *Experimental Cell Research*, **86**, 365–73.

Moran, D. J. (1976). *Journal of Experimental Zoology*, **198**, 409–16.

Moran, D. J. (1978*a*). *Bioelectrochemistry and Bioenergetics*, **5**, 373–7.

Moran, D. J. (1978*b*). *Zoon, a Journal of Zoology*, **6**, 81–6.

Moran, D. J. & Mouradian, W. E. (1975). *Developmental Biology*, **46**, 422–9.

Moran, D. J., Palmer, J. D. & Model, P. G. (1973). *Developmental Biology*, **32**, 15–27.

Moran, D. J. & Rice, R. W. (1975). *Journal of Cell Biology*, **64**, 172–81.

Moran, D. J. & Rice, R. W. (1976). *Nature* (London), **261**, 497–9.

Moreau, M., Dorée, M. & Guerrier, P. (1976). *Journal of Experimental Zoology*, **197**, 443–9.

Moreau, M., Guerrier, P. & Dorée, M. (1976). *Compte Rendu de l'Académie des Sciences*, Paris, Sér. D, **282**, 1309–12.

Morgan, T. H. (1934*). *Embryology and Genetics*, 766 p. New York: Columbia University Press.

Morrill, G. A., Kostellow, A. B. & Murphy, J. B. (1971). *Experimental Cell Research*, **66**, 289–98.

Morrill, G. A., Kostellow, A. B. & Murphy, J. B. (1974). *Annals of the New York Academy of Sciences*, **242**, 543–59.

Morrill, G. A., Kostellow, A. B. & Watson, D. E. (1966). *Life Sciences*, **5**, 705–9.
Morrill, G. A. & Watson, D. E. (1966). *Journal of Cellular Physiology*, **67**, 85–92.
Moscona, A. A. (1959). *Developmental Biology*, **1**, 1–23.
Moscona, A. A. (1960). *Transplantation Bulletin*, **26**, 120–4.
Moscona, A. A. & Degenstein, L. (1982*). In *Stability and Switching in Cellular Differentiation*, eds R. M. Clayton & D. E. S. Truman, pp. 187–98. New York: Plenum Press.
Moscona, A. A., Moscona, M. H. & Saenz, N. (1968). *Proceedings of the National Academy of Sciences of the United States of America*, **61**, 160–7.
Moscona, A. A., Saenz, N. & Moscona, M. H. (1967). *Experimental Cell Research*, **48**, 646–9.
Moss, M. L. (1969). *Journal of Dental Research*, **48**, 732–7.
Motomura, I. (1960). *Science Reports of the Tohoku University*, **26**, 53–8.
Motomura, I. (1967). *Science Reports of the Tohoku University*, **33**, 143–8.
Muchmore, W. B. (1951). *Journal of Experimental Zoology*, **118**, 137–86.
Muchmore, W. B. (1957*a*). *Journal of Experimental Zoology*, **134**, 293–313.
Muchmore, W. B. (1957*b*). *Proceedings of the National Academy of Sciences of the United States of America*, **43**, 435–9.
Muchmore, W. B. (1958). *Journal of Experimental Zoology*, **139**, 181–8.
Muchmore, W. B. (1964*a*). *The American Zoologist*, **4**, 387, Abstr. no. 42.
Muchmore, W. B. (1964*b*). *Journal of Embryology and Experimental Morphology*, **12**, 587–96.
Mulnard, J. G. (1967). *Archives de Biologie*, **78**, 107–38.
Mulnard, J., Creteur, V. & Verbruggen, J. L. (1977). *Archives de Biologie*, **88**, 15–23.
Murakawa, Sh. (1960). *Mie Medical Journal*, **10**, 377–86.
Murillo-Ferrol, N. L. (1965). *Acta Anatomica*, **62**, 80–103.
Muthakkaruppan, V. (1965). *Journal of Experimental Zoology*, **159**, 269–88.
Nagele, R. G. & Lee, H. Y. (1980). *Journal of Experimental Zoology*, **213**, 391–8.
Nagele, R. G., Pietrolungo, J. F. & Lee, H. Y. (1981). *Experientia*, **37**, 304–6.
Nakamura, O. (1942). *Annotationes Zoologicae Japonenses*, **21**, 169–236.
Nakamura, O. (1961). *Embryologia*, **6**, 99–109.
Nakamura, O. (1978*). In *Organiser – a Milestone of a Half-Century from Spemann*, eds O. Nakamura & S. Toivonen, pp. 179–220. Amsterdam: Elsevier/North Holland Biomedical Press.
Nakamura, O. & Aochi, M. (1970). *Proceedings of the Japanese Academy*, **46**, 852–7.
Nakamura, O., Hayakawa, H. & Yamamoto, K. (1966). *Proceedings of the 6th International Congress of Electron Microscopy*, Kyoto, 651–2.
Nakamura, O., Hayashi, Y. & Asashima, M. (1978*). In *Organiser – a Milestone of a Half-Century from Spemann*, eds O. Nakamura & S. Toivonen, pp. 1–47. Amsterdam: Elsevier/North Holland Biomedical Press.
Nakamura, O. & Kishiyama, K. (1971). *Proceedings of the Japanese Academy*, **47**, 407–12.
Nakamura, O. & Matsuzawa, T. (1967). *Embryologia*, **9**, 223–37.
Nakamura, O. & Takasaki, H. (1970). *Proceedings of the Japanese Academy*, **46**, 546–51.
Nakamura, O. & Takasaki, H. (1971*a*). *Proceedings of the Japanese Academy*, **47**, 92–7.
Nakamura, O. & Takasaki, H. (1971*b*). *Proceedings of the Japanese Academy*, **47**, 499–504.
Nakamura, O., Takasaki, H. & Ishihara, M. (1971*a*). *Proceedings of the Japanese Academy*, **47**, 313–18.
Nakamura, O., Takasaki, H. & Mizohata, T. (1970*a*). *Proceedings of the Japanese Academy*, **46**, 694–9.
Nakamura, O., Takasaki, H., Okumoto, T. & Iida, H. (1971*b*). *Proceedings of the Japanese Academy*, **47**, 203–8.

Nakamura, O., Takasaki, H., Yamane, H., Obayashi, N., Kono, S., Akamoto, H. & Okumoto, T. (1970*b*). *Proceedings of the Japanese Academy*, **46**, 700–5.

Nakamura, O. & Toivonen, S. (1978*). *Organiser – a Milestone of a Half-Century from Spemann*, 379 p. Amsterdam: Elsevier/North Holland Biomedical Press.

Nakamura, O. & Yamada, K. (1971). *Development, Growth and Differentiation*, **13**, 303–21.

Nakatsuji, N. (1974*a*). *Development, Growth and Differentiation*, **16**, 257–65.

Nakatsuji, N. (1974*b*). *Journal of Embryology and Experimental Morphology*, **32**, 795–804.

Nakatsuji, N. (1975*a*). *Journal of Embryology and Experimental Morphology*, **34**, 669–85.

Nakatsuji, N. (1975*b*). *W. Roux's Archives of Developmental Biology*, **178**, 1–14.

Nakatsuji, N. (1976). *W. Roux's Archives of Developmental Biology*, **180**, 229–40.

Nakatsuji, N. (1979). *Developmental Biology*, **68**, 140–50.

Nakatsuji, N. (1985). *The American Zoologist*, (in press).

Nakatsuji, N., Gould, A. C. & Johnson, K. E. (1982). *Journal of Cell Science*, **56**, 207–22.

Nakatsuji, N. & Johnson, K. E. (1982*a*). *Cell Motility*, **2**, 149–61.

Nakatsuji, N. & Johnson, K. E. (1982*b*). *Development, Growth and Differentiation*, **24**, 415 (abstract no. C 51).

Nakatsuji, N. & Johnson, K. E. (1983). *Journal of Cell Science*, **59**, 43–60.

Nakatsuji, N. & Johnson, K. E. (1984). *Nature* (London), **307**, 453–5.

Nakauchi, M. & Takeshita, T. (1983). *Journal of Experimental Zoology*, **227**, 155–8.

Nanjundiah, V. (1974). *Experimental Cell Research*, **86**, 408–11.

Neff, A. W., Wakahara, M., Jurand, A. & Malacinski, G. M. (1984). *Journal of Embryology and Experimental Morphology*, **80**, 197–224.

Neufang, O., Born, J., Tiedemann, H. & Tiedemann, H. (1978). *Medical Biology*, **56**, 361–5.

Neumann, T. (1983). *Experientia*, **39**, 96–7.

Neumann, T., Laasberg, T. & Kärner, J. (1983). *W. Roux's Archives of Developmental Biology*, **192**, 42–4.

Newgreen, D. F. & Gibbins, J. L. (1982). *Cell and Tissue Research*, **224**, 145–60.

Newgreen, D. F., Gibbins, J. L., Sauter, J., Wallenfels, B. & Würtz, R. (1982). *Cell and Tissue Research*, **221**, 521–49.

Newgreen, D. F. & Thiery, J. P. (1980). *Cell and Tissue Research*, **211**, 269–91.

Newport, J. & Kirschner, M. (1982*a*). *Cell*, **30**, 675–86.

Newport, J. & Kirschner, M. (1982*b*). *Cell*, **30**, 687–96.

Newsome, D. A. (1972). *Developmental Biology*, **27**, 575–9.

Neyfakh, A. A. (1959). *Journal of Embryology and Experimental Morphology*, **7**, 173–92.

Niazi, I. A. (1969). *Journal of Embryology and Experimental Morphology*, **22**, 1–14.

Nicholson, G. L. (1974*). *International Review of Cytology*, **39**, 89–190.

Nicolet, G. (1965), *Acta Embryologiae et Morphologiae Experimentalis*, **8**, 213–20.

Nicolet, G. (1968). *Experientia*, **24**, 263–4.

Nicolet, G. (1970*a*). *Journal of Embryology and Experimental Morphology*, **23**, 79–108.

Nicolet, G. (1970*b**). *Médicine et Hygiène*, **28**, 1433–7.

Nicolet, G. (1970*c*). *Journal of Embryology and Experimental Morphology*, **24**, 467–78.

Nicolet, G. (1971*). *Advances in Morphogenesis*, **9**, 231–62.

Nieuwkoop, P. D. (1946). *Archives Néerlandaises de Zoologie*, **8**, 1–205.

Nieuwkoop, P. D. (1947). *Journal of Experimental Biology*, **24**, 145–83.

Nieuwkoop, P. D. (1955*). *Experimental Cell Research*, suppl. 3, 262–73.

Nieuwkoop, P. D. (1958). *Acta Embryologiae et Morphologiae Experimentalis*, **2**, 13–53.

Nieuwkoop, P. D. (1960). *Archives Néerlandaises de Zoologie*, **13**, 588–90.

Nieuwkoop, P. D. (1962). *Acta Biotheoretica*, **16**, 57–68.

Nieuwkoop, P. D. (1963). *Developmental Biology*, **7**, 255–79.

Nieuwkoop, P. D. (1966*). In *Cell Differentiation and Morphogenesis*, eds W. Beerman,

P. D. Nieuwkoop & E. Wolff, pp. 120–43. Amsterdam: North Holland Publishing Company.

Nieuwkoop, P. D. (1967*a*). *Acta Biotheoretica*, **17**, 151–77.

Nieuwkoop, P. D. (1967*b*). *Acta Biotheoretica*, **17**, 178–94.

Nieuwkoop, P. D. (1967*c**). Informal round-table Conference, 1966, *Morphological and Biochemical Aspects of Cytodifferentiation*, eds E. Hagen, W. Wechsler & P. Zilliken. Basel: Karger Verlag, *Experimental Biology and Medicine*, **1**, 22–36.

Nieuwkoop, P. D. (1968). In *Le Basi Moleculari del Differenziamento*, Milano, 1967, ed. S. Ranzi, Academia Nazionale Lincei, Quaderno, **104**, 203–8.

Nieuwkoop, P. D. (1969*a*). *W. Roux' Archiv für Entwicklungsmechanik der Organismen*, **162**, 341–73.

Nieuwkoop, P. D. (1969*b*). *W. Roux' Archiv für Entwicklungsmechanik der Organismen*, **163**, 298–315.

Nieuwkoop, P. D. (1970). *W. Roux' Archiv für Entwicklungsmechanik der Organismen*, **166**, 105–23.

Nieuwkoop, P. D. (1973*). *Advances in Morphogenesis*, **10**, 1–39.

Nieuwkoop, P. D. (1977*). *Current Topics in Developmental Biology*, **11**, 115–32.

Nieuwkoop, P. D., Boterenbrood, E. C., Kremer, A., Bloemsma, F. F. S. N., Hoessels, E. L. M. J., Meyer, G. & Verheyen, F. J. (1952). *Journal of Experimental Zoology*, **120**, 1–108.

Nieuwkoop, P. D. & Faber, J. (1975). *Normal Table of Xenopus laevis* (Daudin), 252 p. Second reprinted edition, Amsterdam: North Holland Publishing Company.

Nieuwkoop, P. D. & Florschütz, P. A. (1950). *Archives de Biologie*, **61**, 113–50.

Nieuwkoop, P. D., Niermeyer, E. K. & Jansen, W. F. (1964). *Journal of Animal Morphology and Physiology*, **11**, 21–44.

Nieuwkoop, P. D., Oikawa, I. & Boddingius, J. (1958). *Archives Néerlandaises de Zoologie*, **13**, 167–84.

Nieuwkoop, P. D. & Sutasurya, L. A. (1979*). *Primordial Germ Cells in the Chordates, Embryogenesis and Phylogenesis*, 187 p. Cambridge University Press.

Nieuwkoop, P. D. & Sutasurya, L. A. (1983). In *Development and Evolution*, eds B. C. Goodwin, N. Holder & C. C. Wylie, pp. 123–35. Cambridge University Press.

Nieuwkoop, P. D. & Ubbels, G. A. (1972). *W. Roux' Archiv für Entwicklungsmechanik der Organismen*, **169**, 185–99.

Nieuwkoop, P. D. & van der Grinten, S. J. (1961). *Embryologia*, **6**, 51–66.

Nieuwkoop, P. D. & Weijer, C. J. (1978). *Medical Biology*, **56**, 366–71.

Nishijima, K., Noda, S., Kurihara, K. & Sasaki, N. (1978). *Development, Growth and Differentiation*, **20**, 275–81.

Nishimura, K. (1967). *Mie Medical Journal*, **16**, 269–76.

Niu, M. C. (1958). *The Anatomical Record*, **131**, 585, Abstr. no. 78.

Niu, M. C. (1959*). In *Evolution of Nervous Control*, American Association for the Advancement of Science, ed. A. D. Bass, pp. 7–30. London: Bailey & Swinfen.

Niu, M. C. (1963). *Developmental Biology*, **7**, 379–93.

Niu, M. C. (1964). *National Cancer Institute Monographs*, **13**, 167–77.

Niu, M. C. & Deshpande, A. K. (1973). *Journal of Embryology and Experimental Morphology*, **29**, 485–501.

Niu, M. C. & Sasaki, N. (1971). *Experimental Cell Research*, **64**, 57–64.

Noda, S. & Kawakami, I. (1976). *Journal of Embryology and Experimental Morphology*, **36**, 55–66.

Noda, S., Sasaki, N. & Iyeiri, S. (1972*a*). *The Zoological Magazine*, **81**, 59–62.

Noda, S., Sasaki, N. & Iyeiri, S. (1972*b*). *The Zoological Magazine*, **81**, 63–6.

Noda, S., Sasaki, N. & Iyeiri, S. (1972*c*). *The Zoological Magazine*, **81**, 150–3.

Noden, D. M. (1975). *Developmental Biology*, **42**, 106–30.

Noden, D. M. (1978*a*). *Developmental Biology*, **67**, 296–312.

Noden, D. M. (1978*b*). *Developmental Biology*, **67**, 313–29.

Noden, D. M. (1978*c**). In *Specificity of Embryological Interactions; Receptors and Recognition* (B 4), ed. D. R. Garred, pp. 3–49. London: Chapman & Hall.

Nomura, K. & Okada, T. S. (1979). *Development, Growth and Differentiation*, **21**, 161–8.

Nordling, S., Miettinen, H., Wartiovaara, J. & Saxén, L. (1971). *Journal of Embryology and Experimental Morphology*, **26**, 231–52.

Norr, S. C. (1973). *Developmental Biology*, **34**, 16–38.

Nosek, J. (1978). *W. Roux's Archives of Developmental Biology*, **184**, 181–93.

Noto, T. (1967). *Science Report Tohoku University*, Ser. IV (Biology), **33**, 51–7.

Novel, G. (1973). *Journal of Embryology and Experimental Morphology*, **30**, 605–33.

Nyholm, M., Saxén, L., Toivonen, S. & Vainio, T. (1962). *Experimental Cell Research*, **28**, 209–12.

O'Dell, D. S., Tencer, R., Monroy, A. & Brachet, J. (1974). *Cell Differentiation*, **3**, 193–8.

Odell, G. M., Oster, G., Alberch, P. & Burnside, B. (1981). *Development Biology*, **85**, 446–62.

Ogi, K. I. (1961). *Embryologia*, **5**, 384–96.

Ogi, K. I. (1967). *Science Report Tohoku University*, Ser. IV (Biology), **33**, 239–47.

Ogi, K. I. (1969). *Research Bulletin of the Department of General Education, Nagoya University*, **13**, 31–40.

Ohara, A. (1980). *Development, Growth and Differentiation*, **22**, 805–12.

Ohara, A. (1981). *Development, Growth and Differentiation*, **23**, 51–8.

Ohara, A. & Hama, T. (1979*a*). *W. Roux's Archives of Developmental Biology*, **187**, 13–23.

Ohara, A. & Hama, T. (1979*b*). *Development, Growth and Differentiation*, **21**, 509–17.

O'Hare, M. J. (1972). *Journal of Embryology and Experimental Morphology*, **27**, 119–34.

Ohnishi, T. & Sugiyama, M. (1963). *Embryologia*, **8**, 79–88.

Okada, T. S. (1953*a*). *The Zoological Magazine*, **62**, 288–91.

Okada, T. S. (1953*b*). *Memoirs of the College of Science, Kyoto University*, Ser. B, **20**, 157–62.

Okada, T. S. (1957). *Journal of Embryology and Experimental Morphology*, **5**, 438–48.

Okada, T. S. (1960*a*). *W. Roux' Archiv für Entwicklungsmechanik der Organismen*, **151**, 559–71.

Okada, T. S. (1960*b*). *W. Roux' Archiv für Entwicklungsmechanik der Organismen*, **152**, 1–21.

Okada, T. S., Itoh, Y., Watanabe, K. & Eguchi, G. (1975). *Developmental Biology*, **45**, 318–29.

Okada, T. S. & Sirlin, J. L. (1960). *Journal of Embryology and Experimental Morphology*, **8**, 54–9.

Okada, T. S., Yasuda, K., Araki, M. & Eguchi, G. (1979*a*). *Developmental Biology*, **68**, 600–17.

Okada, T. S., Yasuda, K. & Nomura, K. (1979*b*). In *Cell Lineage, Stem Cells and Cell Determination*, ed. N. le Douarin, pp. 335–46. Amsterdam: Elsevier/North Holland Biomedical Press.

Okada, T. S. & Ichikawa, M. (1947). *Japanese Journal of Experimental Morphology*, **3**, 1–6.

Okamoto, M. & Eguchi, G. (1975). *Development, Growth and Differentiation*, **17**, 209–19.

Olson, G. E. & Hamilton, D. W. (1978). *Biology of Reproduction*, **19**, 26–35.

Olszanska, B. & Kludkiewicz, B. (1983). *Cell Differentiation*, **12**, 115–20.

Oppenheimer, J. M. (1947*). *Quarterly Review of Biology*, **22**, 105–18.

Oppenheimer, J. M. (1966). *Archivio Zoologico Italiano*, **51**, 667–82.

Ortolani, G. (1958). *Acta Embryologiae et Morphologiae Experimentalis*, **1**, 247–72.

Ortolani, G. (1961). *La Ricerca Scientifica*, II B, **31**, 157–62.

Orts-Llorca, F. & Domenech Mateu, J. M. (1980). *Acta Anatomica*, **106**, 415–23.

Orts-Llorca, F. & Jimenez Collado, J. (1967). *W. Roux' Archiv für Entwicklungsmechanik der Organismen*, **158**, 147–63.

Orts-Llorca, F. & Jimenez Collado, J. (1969). *Developmental Biology*, **19**, 213–27.

Orts-Llorca, F. & Jimenez Collado, J. (1970). *Archives d'Anatomie, d'Histologie et d'Embryologie Expérimentale*, **53**, 115–24.

Orts-Llorca, F. & Murillo-Ferrol, N. L. (1965). *W. Roux' Archiv für Entwicklungsmechanik der Organismen*, **156**, 363–7.

Orts-Llorca, F. & Ruano-Gil, D. (1965). *W. Roux' Archiv für Entwicklungsmechanik der Organismen*, **156**, 368–70.

Osborn, J. C., Duncan, C. J. & Smith, J. L. (1979). *Journal of Cell Biology*, **80**, 589–604.

Overton, J. (1974*). *Progress in Surface and Membrane Science*, **8**, 161–208.

Packard, D. S. (1978). *Journal of Experimental Zoology*, **203**, 295–306.

Packard, D. S. (1980). *American Journal of Anatomy*, **158**, 83–91.

Packard, D. S. & Jacobson, A. G. (1976). *Developmental Biology*, **53**, 36–48.

Páleček, J., Ubbels, G. A. & Rzehak, K. (1978). *Journal of Embryology and Experimental Morphology*, **45**, 203–14.

Palmer, J. F. & Slack, C. (1969). *Nature* (London), **223**, 1286–7.

Palmer, J. F. & Slack, C. (1970). *Journal of Embryology and Experimental Morphology*, **24**, 535–53.

Papaioannou, V. E. (1982). *Journal of Embryology and Experimental Morphology*, **68**, 199–209.

Pasteels, J. (1937*a*). *Archives de Biologie*, **48**, 107–84.

Pasteels, J. (1937*b*). *Archives de Biologie*, **48**, 381–488.

Pasteels, J. (1937*c*). *Archives d'Anatomie Microscopique*, **33**, 279–300.

Pasteels, J. (1938). *Archives de Biologie*, **49**, 629–67.

Pasteels, J. (1939). *Archives de Biologie*, **50**, 291–320.

Pasteels, J. (1940*a*). *Archives de Biologie*, **51**, 103–49.

Pasteels, J. (1940*b*). *Archives de Biologie*, **51**, 335–86.

Pasteels, J. (1941*a*). *Archives de Biologie*, **52**, 321–39.

Pasteels, J. (1941*b*). *Archives de Biologie*, **52**, 341–60.

Pasteels, J. (1942). *Journal of Experimental Zoology*, **89**, 255–81.

Pasteels, J. (1946). *Acta Anatomica*, **2**, 1–16.

Pasteels, J. (1948). *Folia Biotheoretica*, Ser. B, **3**, 83–108.

Pasteels, J. (1953*a*). *Journal of Embryology and Experimental Morphology*, **1**, 5–24.

Pasteels, J. (1953*b*). *Journal of Embryology and Experimental Morphology*, **1**, 125–45.

Pasteels, J. (1954*a*). *Journal of Embryology and Experimental Morphology*, **2**, 122–48.

Pasteels, J. (1954*b*). *Archives d'Anatomie, Strasbourg*, **37**, 125–30.

Pasteels, J. (1957*a*). *Annales de la Société Royale Zoologique de Belgique*, **87**, 217–41.

Pasteels, J. (1957*b*). *Acta Anatomica*, **30**, 601–12.

Pasteels, J. (1964*). *Advances in Morphogenesis*, **3**, 363–88.

Pasternak, L. & McCallion, D. J. (1962). *Canadian Journal of Zoology*, **40**, 585–91.

Patton, G. W. & Villee, C. A. (1968*). *American Journal of Obstetrics and Gynecology*, **101**, 424–37.

Paul, J. (1968*). *Advances in Comparative Physiology and Biochemistry*, **3**, 115–72.

Pautou, M. P. (1977*). In *Vertebrate Limb and Somite Morphogenesis*, eds D. A. Ede, J. R. Hinchliffe & M. Balls, pp. 257–66. Cambridge University Press.

Pautou, M. P. (1978). *Archives de Biologie*, **89**, 27–66.

Pays-de Schutter, A., Kram, R., Hubert, K. E. & Brachet, J. (1975). *Experimental Cell Research*, **96**, 7–14.

Pearson, M. J. & Elsdale, T. (1979). *Journal of Embryology and Experimental Morphology*, **51**, 27–50.

Pedersen, R. A., Spindle, A. I. & Wiley, L. M. (1977). *Nature* (London), **270**, 435–7.

Pehlemann, F. W. (1961). *Verhandlungen der Deutschen Zoologischen Gessellschaft, Zoologischer Anzeiger*, Suppl. **25**, 274–83.

Penners, A. & Schleip, W. (1928 a). *Zeitschrift für Wissenschaftliche Zoologie*, **130**, 305–454.

Penners, A. & Schleip, W. (1928 b). *Zeitschrift für Wissenschaftliche Zoologie*, **131**, 1–156.

Perlmann, P. & de Vincentiis, M. (1961). *Experimental Cell Research*, **23**, 612–16.

Perry, M. M. (1975). *Journal of Embryology and Experimental Morphology*, **33**, 127–46.

Perry, M. M., John, H. A. & Thomas, N. S. T. (1971). *Experimental Cell Research*, **65**, 249–53.

Perry, M. M., Selman, G. G. & Jacob, J. (1976). *Journal of Embryology and Experimental Morphology*, **36**, 209–23.

Perry, M. M. & Waddington, C. H. (1966). *Journal of Embryology and Experimental Morphology*, **15**, 317–30.

Peters, H., Borum, K., Butler, H., Ioannou, J. M., Pedersen, T., Ruby, J. B., Byskov, A. G. S., Anderson, E. & Pool, W. R. (1972*). *Mammalian Oogenesis* **1**, 169 p. New York: MSS Information Cooperation.

Peterson, A. W. (1971). *Canadian Journal of Genetics and Cytology*, **13**, 898–901.

Petriconi, V. (1964). *W. Roux' Archiv für Entwicklungsmechanik der Organismen*, **155**, 358–90.

Pfautsch, M. E. (1960). *Embryologia*, **5**, 139–69.

Phillips, C. R. (1982). *Journal of Experimental Zoology*, **223**, 265–75.

Phillips, H. M. & Davis, G. S. (1978). *The American Zoologist*, **18**, 81–93.

Piatt, J. (1969). *Developmental Biology*, **19**, 608–16.

Picard, J. J. (1975 a). *Journal of Embryology and Experimental Morphology*, **33**, 957–67.

Picard, J. J. (1975 b). *Journal of Embryology and Experimental Morphology*, **33**, 969–78.

Picheral, B. & Charbonneau, M. (1982). *Journal of Ultrastructure Research*, **81**, 306–21.

Pinot, M. (1970 a). *Journal of Embryology and Experimental Morphology*, **23**, 109–51.

Pinot, M. (1970 b*). *Année de Biologie*, **9**, 277–84.

Poelmann, R. E. (1980). *Journal of Embryology and Experimental Morphology*, **55**, 33–51.

Poole, T. J. & Steinberg, M. S. (1981). *Journal of Embryology and Experimental Morphology*, **63**, 1–16.

Poole, T. J. & Steinberg, M. S. (1982). *Developmental Biology*, **92**, 144–58.

Poste, G. & Allison, A. C. (1973*). *Biochimica et Biophysica Acta*, **300**, 421–65.

Powell, J. A. & Segil, N. (1976). *Developmental Biology*, **52**, 128–40.

Pratt, R. M., Larsen, M. A. & Johnston, M. C. (1975). *Developmental Biology*, **44**, 298–305.

Pritchard, D. J., Clayton, R. M. & de Pomerai, D. I. (1978). *Journal of Embryology and Experimental Morphology*, **48**, 1–21.

Propper, A. Y. (1970). *Année de Biologie*, **9**, 267–75.

Propper, A. Y. (1975). In *Extracellular Matrix Influences on Gene Expression*, eds H. C. Slavkin & R. C. Greulich, pp. 541–7. San Francisco: Academic Press.

Propper, A. Y. & Gomot, L. (1973). *Experientia*, **29**, 1543–4.

Pugin, E. (1973). *Année de Biologie*, **12**, 497–511.

Raff, E. C. (1977). *Developmental Biology*, **58**, 56–75.

Raff, E. C., Brothers, A. J. & Raff, R. A. (1976). *Nature* (London), **260**, 615–17.

Ranzi, S. (1962*). In *Semaine d'Etude sur le Problème des Macromolécules d'Intérêt Biologique, Pontificiae Academia Scientifica, Scripta Varia*, **22**, 255–68.

Ranzi, S. (1975). *Revue Suisse de Zoologie*, **82**, 91–100.

Ranzi, S. (1981). In *Results and Problems in Cell Differentiation*, vol. 11. *Differentiation and Neoplasia*, eds R. G. McKinnell, M. A. di Bernardino, M. Blumenfield & R. D. Bergad, pp. 191–5. Berlin: Springer Verlag.

Ranzi, S., Vailati, G. & Vitali, P. (1972). *Rendiconti dell'Istituto Lombardo, Accademia della Scienza e Littera*, B, **106**, 218–20.

Rao, B. R. (1968). *W. Roux' Archiv für Entwicklungsmechanik der Organismen*, **160**, 187–236.

Rao, K. V. (1969). *W. Roux' Archiv für Entwicklungsmechanik der Organismen*, **163**, 161–5.

Rao, K. V. (1973). *Current Science*, **42**, 826–7.

Rappaport, R. (1971*). *International Review of Cytology*, **31**, 169–213.

Rappaport, R. & Rappaport, B. N. (1974). *Journal of Experimental Zoology*, **189**, 189–96.

Rasilo, M. L. & Leikola, A. (1976). *Differentiation*, **5**, 1–7.

Raveh, D., Friedländer, M. & Eyal-Giladi, H. (1971). *W. Roux' Archiv für Entwicklungsmechanik der Organismen*, **166**, 287–99.

Raven, C. P. (1931). *W. Roux' Archiv für Entwicklungsmechanik der Organismen*, **125**, 211–92.

Raven, C. P. (1936). *W. Roux' Archiv für Entwicklungsmechanik der Organismen*, **134**, 122–46.

Raven, C. P. (1937). *Journal of Comparative Neurology*, **67**, 221–40.

Raven, C. P. (1938). *Acta Biotheoretica*, Ser. A, **4**, 51–64.

Raven, C. P. (1961*). *Oogenesis: The Storage of Developmental Information*, 274 p. Oxford: Pergamon Press.

Rawles, M. (1963). *Journal of Embryology and Experimental Morphology*, **11**, 765–89.

Raynaud, A. & Brachet, J. (1979). *Compte Rendu de l'Académie des Sciences*, Paris, **288**, 1675–7.

Reed, P. W. & Lardy, H. A. (1972). *Journal of Biological Chemistry*, **247**, 6970–7.

Reinbold, R. (1968). *Journal of Embryology and Experimental Morphology*, **19**, 43–7.

Reuss, C. & Saunders, J. W. (1965). *The American Zoologist*, **5**, 214, Abstr. no. 94.

Reverberi, G. (1961). *Rendiconti dell'Istituto Scientifica Universita del Camerino*, **2**, 167–209.

Reverberi, G. & Farinella-Ferruzza, N. (1961). *Acta Embryologiae et Morphologiae Experimentalis*, **4**, 139–49.

Reverberi, G. & Ortolani, G. (1962). *Developmental Biology*, **5**, 84–100.

Reverberi, G., Ortolani, G. & Farinella-Ferruzza, N. (1960). *Acta Embryologiae et Morphologiae Experimentalis*, **3**, 296–336.

Reyer, R. W. (1962). *Journal of Experimental Zoology*, **151**, 123–54.

Reyer, R. W. (1966). *Journal of Experimental Zoology*, **162**, 99–132.

Reyer, R. W. (1977*). In *Handbook of Sensory Physiology*, vol. VII, *The Visual System in Vertebrates*, ed. F. Crescitelli, pp. 309–90. Berlin: Springer Verlag.

Reynhout, J. K. & Smith, L. D. (1974). *Developmental Biology*, **38**, 394–400.

Reyss-Brion, M. (1963). *Journal of Embryology and Experimental Morphology*, **11**, 649–57.

Reyss-Brion, M. (1964). *Archives d'Anatomie Microscopique et de Morphologie Expérimentale*, **53**, 397–465.

Reyss-Brion, M. (1973). *Année de Biologie*, **12**, 467–80.

Reyss-Brion, M. (1974). *Année de Biologie*, **13**, 51–6.

Reyss-Brion, M. (1976). *Archives de Biologie*, **87**, 69–77.

Rice, R. W. & Moran, D. J. (1977). *Journal of Experimental Zoology*, **201**, 471–8.

Richter, J. D., Wasserman, W. J. & Smith, L. D. (1982). *Developmental Biology*, **89**, 159–67.

Ridgway, E. B., Gilkey, J. C. & Jaffe, L. F. (1977). *Proceedings of the National Academy of Sciences of the United States of America*, **74**, 623–7.

Roach, F. C. (1945). *Journal of Experimental Zoology*, **99**, 53–77.

Roberson, M., Armstrong, J. & Armstrong, P. (1980). *Journal of Cell Science*, **44**, 19–31.

Roberts, H. S. (1961*). *Quarterly Review of Biology*, **36**, 155–77.

Rollhäuser-ter Horst, J. (1977*a*). *Anatomy and Embryology*, **151**, 309–16.

Rollhäuser-ter Horst, J. (1977*b*). *Anatomy and Embryology*, **151**, 317–24.

Rollhäuser-ter Horst, J. (1979). *Anatomy and Embryology*, **157**, 113–20.
Rollhäuser-ter Horst, J. (1981). *Anatomy and Embryology*, **162**, 69–80.
Roosen-Runge, E. C. (1977*). *The Process of Spermatogenesis in Animals*, 214 p. Cambridge University Press.
Rosenquist, G. C. (1966). *Contributions to Embryology*, **38**, 71–110, Carnegie Institute of Washington Publication no. 625.
Rosenquist, G. C. (1971). *Developmental Biology*, **26**, 323–35.
Rosenquist, G. C. (1972). *Journal of Experimental Zoology*, **180**, 95–104.
Rostedt, I. (1968). *Scandinavian Journal of Clinical and Laboratory Investigation*, **21**, suppl., 101, 49 (abstract).
Rostedt, I. (1971). *Annales Medicinae Experimentalis et Biologiae Fenniae*, **49**, 186–203.
Rounds, D. E. & Flickinger, R. A. (1958). *Journal of Experimental Zoology*, **137**, 479–500.
Rubin, L. & Saunders, J. W. (1972). *Developmental Biology*, **28**, 94–112.
Rudnick, D. (1945). *Transactions of the Connecticut Academy of Arts and Sciences*, **36**, 353–77.
Rudnick, D. (1948). *Annals of the New York Academy of Sciences*, **49**, 761–73.
Runnström, J. (1967). *Experimental Biology and Medicine*, **1**, 52–62.
Rutter, W. J., Ingles, C. J., Weaver, R. F., Blatti, S. P. & Morris, P. W. (1972*). In *Molecular Genetics and Developmental Biology*, ed. M. Sussman, pp. 143–62. New Jersey: Prentice Hall.
Rutter, W. J., Pictet, R. L. & Morris, P. W. (1973*). *Annual Review of Biochemistry*, **42**, 601–46.
Ryabova, L. V. (1983). *Cell Differentiation*, **13**, 171–5.
Rzehak, K. (1972). *Folio Biologica*, **20**, 409–16.
Sabbadin, A. & Zaniolo, G. (1979). *Journal of Experimental Zoology*, **207**, 289–304.
Sakai, M. & Kubota, H. Y. (1981). *Development, Growth and Differentiation*, **23**, 41–9.
Sakakura, T., Sakagami, Y. & Nishizuka, Y. (1979). *Developmental Biology*, **72**, 201–10.
Sala, M. (1955). *Proceedings of the Royal Academy of Sciences*, Amsterdam, Ser. C, **58**, 635–47.
Sala, M. (1956). *Proceedings of the Royal Academy of Sciences*, Amsterdam, Ser. C, **59**, 661–7.
Sanchez, S. S. & Barbieri, F. D. (1983). *W. Roux's Archives of Developmental Biology*, **192**, 37–41.
Sanders, E. J. (1980). *Journal of Cell Science*, **44**, 225–42.
Sanders, E. J. & Dicaprio, R. A. (1976*a*). *Differentiation*, **7**, 13–21.
Sanders, E. J. & Dicaprio, R. A. (1976*b*). *Journal of Experimental Zoology*, **197**, 415–21.
Sanders, E. J. & Singal, P. K. (1973). *Micron*, **4**, 156–62.
Sanders, E. J. & Singal, P. K. (1975). *Experimental Cell Research*, **93**, 219–24.
Sanders, E. J. & Zalik, S. E. (1972). *W. Roux' Archiv für Entwicklungsmechanik der Organismen*, **171**, 181–94.
Sandor, S. (1972). *Revue Roumaine d'Embryologie et de Cytologie, Sér. Embryologie*, **9**, 113–21.
Sandor, S. & Amels, D. (1970). *Revue Roumaine d'Embryologie et de Cytologie, Sér. Embryologie*, **7**, 49–57.
Sandor, S. & Amels, D. (1971). *Revue Roumaine d'Embryologie et de Cytologie, Sér. Embryologie*, **8**, 37–42.
Sanger, J. W. & Sanger, J. M. (1979*). *Methods and Achievements in Experimental Pathology*, **8**, 110–42.
Sanyal, S. & Niu, M. C. (1966). *Proceedings of the National Academy of Sciences of the United States of America*, **55**, 743–50.
Sariola, H., Timpl, R., von der Mark, K., Mayne, R., Fitch, J. M., Linsenmayer, T. F. & Ekblom, P. (1983). *Developmental Biology*, **101**, 86–96.

Sasaki, N. (1961). *Kumamoto Journal of Sciences*, Ser. B 2, *Biology*, **5**, 173–84.

Sasaki, N. & Iyeiri, S. (1972*a*). *The Zoological Magazine*, **81**, 63–6.

Sasaki, N. & Iyeiri, S. (1972*b*). *The Zoological Magazine*, **81**, 150–3.

Sasaki, N., Iyeiri, S. & Kurihara, K. (1976*a*). *W. Roux's Archives of Developmental Biology*, **179**, 237–41.

Sasaki, N., Iyeiri, S. & Tadokoro, T. (1975*a*). *The Zoological Magazine*, **84**, 64–6.

Sasaki, N., Iyeiri, S. & Tadokoro, T. (1975*b*). *The Zoological Magazine*, **84**, 148–50.

Sasaki, N., Iwamoto, K., Noda, S. & Kawakami, I. (1976*b*). *Development, Growth and Differentiation*, **18**, 457–65.

Sato, T. (1953). *W. Roux' Archiv für Entwicklungsmechanik der Organismen*, **146**, 487–514.

Satoh, N. (1977). *Development, Growth and Differentiation*, **19**, 111–17.

Satoh, N. (1979). *Journal of Embryology and Experimental Morphology*, **54**, 131–9.

Satoh, N., Kageyama, T. & Sirakami, K. I. (1976). *Development, Growth and Differentiation*, **18**, 55–67.

Saunders, J. W. (1972*). *Annals of the New York Academy of Sciences*, **193**, 29–42.

Saunders, J. W. & Gasseling, M. T. (1968*). In *Epithelial–Mesenchymal Interactions*, eds R. Fleischmajer & R. E. Billingham, pp. 78–97. Baltimore: Williams & Wilkins.

Saunders, J. W., Gasseling, M. T. & Errick, J. E. (1976). *Developmental Biology*, **50**, 16–25.

Saunders, J. W. & Reuss, C. (1974). *Developmental Biology*, **38**, 41–50.

Sawai, T. (1972). *Journal of Cell Science*, **11**, 543–56.

Sawai, T. (1974). *Journal of Cell Science*, **15**, 259–67.

Sawai, T. (1976*a*). *Development, Growth and Differentiation*, **18**. 357–61.

Sawai, T. (1976*b*). *Journal of Cell Science*, **21**, 537–51.

Sawai, T. (1979). *Journal of Embryology and Experimental Morphology*, **51**, 183–93.

Sawai, T. (1983). *Journal of Embryology and Experimental Morphology*, **77**, 243–54.

Sawai, T., Kubota, T. & Kojima, M. K. (1969). *Development, Growth and Differentiation*, **11**, 246–54.

Sawai, T. & Yoneda, M. (1974). *The Journal of Cell Biology*, **60**, 1–7.

Sawyer, R. H. (1975). *Journal of Experimental Zoology*, **191**, 133–9.

Saxén, L. (1961). *Developmental Biology*, **3**, 140–52.

Saxén, L. (1963*). In *Biological Organization*, Symposium Varenna, September 1962, ed. R. J. C. Harris, pp. 211–27. London: Academic Press.

Saxén, L. (1970). *Developmental Biology*, **23**, 511–23.

Saxén, L. (1975). *Clinical Obstetrics and Gynecology*, **18**, 149–75.

Saxén, L. (1977*). In *Cell and Tissue Interaction*, eds J. W. Lash & M. M. Burger, pp. 1–9. New York: Raven Press.

Saxén, L. (1978). *Medical Biology*, **56**, 293–8.

Saxén, L. (1980*a*). *Current Topics in Developmental Biology*, **15**, 409–18.

Saxén, L. (1980*b*). *In Differentiation and Neoplasia*, eds R. G. McKinnell, M. A. Diberardino, M. Blumenfeld & R. D. Bergad, pp. 147–54. Berlin: Springer Verlag.

Saxén, L., Karkinen-Jääskeläinen, M., Lehtonen, E., Nordling, S. & Wartiovaara, J. (1976*). In *The Cell Surface in Animal Embryogenesis and Development*, eds G. Poste & G. L. Nicolson, pp. 331–407. Amsterdam: Elsevier/North Holland Biomedical Press.

Saxén, L. & Kohonen, J. (1968*). *International Reviews in Experimental Pathology*, **8**, 57–128.

Saxén, L. & Lehtonen, E. (1977*). In *Third Department of Pathology, University of Helsinki*, 1967–1977. Eds S. Stenman, L. Saxén & E. Saxén, pp. 79–82.

Saxén, L. & Saksela, E. (1971). *Experimental Cell Research*, **66**, 369–77.

Saxén, L. & Toivonen, S. (1957). *Embryologia*, **3**, 353–60.

Saxén, L. & Toivonen, S. (1961). *Journal of Embryology and Experimental Morphology*, **9**, 514–33.

Saxén, L. & Toivonen, S. (1962*). *Primary Embryonic Induction*, 271 p. London: Logos Press.

Saxén, L., Toivonen, S. & Nakamura, O. (1978*). In *Organiser – A Milestone of a Half-Century from Spemann*, eds O. Nakamura & S. Toivonen, pp. 315–20. Amsterdam: Elsevier/North Holland Biomedical Press.

Saxén, L., Toivonen, S. & Vainio, T. (1961). *Acta Pathologica et Microbiologica Scandinavica*, **51**, suppl. 144, 163–4.

Saxén, L. & Wartiovaara, J. (1976*). In *Developmental Biology of Plants and Animals*, eds C. F. Graham & P. F. Wareing, pp. 127–40. Oxford: Blackwell.

Saxén, L., Wartiovaara, J., Häyry, P. & Vainio, T. (1965*). *Fourth Scandinavian Congress of Cell Research*, 21–36.

Schaeffer, B. E., Schaeffer, H. E. & Brick, I. (1973*a*). *Developmental Biology*, **34**, 66–76.

Schaeffer, H. E., Schaeffer, B. E. & Brick, I. (1973*b*). *Developmental Biology*, **34**, 163–8.

Schaeffer, H. E., Schaeffer, B. E. & Brick, I. (1973*c*). *Developmental Biology*, **35**, 376–81.

Scharf, S. R. & Gerhart, J. C. (1980). *Developmental Biology*, **79**, 181–98.

Schatten, G. & Mazia, D. (1977*). In *Cell Shape and Surface Architecture*, eds J. P. Revel, U. Henning & C. F. Fox. New York: A. R. Liss: and *Journal of Supramolecular Structure*, **5**, 343–69; and *Progress in Clinical and Biological Research*, **17**, 295–321.

Schechtman, A. M. (1934). *University of California Publications in Zoology*, **39**, 303–10.

Schlichter, L. C. & Elinson, R. P. (1981). *Developmental Biology*, **83**, 33–41.

Schmidt, B. (1979). PhD thesis, University of Cologne, 113 p.

Schoenwolf, G. C. (1977). *Journal of Experimental Zoology*, **201**, 227–46.

Schoenwolf, G. C. & Delongo, J. (1980). *American Journal of Anatomy*, **158**, 43–63.

Schoenwolf, G. C. & Kelley, R. O. (1980). *American Journal of Anatomy*, **158**, 29–41.

Schoots, A. F. M., Stikkelbroeck, J. J. M., Bekhuis, J. F. & Denucé, J. M. (1982). *Journal of Ultrastructure Research*, **80**, 185–96.

Schowing, J. (1968*a*). *Journal of Embryology and Experimental Morphology*, **19**, 9–22.

Schowing, J. (1968*b*). *Journal of Embryology and Experimental Morphology*, **19**, 23–32.

Schowing, J. (1968*c*). *Journal of Embryology and Experimental Morphology*, **19**, 88–93.

Schowing, J. (1974). *Année de Biologie*, **13**, 69–76.

Schroeder, T. E. (1970). *Journal of Embryology and Experimental Morphology*, **23**, 427–62.

Schroeder, T. E. (1971). *International Journal of Neuroscience*, **2**, 183–98.

Schroeder, T. E. (1973). *The American Zoologist*, **13**, 949–60.

Schroeder, T. E. & Strickland, D. L. (1974). *Experimental Cell Research*, **83**, 139–42.

Schubert, L. & Lohmann, K. (1982). *Development, Growth and Differentiation*, **24**, 25–38.

Schuetz, A. W. (1967). *Journal of Experimental Zoology*, **166**, 347–54.

Schuetz, A. W. (1974). *Biology and Reproduction*, **10**, 150–78.

Schuetz, A. W. (1978). *Biologie Cellulaire*, **32**, 89–95. Symposium *Membrane Cell Regulation*, Roscoff, 1977, eds P. Querrier & M. Moreau. Paris: Société Française de Microscopie Electronique.

Schultze, O. (1894). *W. Roux' Archiv für Entwicklungsmechanik der Organismen*, **1**, 269–305.

Searle, R. F. & Jenkinson, E. J. (1978). *Journal of Embryology and Experimental Morphology*, **43**, 147–56.

Seidel, F. (1969*). *Entwicklungspotenzen des frühen Säugetierkeimes*, 91 p. Köln: Westdeutscher Verlag.

Selman, G. G. (1982). *Development, Growth and Differentiation*, **24**, 1–6.

Selman, G. G., Jacob, J. & Perry, M. M. (1976). *Journal of Embryology and Experimental Morphology*, **36**, 321–41.

Selman, G. G. & Perry, M. M. (1970). *Journal of Cell Science*, **6**, 207–27.

Sengel, P. (1958*). *Année de Biologie*, **34**, 29–52.

Sengel, P. (1964*). In *The Epidermis*, eds W. Montagna & W. C. Lobitz, pp. 15–34. New York: Academic Press.

Sengel, P. (1972). *Bulletin de la Société Zoologique de France*, **97**, 485–95.

Sengel, P. (1975*). In *Cell Patterning*, eds R. Porter & J. Rivers, pp. 51–70. Amsterdam: Elsevier/Excerpta Medica/North Holland Publishing Company.

Sengel, P., Dhouailly, D. & Kieny, M. (1969). *Developmental Biology*, **19**, 436–46.

Sengel, P. & Pautou, M. P. (1969). *Nature* (London), **222**, 673–94.

Sengel, P. & Rusaouën, M. (1969). *Archives d'Anatomie Microscopique et de Morphologie Expérimentale*, **58**, 77–96.

Sentein, P. (1961). *Bulletin de l'Association des Anatomistes*, 47ième Réunion, 737–46.

Sentein, P. (1967). *Chromosoma*, **21**, 51–71.

Sentein, P. (1968). *Archives d'Anatomie, d'Histologie et d'Embryologie*, **51**, 641–52.

Setchell, B. P. (1982*). In *Reproduction in Mammals*, vol. 1, *Germ Cells and Fertilization*, eds R. Austin & R. V. Short, pp. 63–101. Cambridge University Press.

Shapiro, B. M. (1981*). In *Fertilization and Embryonic Development in Vitro*, eds L. Mastroianni & J. D. Biggers, pp. 233–55. New York: Plenum Press.

Shapiro, B. M. & Eddy, E. M. (1980*). *International Review of Cytology*, **66**, 257–302.

Shapiro, B. M., Schackmann, R. W. & Gabel, C. A. (1981*). *Annual Review of Biochemistry*, **50**, 815–44.

Sherbet, G. V. (1962). *Naturwissenschaften*, **20**, 471–2.

Sherbet, G. V. (1963). *Journal of Embryology and Experimental Morphology*, **11**, 227–37.

Sherbet, G. V. (1966a). *Progress in Biophysics and Molecular Biology*, **16**, 91–106.

Sherbet, G. V. (1966b). *Journal of Embryology and Experimental Morphology*, **16**, 159–70.

Sherbet, G. V. & Lakshmi, M. S. (1967). *Nature* (London), **215**, 1089–90.

Sherbet, G. V. & Lakshmi, M. S. (1969a). *Experientia*, **25**, 481–2.

Sherbet, G. V. & Lakshmi, M. S. (1969b). *Experientia*, **25**, 1130–1.

Sherbet, G. V. & Lakshmi, M. S. (1974*). *Differentiation*, **2**, 51–63.

Sherbet, G. V. & Mulherkar, L. (1963). *W. Roux' Archiv für Entwicklungsmechanik der Organismen*, **154**, 506–12.

Sherbet, G. V. & Mulherkar, L. (1965). *W. Roux' Archiv für Entwicklungsmechanik der Organismen*, **155**, 701–8.

Sheridan, J. D. (1976*). In *The Cell Surface in Animal Embryogenesis and Development*, **1**., eds G. Poste & G. L. Nicolson, pp. 409–47. Amsterdam: Elsevier/North Holland Biomedical Press.

Sherman, J. E. (1960). *Wisconsin Academy of Sciences, Arts and Letters*, **49**, 171–89.

Sherman, M. I. (1975*). In *The Early Development of Mammals*, eds M. Balls & A. E. Wild, pp. 145–65. Cambridge University Press.

Sherman, M. I., McLaren, A. & Walker, P. M. B. (1972). *Nature New Biology*, **238**, 175–6.

Shieh, S.-P., Ning, I.-L. & Tsung, S.-D. (1963). *Acta Biologiae Experimentalis Sinica*, **8**, 1–9.

Shieh, S.-P., Ning, I.-L. & Tsung, S.-D. (1965). *Acta Anatomica Sinica*, **8**, 1–12.

Shimada, Y. (1965). *Okajimas Folio Anatomica Japonica*, **41**, 179–97.

Shinagawa, A. (1983). *Journal of Cell Science*, **64**, 147–62.

Shinohara, T. & Piatigorsky, J. (1976). *Proceedings of the National Academy of Sciences of the United States of America*, **73**, 2808–12.

Shiokawa, K. Tashiro, K., Oka, T. & Yamana, K. (1982). *Development, Growth and Differentiation*, **24**, 402 (abstract no. B61).

Shiokawa, K. & Yamana, K. (1979). *Development, Growth and Differentiation*, **21**, 501–7.

Shoger, R. L. (1960). *Journal of Experimental Zoology*, **143**, 221–38.

Shumway, W. (1940). *The Anatomical Record*, **78**, 139–47.

Shumway, W. (1942). *The Anatomical Record*, **83**, 309–15.

Sieber-Blum, M. & Cohen, A. M. (1980). *Developmental Biology*, **80**, 96–106.

Signoret, J. (1977*a*). *Mémoires de la Société Zoologique de France*, **41**, 123–32 (Symposium L. Gallien).

Signoret, J. (1977*b*). *Journal of Embryology and Experimental Morphology*, **42**, 5–14.

Signoret, J. & Lefresne, J. (1971). *Annales d'Embryologie et de Morphogenèse*, **4**, 113–23.

Signoret, J. & Lefresne, J. (1973). *Annales d'Embryologie et de Morphogenèse*, **6**, 299–307.

Sigot, M. (1971). *Archives d'Anatomie microscopique et de Morphologie Expérimentale*, **60**, 169–204.

Sigot, M. & Marin, L. (1970). *Journal of Embryology and Experimental Morphology*, **24**, 43–63.

Singal, P. K. (1975*a*). *Cell and Tissue Research*, **163**, 215–21.

Singal, P. K. (1975*b*). *Cytobios*, **14**, 29–37.

Singal, P. K. & Sanders, E. J. (1974*a*). *Journal of Ultrastructure Research*, **47**, 433–51.

Singal, P. K. & Sanders, E. J. (1974*b*). *Cell and Tissue Research*, **154**, 189–209.

Sirakami, K. I. (1963). *Memoirs of the Faculty of Liberal Arts and Education, Yamanashi University*, **14**, 132–40.

Sirakami, K. I., Gejo, M. & Hirose, N. (1962). *Memoirs of the Faculty of Liberal Arts and Education, Yamanashi University*, **13**, 151–7.

Slack, C. & Palmer, J. F. (1969). *Experimental Cell Research*, **55**, 416–19.

Slack, C. & Warner, A. E. (1973). *Journal of Physiology*, **232**, 313–30.

Slack, C., Warner, A. E. & Warren, R. L. (1973). *Journal of Physiology*, **232**, 297–312.

Slack, J. M. W. (1977*a*). *Journal of Embryology and Experimental Morphology*, **39**, 151–68.

Slack, J. M. W. (1977*b*). *Journal of Embryology and Experimental Morphology*, **39**, 169–82.

Slack, J. M. W. (1979). *Nature* (London), **279**, 583–4.

Slack, J. M. W. (1984). *Nature*, **311**, 107–81.

Slack, J. M. W. & Forman, D. (1980). *Journal of Embryology and Experimental Morphology*, **56**, 283–99.

Slàdeček, F. (1952). *Věstník Československé Zoologické Společnosti*, **16**, 322–33.

Slavkin, H. C. (1982). *Journal of Craniofacial Genetics and Developmental Biology*, **2**, 179–89.

Slavkin, H. C., Beierle, J. & Bavetta, L. A. (1968). *Nature* (London), **217**, 269–70.

Smith, J. C. & Malacinski, G. M. (1983). *Developmental Biology*, **98**, 250–4.

Smith, J. C. & Slack, J. M. W. (1983). *Journal of Embryology and Experimental Morphology*, **78**, 299–317.

Smith, J. L., Osborn, J. C. & Stanisstreet, M. (1976). *Journal of Embryology and Experimental Morphology*, **36**, 513–22.

Smith, L. & Thorogood, P. (1983). *Journal of Embryology and Experimental Morphology*, **75**, 165–88.

Smith, L. D. (1975*). In *The Biochemistry of Animal Development*, vol. 3, *Molecular Aspects of Animal Development*, ed. R. Weber, pp. 1–46. New York: Academic Press.

Smith, L. D. & Ecker, R. E. (1968*). *Eighth Canadian Cancer Conference*, pp. 103–29. Pergamon of Canada Ltd.

Smith, L. D. & Ecker, R. E. (1969). *Developmental Biology*, **19**, 281–309.

Smith, L. D. & Ecker, R. E. (1970*a**). In *Problems in Biology: RNA in Development*, 1969, ed. E. W. Hanly, pp. 355–79. Salt Lake City: University of Utah Press.

Smith, L. D. & Ecker, R. E. (1970*b**). *Current Topics in Developmental Biology*, **5**, 1–38.

Smith, L. D. & Ecker, R. E. (1970*c*). *Developmental Biology*, **22**, 622–37.

Smith, L. D. & Ecker, R. E. (1971). *Developmental Biology*, **25**, 232–47.

Smith, L. D., Ecker, R. E. & Subtelny, S. (1968). *Developmental Biology*, **17**, 627–43.

Smith, L. D. & Williams, M. A. (1975*). In *The Developmental Biology of Reproduction*, ed. C. L. Markert, pp. 3–24. New York: Academic Press.

Smith, L. J. (1980). *Journal of Embryology and Experimental Morphology*, **55**, 257–77.

Smith, S. D. (1965). *Journal of Experimental Zoology*, **159**, 149–66.
Solursh, M. & Revel, J. P. (1978). *Differentiation*, **11**, 185–90.
Sonneborn, T. M. (1970). *Proceedings of the Royal Society, London*, **176**, 347–66.
Spelsberg, T. C. (1974*). In *Acid Proteins of the Nucleus*, eds I. L. Cameron & J. Jeter, pp. 248–90. New York: Academic Press.
Spemann, H. (1931). *W. Roux' Archiv für Entwicklungsmechanik der Organismen*, **123**, 389–517.
Spemann, H. (1938*, reprinted in 1962*). *Embryonic Development and Induction*, 401 p. New York: Hafner Publishing Company.
Spemann, H. (1936*, reprinted in 1968*). *Experimentelle Beiträge zu einer Theorie der Entwicklung*, 206 p. Berlin: Springer Verlag.
Spemann, H. & Falkenberg, H. (1919). *Archiv für Entwicklungsmechanik der Organismen*, **45**, 371–422.
Spemann, H. & Mangold, H. (1924). *Archiv für mikroskopische Anatomie und Entwicklungsmechanik der Organismen*, **100**, 599–638.
Spieth, J. & Keller, R. E. (1984). *Journal of Experimental Zoology*, **229**, 91–107.
Spofford, W. R. (1945). *Journal of Experimental Zoology*, **99**, 35–52.
Spofford, W. R. (1948). *Journal of Experimental Zoology*, **107**, 123–64.
Spofford, W. R. (1953). *Archives de Biologie*, **64**, 439–93.
Spratt, N. T. (1952). *Journal of Experimental Zoology*, **120**, 109–30.
Spratt, N. T. (1955). *Journal of Experimental Zoology*, **128**, 121–63.
Spratt, N. T. (1957*a*). *Journal of Experimental Zoology*, **134**, 577–612.
Spratt, N. T. (1957*b*). *Journal of Experimental Zoology*, **135**, 319–53.
Spratt, N. T. & Haas, H. (1960*a*). *Journal of Experimental Zoology*, **144**, 139–57.
Spratt, N. T. & Haas, H. (1960*b*). *Journal of Experimental Zoology*, **144**, 257–75.
Spratt, N. T. & Haas, H. (1960*c*). *Journal of Experimental Zoology*, **145**, 97–137.
Spratt, N. T. & Haas, H. (1961*a*). *Journal of Experimental Zoology*, **147**, 57–93.
Spratt, N. T. & Haas, H. (1961*b*). *Journal of Experimental Zoology*, **147**, 271–93.
Spratt, N. T. & Haas, H. (1962). *Journal of Experimental Zoology*, **149**, 75–102.
Spray, D. C., Harris, A. L. & Bennett, M. V. L. (1979). *Science*, **204**, 432–4.
Stanisstreet, M. (1975). *Acta Embryologiae et Morphologiae Experimentalis*, 33–7.
Stanisstreet, M. (1982). *Journal of Embryology and Experimental Morphology*, **67**, 195–205.
Stanisstreet, M. & Deuchar, E. M. (1972). *Cell Differentiation*, **1**, 15–18.
Stanisstreet, M. & Jumah, H. (1982). *Acta Embryologiae et Morphologiae Experimentalis*, 3–13.
Stanisstreet, M. & Panayi, M. (1980). *Experientia*, **36**, 1110–12.
Stanisstreet, M. & Smith, J. L. (1978). *Acta Embryologiae et Morphologiae Experimentalis*, 3–12.
Stanisstreet, M., Wakely, J. & England, M. A. (1980). *Journal of Embryology and Experimental Morphology*, **59**, 341–53.
Starck, D. (1982*). *Vergleichende Anatomie der Wirbeltiere*, Bd. 3, *Zentralnervensystem*, pp. 281–7. Berlin: Springer Verlag.
Stark, R. J. & Searls, R. L. (1974). *Developmental Biology*, **38**, 51–63.
Steding, G. (1967). Dr. Sc. thesis, University of Göttingen, 108 p.
Steinberg, M. S. (1964*). In *Cell Membranes in Development*, 22nd Growth Symposium, ed. M. Locke, pp. 321–66. New York: Academic Press.
Steinberg, M. S. (1970). *Journal of Experimental Zoology*, **173**, 395–433.
Steinberg, M. S. & Poole, T. J. (1981). *Philosophical Transactions of the Royal Society, London, B*, **295**, 451–60.
Steinberg, M. S. & Poole, T. J. (1982*). In *Developmental Order: its Origin and Regulation*, eds S. Subtelny & P. B. Green, pp. 351–78. New York: Alan R. Liss Inc.
Steinhardt, R. A. & Mazia, D. (1973). *Nature* (London), **241**, 400–1.

Stern, C. D. & Goodwin, B. C. (1977). *Journal of Embryology and Experimental Morphology*, **41**, 15–22.

Stewart, P. A. & McCallion, D. J. (1975). *Developmental Biology*, **46**, 383–9.

Stinnakre, M. G., Evans, M. J., Willison, K. R. & Stern, P. L. (1981). *Journal of Embryology and Experimental Morphology*, **61**, 117–31.

Stone, L. S. (1950). *Journal of Experimental Zoology*, **113**, 9–26.

Stroeva, O. G. (1963). *Doklady Akademii Nauk SSSR*, **151**, 464–7.

Strudel, G. (1953). *Annales des Sciences Naturelles, Zoologie*, 11ième Sér., **15**, 253–329.

Strudel, G. (1963). *Journal of Embryology and Experimental Morphology*, **11**, 399–412.

Strudel, G. (1970). *Compte Rendu de l'Académie des Sciences*, Paris, **270**, 128–30.

Strudel, G. (1971). *Compte Rendu de l'Académie des Sciences*, Paris, **272**, 473–6.

Subtelny, S. & Wessels, N. K. (1980*). *The Cell Surface: Mediator of Developmental Processes*, 374 p. New York: Academic Press.

Sudarwati, S. & Nieuwkoop, P. D. (1971). *W. Roux' Archiv für Entwicklungsmechanik der Organismen*, **166**, 189–204.

Sumiya, M. (1976a). *W. Roux's Archives of Developmental Biology*, **179**, 1–17.

Sumiya, M. (1976b). *Journal of the Faculty of Science of the University of Tokyo*, Ser IV, **13**, 363–81.

Sumiya, M. & Mizuno, T. (1976). *Proceedings of Japanese Academy*, **52**, 587–90.

Summerbell, D. (1979). *Journal of Embryology and Experimental Morphology*, **50**, 217–33.

Summerbell, D. & Honig, L. S. (1982). *The American Zoologist*, **22**, 105–16.

Sunkara, P. S., Wright, D. A. & Rao, P. N. (1979). *Proceedings of the National Academy of Sciences of the United States of America*, **76**, 2799–802.

Sutasurya, L. A. & Nieuwkoop, P. D. (1974). *W. Roux' Archiv für Entwicklungsmechanik der Organismen*, **175**, 199–220.

Suzuki, A. (1966). *Kumamoto Journal of Science*, B2, **8**, 13–19.

Suzuki, A. (1968a). *Kumamoto Journal of Science*, B2, **9**, 1–8.

Suzuki, A. (1968b). *Kumamoto Journal of Science*, B2, **9**, 9–16.

Suzuki, A. (1968c). *Kumamoto Journal of Science*, B2, **9**, 17–22.

Suzuki, A., Hakatake, H. & Hidaka, T. (1984). *Differentiation*, **28**, 73–7.

Suzuki, A. & Ikeda, K. (1979). *Development, Growth and Differentiation*, **21**, 175–88.

Suzuki, A. & Kawakami, I. (1963a). *The Zoological Magazine*, **72**, 110–14.

Suzuki, A. & Kawakami, I. (1963b). *Embryologia*, **8**, 75–8.

Suzuki, A. & Kuwabara, K. (1974). *Development, Growth and Differentiation*, **16**, 29–40.

Suzuki, A., Kuwabara, K. & Kuwabara, Y. (1975). *Development, Growth and Differentiation*, **17**, 343–52.

Suzuki, A., Kuwabara, Y. & Kuwana, T. (1976). *Development, Growth and Differentiation*, **18**, 447–55.

Suzuki, A., Mifune, Y. & Kanéda, T. (1984). *Development, Growth and Differentiation*, **26**, 81–94.

Suzuki, A. & Miki, K. (1983). *Development, Growth and Differentiation*, **25**, 289–97.

Sze, L. C. (1953). *Physiological Zoology*, **26**, 212–23.

Tahara, Y. (1959). *Japanese Journal of Experimental Morphology*, **13**, 49–60.

Tahara, Y. (1975). *Memoirs of the Osaka Kyoiku University*, **23**, 33–53.

Takahashi, M. & Ito, S. (1968). *The Zoological Magazine*, **77**, 307–16.

Takahashi, Y. M. & Sugiyama, M. (1973). *Development, Growth and Differentiation*, **15**, 261–7.

Takata, C. (1960a). *Embryologia*, **5**, 38–70.

Takata, C. (1960b). *Embryologia*, **5**, 194–205.

Takata, C. & Yamada, T. (1960). *Embryologia*, **5**, 8–20.

Takata, K. & Hama, T. (1978*). In *Organiser – A Milestone of a Half-Century from Spemann*, eds O. Nakamura & S. Toivonen, pp. 267–82. Amsterdam: Elsevier/North Holland Biomedical Press.

Takata, K., Yamamoto, K. Y. & Ozawa, R. (1981). *W. Roux's Archives of Developmental Biology*, **190**, 92–6.

Takata, K., Yamamoto, K. Y. & Takahashi, N. (1983). *Development, Growth and Differentiation*, **25**, 418 (abstract).

Takaya, H. (1956*a*). *Proceedings of the Japanese Academy*, **32**, 282–6.

Takaya, H. (1956*b*). *Proceedings of the Japanese Academy*, **32**, 287–92.

Takaya, H. (1961). *Embryologia*, **6**, 123–34.

Takaya, H. (1973). *Development, Growth and Differentiation*, **15**, 39–45.

Takaya, H. (1977). *Differentiation*, **7**, 187–92.

Takaya, H. (1978*). In *Organiser – A Milestone of a Half-Century from Spemann*, eds O. Nakamura & S. Toivonen, pp. 49–70. Amsterdam: Elsevier/North Holland Biomedical Press.

Takaya, H. & Kayahara, T. (1978). *Development, Growth and Differentiation*, **20**, 143–50.

Takeuchi, S. (1963). *Embryologia*, **8**, 21–44.

Takeuchi, Y. K. & Takeuchi, I. K. (1980). *Development, Growth and Differentiation*, **22**, 627–37.

Takor Takor, T. & Pearse, A. G. E. (1975). *Journal of Embryology and Experimental Morphology*, **34**, 311–25.

Tanaka, T., Noda, S., Kawakami, I. & Sato, A. (1976). *Development, Growth and Differentiation*, **18**, 267–72.

Tarin, D. (1971*a*). *Journal of Embryology and Experimental Morphology*, **26**, 543–70.

Tarin, D. (1971*b*). *Journal of Anatomy*, **109**, 535–47.

Tarin, D. (1972). *Journal of Anatomy*, **111**, 1–28.

Tarin, D. (1973). *Differentiation*, **1**, 109–26.

Tarin, D. (1978). *Medical Biology*, **56**, 380–5.

Tarkowsky, A. K. & Wróblewska, J. (1967). *Journal of Embryology and Experimental Morphology*, **18**, 155–80.

Taylor, A. C. & Kollros, J. J. (1946). *The Anatomical Record*, **94**, 7–24.

Teillet, M. A. (1971). *Bulletin de l'Association des anatomists*, 56ième Congrès, Nantes, 734–43.

Teillet, M. A., Cochard, P. & le Douarin, N. (1978). *Zoon, a Journal of Zoology*, **6**, 115–22.

Teillet, M. A. & le Douarin, N. (1974). *Archives d'Anatomie Microscopique et de Morphologie Expérimentale*, **63**, 51–62.

Teillet, M. A. & le Douarin, N. (1983). *Developmental Biology*, **98**, 192–211.

Tencer, R. (1978*a*). *Experimental Cell Research*, **116**, 253–60.

Tencer, R. (1978*b*). *Biologie Cellulaire*, **32**, 109–14.

Thesleff, I., Lehtonen, E., Wartiovaara, J. & Saxén, L. (1977). *Developmental Biology*, **58**, 197–203.

Thiébaud, C. H. (1983). *Developmental Biology*, **98**, 245–9.

Thiery, J. P., Duband, J. L., Rutishauser, U. & Edelman, G. M. (1982). *Proceedings of the National Academy of Sciences of the United States of America*, **79**, 6737–41.

Thoman, M. & Gerhart, J. C. (1979). *Developmental Biology*, **68**, 191–202.

Thomas, N. & Deuchar, E. M. (1971). *Acta Embryologiae et Morphologiae Experimentalis*, 195–200.

Thomas, W. A., Edelman, B. A., Lobel, S. M., Breitbart, A. S. & Steinberg, M. S. (1981). *Journal of Supramolecular Structure and Cellular Biochemistry*, **16**, 15–27 and *Cellular Recognition*, 77–89.

Thomson, I., Wilkinson, C. E., Jackson, J. F., de Pomerai, D. I., Clayton, R. M., Truman, D. E. E. & Williamson, R. (1978). *Developmental Biology*, **65**, 372–82.

Tiedemann, H. (1966*a**). In *Morphological and Biochemical Aspects of Cytodifferentiation*, First Round Table Conference, ed. F. Zilliken, 27 p. Basel: Karger Verlag (Sponsor).

Tiedemann, H. (1966*b**). *Current Topics in Developmental Biology*, **1**, 85–112.
Tiedemann, H. (1968*a*). *Verhandlungen der Deutschen Zoologischen Gesellschaft, Zoologischer Anzeiger*, **31**, 59–71.
Tiedemann, H. (1968*b*). *Journal of Cellular and Comparative Physiology*, **72**, 129–44.
Tiedemann, H. (1969*). *Verhandlungen der Schweizeren Naturforschenden Gesselschaft*, 32–55.
Tiedemann, H. (1975*). In *Biochemistry of Animal Development*, ed. R. Weber, vol. 3, 257–92. San Francisco: Academic Press.
Tiedemann, H. (1976). *Journal of Embryology and Experimental Morphology*, **35**, 437–44.
Tiedemann, H. (1978*). In *Organiser – A Milestone of a Half-Century from Spemann*, eds O. Nakamura & S. Toivonen, pp. 91–117. Amsterdam: Elsevier/North Holland Biomedical Press.
Tiedemann, H. (1982*). In *Biochemistry of Differentiation and Morphogenesis*, ed. L. Jaenicke, pp. 275–87. Berlin: Springer Verlag.
Tiedemann, H. & Born, J. (1978). *W. Roux's Archives of Developmental Biology*, **184**, 285–99.
Tiedemann, H., Born, J. & Kocher-Becker, U. (1965). *Zeitschrift für Naturforschung*, **20 B**, 997–1004.
Tiedemann, H., Born, J. & Tiedemann, H. (1967). *Zeitschrift für Naturforschung*, **22 B**, 649–59.
Tiedemann, H., Born, J. & Tiedemann, H. (1972). *W. Roux' Archiv für Entwicklungsmechanik der Organismen*, **171**, 160–9.
Tiedemann, H., Tiedemann, H., Born, J. & Kocher-Becker, U. (1969). *W. Roux' Archiv für Entwicklungsmechanik der Organismen*, **163**, 316–24.
Toivonen, S. (1938). *Annales Societatis Zoologicae-Botanicae Fennicae Vanamo*, **6**, 1–12.
Toivonen, S. (1940). *Annales Academiae Scientiarum Fennicae*, Ser. A, **55**, 1–150.
Toivonen, S. (1953). *Journal of Embryology and Experimental Morphology*, **1**, 97–104.
Toivonen, S. (1961). *Experientia*, **17**, 87–90.
Toivonen, S. (1967). *Karger Gazette*, **16**, 1–4; and *Experimental Biology and Medicine*, **1**, 1–7.
Toivonen, S. (1970). In *Intercellular Interactions in Differentiation and Growth*, eds G. V. Lopashov, N. N. Rott & G. D. Tumanishvili, pp. 140–9. Moscow: Nauka.
Toivonen, S. (1972). In *Cell Differentiation*, eds R. Harris, P. Allin & D. Viza, pp. 30–4. Copenhagen: Munksgaard.
Toivonen, S. (1978*). In *Organiser – A Milestone of a Half-Century from Spemann*, eds O. Nakamura & S. Toivonen, pp. 119–56. Amsterdam: Elsevier/North Holland Biomedical Press.
Toivonen, S. (1979). *Differentiation*, **15**, 177–81.
Toivonen, S., Kohonen, J., Saukkonen, J., Saxén, L. & Vainio, T. (1961*a*). *Embryologia*, **6**, 177–84.
Toivonen, S. & Saxén, L. (1966). *Annales Medicinae Experimentalis et Biologiae Fenniae*, **44**, 128–30.
Toivonen, S. & Saxén, L. (1968). *Science*, **159**, 539–40.
Toivonen, S., Saxén, L. & Vainio, T. (1961*b*). *Experientia*, **17**, 86–9.
Toivonen, S., Saxén, L. & Vainio, T. (1963). *W. Roux' Archiv für Entwicklungsmechanik der Organismen*, **154**, 293–307.
Toivonen, S., Tarin, D. & Saxén, L. (1976). *Differentiation*, **5**, 49–55.
Toivonen, S., Tarin, D., Saxén, L., Tarin, P. J. & Wartiovaara, J. (1975). *Differentiation*, **4**, 1–7.
Toivonen, S., Vainio, T. & Saxén, L. (1964). *Revue Suisse de Zoologie*, **71**, 139–45.
Toivonen, S. & Wartiovaara, J. (1976). *Differentiation*, **5**, 61–6.
Tomasck, J. J., Mazurkiewicz, J. E. & Newman, S. A. (1982). *Developmental Biology*, **90**, 118–26.

Tompkins, R. & Rodman, W. (1971). *Proceedings of the National Academy of Sciences of the United States of America*, **68**, 2921–3.
Tonégawa, Y. (1973). *Development, Growth and Differentiation*, **15**, 57–71.
Townes, P. I. & Holtfreter, J. (1955). *Journal of Experimental Zoology*, **128**, 53–120.
Tremaine, R. & Hall, B. K. (1979). *Acta Anatomica*, **105**, 78–85.
Trinkaus, J. P. (1976*). In *The Cell Surface in Animal Embryogenesis and Development*, eds G. Poste & G. L. Nicolson, vol. 1, pp. 225–329. Amsterdam: Elsevier/North Holland Biomedical Press.
Trinkaus, J. P. (1982*). In *Cell Behaviour; a Tribute to Michael Abercrombie*, eds R. Bellairs, A. Curtis & G. Dunn, pp. 471–98. Cambridge University Press.
Trinkaus, J. P. & Erickson, C. A. (1981). *The American Zoologist*, **21**, 401–11.
Tsafriri, A. (1978*). In *The Vertebrate Ovary; Comparative Biology and Evolution*, ed. R. E. Jones, pp. 409–33. New York: Plenum Press.
Tsafriri, A. & Channing, C. P. (1975). *Journal of Reproduction and Fertility*, **43**, 149–52.
Tsafriri, A., Lindner, H. R., Zor, U. & Lamprecht, S. A. (1972). *Journal of Reproduction and Fertility*, **31**, 39–50.
Tseng, Mi-Pai (1958). *Acta Biologiae Experimentalis Sinica*, **6**, 111–28.
Tseng, Mi-Pai (1960). *Acta Biologiae Experimentalis Sinica*, **7**, 67–80.
Tseng, Mi-Pai (1963a). *Acta Biologiae Experimentalis Sinica*, **8**, 230–42.
Tseng, Mi-Pai (1963b). *Acta Biologiae Experimentalis Sinica*, **8**, 463–75.
Tseng, Mi-Pai (1982). *Scientia Sinica*, **25**, 725–9.
Tseng, Mi-Pai, Mo, Hui-Yin & Chang, Chih-Fu (1965). *Acta Biologiae Experimentalis Sinica*, **131**, 156–67.
Tseng, Mi-Pai, Zhou, Mei-Yun, Zhang, Tie-Feng & Qu, Fu-Jin (1982). *Acta Biologiae Experimentalis Sinica*, **15**, 219–31.
Tsuda, H. (1961). *Acta Anatomica Nipponica*, **36**, 106–32.
Tsung, Shu-Dung, Ning, I-Lan & Shieh, Sher-Pu (1965). *Acta Biologiae Experimentalis Sinica*, **10**, 69–79.
Tuft, P. (1961a). *Nature* (London), **192**, 1049–51.
Tuft, P. (1961b). *Proceedings of the Royal Physical Society of Edinburgh*, **28**, 123–30.
Tuft, P. (1961c). *Nature* (London), **191**, 1072–4.
Tuft, P. (1962). *Journal of Experimental Biology*, **39**, 1–19.
Tung, T. C. (1934). *Archives d'Anatomie Microscopique*, **30**, 381–410.
Tung, T. C., Wu, S. C. & Tung, Y. F. Y. (1958). *Acta Biologiae Experimentalis Sinica*, **6**, 57–90 and *Scientia Sinica*, **7**, 1280–1320.
Tung, T. C., Wu, S. C. & Tung, Y. F. Y. (1959). *Acta Biologiae Experimentalis Sinica*, **6**, 191–210.
Tung, T. C., Wu, S. C. & Tung, Y. F. Y. (1960). *Scientia Sinica*, **9**, 119–41.
Tung, T. C., Wu, S. C. & Tung, Y. F. Y. (1961). *Acta Biologiae Experimentalis Sinica*, **7**, 253–61.
Tung, T. C., Wu, S. C. & Tung, Y. F. Y. (1962a). *Acta Biologiae Experimentalis Sinica*, **7**, 81–92 and *Scientia Sinica*, **11**, 629–44.
Tung, T. C., Wu, S. C. & Tung, Y. F. Y. (1962b). *Acta Biologiae Experimentalis Sinica*, **7**, 263–70 and *Scientia Sinica*, **11**, 805–20.
Tung, T. C., Wu, S. C., Tung, Y. F. Y., Yon, S. Y. & Tu, M. (1965a). *Acta Biologiae Experimentalis Sinica*, **10**, 318–31.
Tung, Y. F. Y., Luh, T. Y. & Tung, S. M. (1965b). *Acta Biologiae Experimentalis Sinica*, **10**, 332–45.
Turin, L. & Warner, A. (1977). *Nature* (London), **270**, 56–7.
Twitty, V. C. (1945). *Journal of Experimental Zoology*, **100**, 141–78.
Twitty, V. C. (1955*). In *Analysis of Development*, eds B. H. Willier, P. A. Weiss & V. Hamburger, pp. 402–14. Philadelphia: W. B. Saunders.

Twitty, V. C. & Bodenstein, D. (1962). In *Experimental Embryology Techniques and Procedures*, ed. R. Rugh, p. 90. Minneapolis: Burgess Publishing Company.

Tyler, A. & Tyler, B. S. (1966*). In *Physiology of Echinodermata*, ed. R. A. Boolootian, pp. 639–82 and 683–741. New York: Wiley & Sons.

Ubbels, G. A. (1977*). *Mémoires de la Société Zoologique de France*, no. 41, 103–16.

Ubbels, G. A., Gerhart, J. C., Kirschner, M. W. & Hara, K. (1979). *Archives d'Anatomie microscopique et de Morphologie Expérimentale*, 68, 211 (abstract).

Ubbels, G. A., Hara, K., Koster, C. H. & Kirschner, M. W. (1983). *Journal of Embryology and Experimental Morphology*, 77, 15–37.

Ubbels, G. A. & Hengst, R. T. M. (1978). *Differentiation*, 10, 109–21.

Ulshafer, R. J. & Clavert, A. (1979). *Journal of Embryology and Experimental Morphology*, 53, 237–43.

Ulshafer, R. J. & Hibbard, E. (1979). *Anatomy and Embryology*, 156, 29–35.

Vacquier, V. D. & Moy, G. W. (1980). *Developmental Biology*, 77, 178–90.

Vacquier, V. D. (1981*). *Developmental Biology*, 84, 1–26.

Vahs, W. (1962). *W. Roux' Archiv für Entwicklungsmechanik der Organismen*, 153, 504–50.

Vahs, W. (1965). *Embryologia*, 8, 308–18.

Vainio, T., Saxén, L. & Toivonen, S. (1960). *Experientia*, 16, 27–32.

Vainio, T., Saxén, L., Toivonen, S. & Rapola, J. (1962). *Experimental Cell Research*, 27, 527–38.

Vakaet, L. (1955). *Archives de Biologie*, 66, 1–73.

Vakaet, L. (1960). *Journal of Embryology and Experimental Morphology*, 8, 321–6.

Vakaet, L. (1962). *Journal of Embryology and Experimental Morphology*, 10, 38–57.

Vakaet, L. (1965). *Compte Rendu de la Société de Biologie*, Paris, 159, 232–3.

Vakaet, L. (1971). *Bulletin de l'Association des Anatomistes*, 56ième Congrès, Nantes, 770–7.

Vakaet, L. (1973). *Compte Rendu de la Société de Biologie*, Paris, 167, 1053–5.

Vakaet, L. & Vanroelen, C. (1982). *Journal of Embryology and Experimental Morphology*, 67, 59–70.

Valough, P., Melichna, J. & Sladeček, F. (1970). *Acta Universitatis Carolinae-Biologica*, 2, 195–205.

Van Beneden, E. (1875). *Bulletin de l'Académie Royale de Belgique*, Sér. 2, 40, 1–53.

Vandebroek, G. (1938). *Vlaamsch Natuurwetenschappelijk Tijdschrift*, 20, 234–9.

Vandebroek, G. (1961). In *Symposium on Germ Cells and Development*, ed. S. Ranzi, pp. 277–82. Institut International d'Embryologie and Fondazione A. Baselli.

Van der Starre, H. (1977). *Acta Morphologica Neerlando-Scandinavica*, 15, 275–86.

Van der Starre, H. (1978). *Acta Morphologica Neerlando-Scandinavica*, 16, 109–20.

Van Gansen, P. & Weber, A. (1972). *Archives de Biologie*, 83, 215–32.

Van Haarlem, R. (1979). *Developmental Biology*, 70, 171–9.

Van Haarlem, R. (1983). *Journal of Morphology*, 176, 31–42.

Van Haarlem, R., van Wijk, R. & Fikkert, A. H. M. (1981). *Cell Tissue Kinetics*, 14, 285–300.

Vanroelen, C., Vakaet, L. & Andries, L. (1980). *Anatomy and Embryology*, 160, 361–7.

Vardy, P. H., Stokes, P. A. & McBride, W. G. (1982). *Development, Growth and Differentiation*, 24, 99–114.

Vasan, N. S. (1981). *Journal of Experimental Zoology*, 215, 229–33.

Vasan, N. S. (1983). *Journal of Embryology and Experimental Morphology*, 73, 263–74.

Veini, M. & Bellairs, R. (1983). *Journal of Embryology and Experimental Morphology*, 74, 1–14.

Veini, M. & Hara, K. (1975). *W. Roux's Archives of Developmental Biology*, 177, 89–100.

Vermey-Keers, C. & Poelmann, R. E. (1980). *Netherlands Journal of Zoology*, 30, 74–81.

Vilain, J. P. (1974). *Compte Rendu de l'Académie des Sciences*, Paris, **278**, 1617–20.

Vintemberger, P. & Clavert, J. (1959). *Compte Rendu de la Société de Biologie*, Paris, **153**, 661–5.

Vogt, W. (1929). *W. Roux' Archiv für Entwicklungsmechanik der Organismen*, **120**, 384–706.

Von Kraft, A. (1968*a*). *W. Roux' Archiv für Entwicklungsmechanik der Organismen*, **160**, 259–97.

Von Kraft, A. (1968*b*). *W. Roux' Archiv für Entwicklungsmechanik der Organismen*, **160**, 255–8.

Von Kraft, A. (1968*c*). *W. Roux' Archiv für Entwicklungsmechanik der Organismen*, **161**, 351–74.

Von Kraft, A. (1969*a*). *W. Roux' Archiv für Entwicklungsmechanik der Organismen*, **163**, 178–83.

Von Kraft, A. (1969*b*). *Zoologische Jahrbücher, Anatomie*, **86**, 615–33.

Von Kraft, A. (1971*a*). *W. Roux' Archiv für Entwicklungsmechanik der Organismen*, **168**, 350–8.

Von Kraft, A. (1971*b*). *Naturwissenschaftliche Rundschau*, **24**, 142–51.

Von Kraft, A. (1971*c*). *Die Naturwissenschaften*, **6**, 326–7.

Von Ubisch, L. (1963). *W. Roux' Archiv für Entwicklungsmechanik der Organismen*, **154**, 466–94.

Von Woellwarth, C. (1952). *W. Roux' Archiv für Entwicklungsmechanik der Organismen*, **145**, 582–668.

Von Woellwarth, C. (1960). *W. Roux' Archiv für Entwicklungsmechanik der Organismen*, **152**, 602–31.

Von Woellwarth, C. (1961). *Embryologia*, **6**, 219–42.

Von Woellwarth, C. (1969). *W. Roux' Archiv für Entwicklungsmechanik der Organismen*, **162**, 336–40.

Von Woellwarth, C. (1970). *W. Roux' Archiv für Entwicklungsmechanik der Organismen*, **165**, 87–90.

Voss, H. (1965). *Acta Histochemica*, **20**, 112–14.

Vyasov, O. E., Averkina, R. F. & Petrosjan, I. L. (1965). In *Cellular Differentiation and Induction Mechanisms, 6th International Embryological Conference*, Helsinki, 1963, pp. 134–46. Moscow: Academia Nauk.

Wachtler, F., Christ, B. & Jacob, H. J. (1982). *Anatomy and Embryology*, **164**, 369–78.

Waddington, C. H. (1930). *Nature* (London), **125**, 924–5.

Waddington, C. H. (1932). *Philosophical Transactions of the Royal Society*, London, Ser. B, **221**, 179–230.

Waddington, C. H. (1933). *Journal of Experimental Biology*, **10**, 38–46.

Waddington, C. H. & Perry, M. M. (1966). *Experimental Cell Research*, **41**, 691–3.

Waddington, C. H. & Schmidt, G. A. (1933). *W. Roux' Archiv für Entwicklungsmechanik der Organismen*, **128**, 522–64.

Waheed, M. A. & McCallion, D. J. (1969). *Annales Zoologici Societatis Zoologico-Botanicae Fennicae*, Vanamo, **6**, 448–51.

Waheed, M. A. & Mulherkar, L. (1967). *Journal of Embryology and Experimental Morphology*, **17**, 161–9.

Wahn, H. L., Lightbody, L. E. & Tchen, T. T. (1975). *Science*, **188**, 336–69.

Wahn, H. L., Lightbody, L. T., Tchen, T. T. & Taylor, J. D. (1976). *Journal of Experimental Zoology*, **196**, 125–30.

Wakely, J. & England, M. A. (1978). *Anatomy and Embryology*, **153**, 167–78.

Wakely, J. & England, M. A. (1979). *Proceedings of the Royal Society*, London, B, **206**, 329–52.

Wall, R. & Faulhaber, I. (1976). *W. Roux's Archives of Developmental Biology*, **180**, 207–12.

Wallace, R. A., Jared, D. W., Dumont, J. N. & Lega, M. W. (1973). *Journal of Experimental Zoology*, **184**, 321–34.

Wallace, R. A., Nichol, J. M., Ho, T. & Jared, D. W. (1972). *Developmental Biology*, **29**, 255–72.

Wallace, R. A. & Selman, K. (1981). *The American Zoologist*, **21**, 325–43.

Wang, Y.-C. (1965). *Acta Biologiae Experimentalis Sinica*, **10**, 233–49.

Wang, Y.-H., Mo, H.-Y. & Shen, J.-Y. (1963). *Acta Biologiae Experimentalis Sinica*, **8**, 356–67.

Warner, A. E., Guthrie, S. C. & Gilula, N. B. (1984). *Nature* **311**, 127–31.

Wartiovaara, J., Lehtonen, E., Nordling, S. & Saxén, L. (1972). *Nature* (London), **238**, 407–8.

Wasserman, P. M. & Letourneau, G. E. (1976). *Journal of Cell Science*, **20**, 549–68.

Wasserman, W. J. (1982). *Developmental Biology* **90**, 445–7.

Wasserman, W. J. & Masui, Y. (1974). *Biology of Reproduction*, **11**, 133–44.

Wasserman, W. J., Richter, J. D. & Smith, L. D. (1982). *Developmental Biology*, **89**, 152–8.

Wasserman, W. J. & Smith, L. D. (1978*). In *The Vertebrate Ovary: Comparative Biology and Evolution*, ed. R. E. Jones, pp. 443–68. New York: Plenum Press.

Watterson, R. L., Fowler, I. & Fowler, B. J. (1954). *American Journal of Anatomy*, **95**, 337–400.

Webb, A. C., LaMarca, M. J. & Smith, L. D. (1975). *Developmental Biology*, **45**, 44–55.

Webb, A. C. & Smith, L. D. (1977). *Developmental Biology*, **56**, 219–25.

Wedlock, D. E. & McCallion, D. J. (1969). *Canadian Journal of Zoology*, **47**, 142–3.

Wehrmaker, A. (1964). *Zoologische Jahrbücher, Abteilung Physiologie*, **70**, 489–512.

Wehrmaker, A. (1967). *Zoologischer Anzeiger*, Suppl. 30, 564–70.

Wehrmaker, A. (1969). *W. Roux' Archiv für Entwicklungsmechanik der Organismen*, **163**, 1–32.

Weijer, C. J., Nieuwkoop, P. D. & Lindenmayer, A. (1977). *Acta Biotheoretica*, **26**, 164–80.

Weismann, A. (1892*). *Das Keimplasma: Eine Theorie der Vererbung*, 628 p. Jena: Gustav Fischer Verlag.

Weiss, L. & Nir, S. (1979). *Journal of Theoretical Biology*, **78**, 11–20.

Weiss, P. (1934). *Journal of Experimental Zoology*, **68**, 393–448.

Weiss, P. (1941*). *Growth*, **5**, 163–203.

Weiss, P. (1959*). *Reviews of Modern Physics*, **31**, 449–54.

Weiss, P. (1961*a*). *Experimental Cell Research*, suppl. **8**, 260–81.

Weiss, P. (1961*b**). In *The Molecular Control of Cellular Activity*, ed. J. M. Allen, pp. 1–72. New York: McGraw-Hill.

Weiss, P. (1962*). In *Biological Interactions in Normal and Neoplastic Growth*, ed. M. J. Brennan, pp. 3–20. Boston: Little, Brown & Co.

Weiss, P. (1969*). *Principles of Development, a Text in Experimental Embryology*, 601 p. New York: Hafner Publishing Company.

Weiss, P. & Taylor, A. C. (1960). *Proceedings of the National Academy of Sciences of the United States of America*, **46**, 1177–85.

Wessels, N. K. (1962). *Developmental Biology*, **4**, 87–107.

Wessels, N. K. (1964). *Journal of Experimental Zoology*, **157**, 139–52.

Wessels, N. K. (1970*a**). *Journal of Investigative Dermatology*, **55**, 221–5.

Wessels, N. K. (1970*b*). *Journal of Experimental Zoology*, **175**, 455–66.

Wessels, N. K. (1973*). In *An Addison-Wesley Module in Biology*, **9**, 1–43. Philippines: Addison-Wesley.

Wessels, N. K. (1977*). *Tissue Interactions and Development*, 276 p. Menlo Park, California: W. A. Benjamin Inc.

Weston, J. A. (1963). *Developmental Biology*, **6**, 279–310.

Weston, J. A. (1970*). *Advances in Morphogenesis*, **8**, 41–114.
Weston, J. A. (1971*). In *Cellular Aspects of Growth and Differentiation in Nervous System*, ed. D. Pease, pp. 1–19. University of California Press.
Weston, J. A. (1980*). In *Current Research in Prenatal Craniofacial Development*, eds R. M. Pratt & R. L. Christiansen, pp. 27–45. Amsterdam: Elsevier Biomedical Press.
Weston, J. A. & Butler, S. L. (1966). *Developmental Biology*, **14**, 246–66.
Wheldon, T. E. & Kirk, J. (1973). *Journal of Theoretical Biology*, **41**, 261–8.
Whitt, G. S. (1975*). In *Isozymes, III Developmental Biology*, ed. C. L. Markert, pp. 1–8. San Francisco: Academic Press.
Whittaker, J. R. (1979*). In *Determinants of Spatial Organisation*, eds S. Subtelny & I. R. Konigsberg, pp. 29–55. 37th Symposium of the Society for Developmental Biology. New York: Academic Press.
Whittaker, J. R. (1982). *Developmental Biology*, **93**, 463–70.
Whittaker, J. R. (1983). *Journal of Embryology and Experimental Morphology*, **76**, 235–50.
Wilde, C. E. (1959*). In *Cell, Organism and Milieu*, ed. D. Rudnick, pp. 3–43. New York: Ronald Press.
Wilde, C. E. & Crawford, R. B. (1968*). In *Epithelial–Mesenchymal Interactions*, eds R. Fleischmajer & R. E. Billingham, pp. 98–113. Baltimore: Williams & Wilkins.
Williams, M. A. & Smith, L. D. (1971). *Developmental Biology*, **25**, 568–80.
Wilson, J. B., Bolton, E. & Cuttler, R. H. (1972). *Journal of Embryology and Experimental Morphology*, **27**, 467–79.
Wintrebert, P. (1922). *Compte Rendu de l'Académie des Sciences*, Paris, **175**, 411–13.
Wischnitzer, S. (1966*). *Advances in Morphogenesis*, **5**, 131–79.
Witschi, E. (1967*). In *Biochemistry of Animal Development*, ed. R. Weber, vol. 2, 193–225. New York: Academic Press.
Wittek, M. (1952). *Archives de Biologie*, **63**, 133–98.
Woerdeman, M. W. (1962*). *Proceedings of the Royal Academy of Sciences*, Amsterdam, Ser. C, **65**, 145–59.
Wolf, D. P. (1974*a*). *Developmental Biology*, **36**, 62–71.
Wolf, D. P. (1974*b*). *Developmental Biology*, **38**, 14–29.
Wolf, D. P. (1974*c*). *Developmental Biology*, **40**, 102–15.
Wolf, D. P. (1981*). In *Fertilization and Embryonic Development in Vitro*, eds L. Mastroianni & J. D. Biggers, pp. 183–97. New York: Plenum Press.
Wolf, D. P., Nishihara, T., West, D. M., Wyrick, R. E. & Hedrick, J. L. (1976). *Biochemistry*, **15**, 3671–8.
Wolff, E. (1961). *Developmental Biology*, **3**, 767–86.
Wolff, E. (1968*). *Current Topics in Developmental Biology*, **3**, 65–94.
Wolff, E. (1969*). In *Theoretical Physics and Biology*, ed. M. Marois, pp. 302–7. Amsterdam: North Holland Biomedical Press.
Wolff, E. (1973*). *Année de Biologie*, **12**, 513–24.
Wolff, E. & Haffen, K. (1952). *Compte Rendu de l'Académie des Sciences*, Paris, **234**, 1396–8.
Wolk, M. & Eyal-Giladi, H. (1977). *Developmental Biology*, **55**, 33–45.
Wolpert, L. (1960*). *International Reviews of Cytology*, **10**, 163–216.
Wolpert, L. (1973). *Developmental Biology*, **30**, f4–f5.
Wolsky, A. & de Issekutz-Wolsky, M. (1968). *Oncology*, **22**, 290–301.
Woo Youn, B., Keller, R. E. & Malacinski, G. M. (1980). *Journal of Embryology and Experimental Morphology*, **59**, 223–47.
Woo Youn, B. & Malacinski, G. M. (1980). *Journal of Experimental Zoology*, **211**, 369–77.
Woodward, D. J. (1968). *Journal of General Physiology*, **52**, 509–31.
Woodland, H. R. & Gurdon, J. B. (1968). *Journal of Embryology and Experimental Morphology*, **19**, 363–85.

Wright, D. A. (1978*). In *Cell Differentiation and Neoplasia*, ed. G. F. Saunders, pp. 391–402. New York: Raven Press.

Wu, S.-C. & Cai, N.-E. (1964). *Acta Biologiae Experimentalis Sinica*, **9**, 119–26.

Wyllie, A. H., Gurdon, J. B. & Price, J. (1977). *Nature* (London), **268**, 150–2.

Wyrick, R. E., Nishihara, T. & Hedrick, J. L. (1974). *Proceedings of the National Academy of Sciences of the United States of America*, **71**, 2067–71.

Yamada, K. M., Olden, K. & Hahn, L. H. E. (1980*). In *The Cell Surface: Mediator of Developmental Processes*, eds S. Subtelny & N. K. Wessels, pp. 43–77. New York: Academic Press.

Yamada, T. (1937). *W. Roux' Archiv für Entwicklungsmechanik der Organismen*, **137**, 151–270.

Yamada, T. (1938). *Okajimas Folia Anatomica Japonica*, **17**, 369–88.

Yamada, T. (1939a). *Okajimas Folia Anatomica Japonica*, **18**, 565–68.

Yamada, T. (1939b). *Okajimas Folia Anatomica Japonica*, **18**, 569–71.

Yamada, T. (1939c). *Japanese Journal of Zoology*, **8**, 265–83.

Yamada, T. (1940). *Okajimas Folia Anatomica Japonica*, **19**, 132–97.

Yamada, T. (1950a). *Biological Bulletin*, **98**, 98–121.

Yamada, T. (1950b). *Embryologia*, **1**, 1–20.

Yamada, T. (1958*). In *A Symposium on Chemical Basis of Development*, eds W. McElroy & B. Glass, pp. 217–38. Baltimore: Johns Hopkins Press.

Yamada, T. (1962). *Journal of Cellular and Comparative Physiology*, **60**, suppl. 1, 49–64.

Yamada, T. (1967*). *Current Topics in Developmental Biology*, **2**, 247–83.

Yamada, T. (1972*). In *Cell Differentiation*, eds R. Harris, P. Allin & D. Viza, pp. 56–60. Copenhagen: Munksgaard.

Yamada, T. (1978*). In *Organiser – A Milestone of a Half-Century from Spemann*, eds O. Nakamura & S. Toivonen, pp. 309–13. Amsterdam: Elsevier/North Holland Biomedical Press.

Yamada, T. (1981*). *Netherlands Journal of Zoology*, **31**, 78–98.

Yamada, T., Dumont, J. N., Moret, R. & Brun, J. P. (1978). *Differentiation*, **11**, 133–47.

Yamada, T. & McDevitt, D. S. (1974). *Developmental Biology*, **38**, 104–18.

Yamada, T. & Takata, K. (1961). *Developmental Biology*, **3**, 411–23.

Yamamoto, K. Y., Ozawa, R., Takata, K. & Kitoh, J. (1981). *W. Roux's Archives of Developmental Biology*, **190**, 313–19.

Yamamoto, K. Y., Takata, K., Ozawa, R. & Kitoh, J. (1982). *Development, Growth and Differentiation*, 24, 399 (abstract no. B50).

Yamamoto, T. (1961a). *Mie Medical Journal*, **11**, 57–65.

Yamamoto, T. (1961b*). *International Reviews of Cytology*, **12**, 361–405.

Yamamoto, Y. (1976). *Development, Growth and Differentiation*, **18**, 273–8.

Yamazaki-Yamamoto, K., Yamazaki, K. & Kato, Y. (1980). *Development, Growth and Differentiation*, **22**, 79–92.

Yanagimachi, R. (1978*). *Current Topics in Developmental Biology*, **12**, 83–105.

Yanagimachi, R. (1981*). In *Fertilization and Embryonic Development in Vitro*, eds L. Mastroianni & J. D. Biggers, pp. 82–182. New York: Plenum Press.

Yasugi, S. (1979). *Development, Growth and Differentiation*, **21**, 343–8.

Yasugi, S. & Mizuno, T. (1974). *W. Roux' Archiv für Entwicklungsmechanik der Organismen*, **174**, 107–16.

Yasugi, S. & Mizuno, T. (1978). *Development, Growth and Differentiation*, **20**, 261–7.

Yntema, C. L. & Hammond, W. S. (1947*). *Biological Reviews*, **22**, 344–59.

Yoshizaki, N. (1976). *Development, Growth and Differentiation*, **18**, 133–43.

Yoshizaki, N. (1979). *Development, Growth and Differentiation*, **21**, 11–18.

Yoshizaki, N. & Katagiri, C. (1975). *Journal of Experimental Zoology*, **192**, 203–12.

Yoshizaki, N., Sackers, R. J., Schoots, A. F. M. & Denucé, J. M. (1980). *Journal of Experimental Zoology*, **213**, 427–9.

Yu, S.-F. & Wolf, D. P. (1981). *Developmental Biology*, **82**, 203–10.
Yvroud, M. (1971). *Annales d'Embryologie et de Morphogenèse*, **4**, 175–88.
Yvroud, M. (1974). *Compte Rendu de l'Académie des Sciences*, Paris, **279**, 81–2.
Zagris, N. & Eyal-Giladi, H. (1982). *Developmental Biology*, **91**, 208–14.
Zalik, S. E., Milos, N. & Ledsham, I. (1983). *Cell Differentiation*, **12**, 121–7.
Zalokar, M. (1964*). In *The Nucleohistones*, eds J. Bonner & P. Tso, pp. 348–51. San Francisco: Holden-Day.
Zalokar, M. (1979). *W. Roux's Archives of Developmental Biology*, **187**, 35–47.
Zamboni, L. (1972*). In *Biology of Mammalian Fertilization and Implantation*, eds K. S. Mighissi & E. S. E. Hafez, pp. 213–62. Springfield: C. C. Thomas.
Zeeman, E. C. (1975). *Symposium on Catastrophe Theory*, Lecture Notes pp. 1–6. Seattle: Springer.
Zeuthen, E. & Hamburger, K. (1972). *Biological Bulletin*, **143**, 699–706.
Ziegler, D. H. & Masui, Y. (1973). *Developmental Biology*, **35**, 282–92.
Ziegler, D. H. & Masui, Y. (1976). *Journal of Cell Biology*, **68**, 620–8.
Ziegler, D. H. & Morrill, G. A. (1977). *Developmental Biology*, **60**, 318–25.
Zilliken, F. (ed.) (1966*). 1st Informal Round Table Conference on *Morphological and Biochemical Aspects of Cytodifferentiation*, 15 p. Karger Press (sponsor).
Zotin, A. I. (1964*). *Journal of Embryology and Experimental Morphology*, **12**, 247–62.
Zotin, A. I. (1965*). In *The State and Movement of Water in Living Organisms*, 19th Symposium, Society for Experimental Biology, Swansea, 1964, pp. 365–84. Cambridge University Press.
Zotin, A. I. & Pagnaeva, R. V. (1963). *Doklady Akademii Nauk SSSR*, **152**, 1236–9.
Zotin, A. I. & Poglazov, B. F. (1962). *Doklady Akademii Nauk SSSR*, **143**, 1233–6.
Züst, B. & Dixon, K. E. (1975). *Journal of Embryology and Experimental Morphology*, **34**, 209–20.
Zwaan, J., Bryan, P. R. & Pearce, T. L. (1969). *Journal of Embryology and Experimental Morphology*, **21**, 71–83.
Zwaan, J. & Hendrix, R. W. (1973). *The American Zoologist*, **13**, 1039–49.
Zwanzig, H. (1938). *Zeitschrift für Wissenschaftliche Zoologie*, **150**, 468–95.
Zwilling, E. (1956*a*). *Journal of Experimental Zoology*, **132**, 157–71.
Zwilling, E. (1956*b*). *Journal of Experimental Zoology*, **132**, 173–87.
Zwilling, E. (1956*c*). *Journal of Experimental Zoology*, **132**, 241–53.
Zwilling, E. (1961*). *Advances in Morphogenesis*, **1**, 301–30.
Zwilling, E. (1968*). Symposium on *The Emergence of Order in Developing Systems* ed. M. Locke, *Developmental Biology*, suppl. 2, 184–207.
Zwilling, E. & Hansborough, L. A. (1956). *Journal of Experimental Zoology*, **132**, 219–39.
Zwirner, R. & Kuhlo, B. (1964). *W. Roux' Archiv für Entwicklungsmechanik der Organismen*, **155**, 511–24.

Subject index

Author index